PRACTICAL STATISTICS AND EXPERIMENTAL DESIGN FOR PLANT AND CROP SCIENCE

PRACTICAL STATISTICS AND EXPERIMENTAL DESIGN FOR PLANT AND CROP SCIENCE

Alan G. Clewer and David H. Scarisbrick

*T. H. Huxley School of Environment,
Earth Sciences and Engineering
Imperial College at Wye,
Ashford, Kent, UK*

JOHN WILEY & SONS, LTD

Chichester • New York • Weinheim • Brisbane • Singapore • Toronto

Copyright © 2001 by John Wiley & Sons, Ltd,
Baffins Lane, Chichester,
West Sussex PO19 1UD, England

National   01243 779777
International (+44) 1243 779777
e-mail (for orders and customer service enquiries):
cs-books@wiley.co.uk
Visit our Home Page on: http://www.wiley.co.uk
or http://www.wiley.com

All Rights Reserved. No part of this publication may be reproduced, stored in a retrieval system, or transmitted, in any form or by any means, electronic, mechanical, photocopying, recording, scanning or otherwise, except under the terms of the Copyright, Designs and Patents Act 1988 or under the terms of a licence issued by the Copyright Licensing Agency, 90 Tottenham Court Road, London W1P 9HE, UK, without the permission in writing of the Publisher.

Templates for many of the figures in this book were created using Microsoft Excel software

Other Wiley Editorial Offices

John Wiley & Sons Inc., 605 Third Avenue,
New York, NY 10158-0012, USA

WILEY-VCH Verlag GmbH, Pappelallee 3,
D-69469 Weinheim, Germany

Jacaranda Wiley Ltd, 33 Park Road, Milton,
Queensland 4064, Australia

John Wiley & Sons (Asia) Pte Ltd, 2 Clementi Loop #02-01,
Jin Xing Distripark, Singapore 129809

John Wiley & Sons (Canada) Ltd, 22 Worcester Road,
Rexdale, Ontario M9W 1L1, Canada

Library of Congress Cataloging-in-Publication Data

Clewer, Alan G.
    Practical statistics and experimental design for plant and crop science / Alan G. Clewer and David H. Scarisbrick.
        p. cm.
    Includes bibliographical references and index.
    ISBN 0-471-89908-9 (hbk : alk. paper) – ISBN 0-471-89909-7 (pbk. : alk. paper)
    1. Botany–Research–Statistical methods. 2. Botany, Economic–Research–Statistical methods. 3. Biometry. I. Scarisbrick, D. H. II. Title.
    QK51. C58 2001
    580'.7'27–dc21   00-047313

British Library Cataloguing in Publication Data

A catalogue record for this book is available from the British Library

ISBN   0 471 89908 9 (PPC)
ISBN 0 471 89909 7 (Pbk)

Typeset in 10/12pt Times by Dobbie Typesetting Ltd, Tavistock, Devon.
Printed and bound in Great Britain by Bookcraft (Bath) Ltd, Midsomer Norton.

This book is printed on acid-free paper responsibily manufactured from sustainable forestry, in which at least two trees are planted for each one used for paper production.

Contents

Preface xi

Chapter 1 **Basic Principles of Experimentation** 1
    1.1   Introduction 1
    1.2   Field and glasshouse experiments 1
    1.3   Choice of site 3
    1.4   Soil testing 4
    1.5   Satellite mapping 5
    1.6   Sampling 6

Chapter 2 **Basic Statistical Calculations** 9
    2.1   Introduction 9
    2.2   Measurements and type of variable 9
    2.3   Samples and populations 10

Chapter 3 **Basic Data Summary** 16
    3.1   Introduction 16
    3.2   Frequency distributions (discrete data) 16
    3.3   Frequency distributions (continuous data) 18
    3.4   Descriptive statistics 22

Chapter 4 **The Normal Distribution, the *t*-Distribution and Confidence Intervals** 24
    4.1   Introduction to the normal distribution 24
    4.2   The standard normal distribution 25
    4.3   Further use of the normal tables 27
    4.4   Use of the percentage points table (Appendix 2) 29
    4.5   The normal distribution in practice 29
    4.6   Introduction to confidence intervals 31
    4.7   Estimation of the population mean, μ 31
    4.8   The sampling distribution of the mean 32
    4.9   Confidence limits for μ when σ is known 32
    4.10   Confidence limits for μ when σ is unknown—use of the *t*-distribution 34
    4.11   Determination of sample size 36
    4.12   Estimation of total crop yield 36

Chapter 5 Introduction to Hypothesis Testing — 38
- 5.1 The standard normal distribution and the t-distribution — 38
- 5.2 The single sample t-test — 39
- 5.3 The P-value — 42
- 5.4 Type I and Type II errors — 43
- 5.5 Choice of level of significance — 44
- 5.6 The usefulness of a test — 45
- 5.7 Estimation versus hypothesis testing — 46
- 5.8 The paired samples t-test — 46

Chapter 6 Comparison of Two Independent Sample Means — 49
- 6.1 Introduction — 49
- 6.2 The Independent Samples t-test — 51
- 6.3 Confidence intervals — 55
- 6.4 The theory behind the t-test — 55
- 6.5 The F-test — 58
- 6.6 Unequal sample variances — 59
- 6.7 Determination of sample size for a given precision — 60

Chapter 7 Linear Regression and Correlation — 63
- 7.1 Basic principles of Simple Linear Regression (SLR) — 63
- 7.2 Experimental versus observational studies — 66
- 7.3 The correlation coefficient — 67
- 7.4 The least squares regression line and its estimation — 67
- 7.5 Calculation of residuals — 71
- 7.6 The goodness of fit — 72
- 7.7 Calculation of the correlation coefficient — 74
- 7.8 Assumptions, hypothesis tests and confidence intervals for simple linear regression — 75
- 7.9 Testing the significance of a correlation coefficient — 83

Chapter 8 Curve Fitting — 87
- 8.1 Introduction — 87
- 8.2 Polynomial fitting — 87
- 8.3 Quadratic regression — 89
- 8.4 Other types of curve — 93
- 8.5 Multiple linear regression — 100

Chapter 9 The Completely Randomised Design — 102
- 9.1 Introduction — 102
- 9.2 Design construction — 103
- 9.3 Preliminary analysis — 105
- 9.4 The one-way analysis of variance model — 108
- 9.5 Analysis of variance — 110
- 9.6 After ANOVA — 118
- 9.7 Reporting results — 123

9.8	The completely randomised design—unequal replication	124
9.9	Determination of number of replicates per treatment	128

Chapter 10 The Randomised Block Design — **132**

10.1	Introduction	132
10.2	The analysis ignoring blocks	135
10.3	The analysis including blocks	136
10.4	Using the computer	136
10.5	The effect of blocking	137
10.6	The randomised blocks model	138
10.7	Using a hand calculator to find the sums of squares	141
10.8	Comparison of treatment means	142
10.9	Reporting the results	144
10.10	Deciding how many blocks to use	144
10.11	Plot sampling	146

Chapter 11 The Latin Square Design — **149**

11.1	Introduction	149
11.2	Randomisation	151
11.3	Interpretation of computer output	153
11.4	The Latin square model	155
11.5	Using your calculator	156

Chapter 12 Factorial Experiments — **159**

12.1	Introduction	159
12.2	Advantages of factorial experiments	160
12.3	Main effects and interactions	163
12.4	Varieties as factors	165
12.5	Analysis of a randomised blocks factorial experiment with two factors	166
12.6	General advice on presentation	176
12.7	Experiments with more than two factors	177
12.8	Confounding	179
12.9	Fractional replication	180

Chapter 13 Comparison of Treatment Means — **182**

13.1	Introduction	182
13.2	Treatments with no structure	182
13.3	Treatments with structure (factorial structure)	191
13.4	Treatments with structure (levels of a quantitative factor)	195
13.5	Treatments with structure (contrasts)	202

Chapter 14 Checking the Assumptions and Transformation of Data — **213**

14.1	The assumptions	213
14.2	Transformations	219

Chapter 15 Missing Values and Incomplete Blocks — 226
  15.1 Introduction — 226
  15.2 Missing values in a completely randomised design — 226
  15.3 Missing values in a randomised block design — 229
  15.4 Other types of experiment — 234
  15.5 Incomplete block designs — 234

Chapter 16 Split Plot Designs — 238
  16.1 Introduction — 238
  16.2 Uses of this design — 238
  16.3 The skeleton analysis of variance tables — 240
  16.4 An example with interpretation of computer output — 242
  16.5 The growth cabinet problem — 250
  16.6 Other types of split plot experiment — 252
  16.7 Repeated measures — 252

Chapter 17 Comparison of Regression Lines and Analysis of Covariance — 256
  17.1 Introduction — 256
  17.2 Comparison of two regression lines — 256
  17.3 Analysis of covariance — 260
  17.4 Analysis of covariance applied to a completely randomised design — 260
  17.5 Comparing several regression lines — 265
  17.6 Conclusion — 270

Chapter 18 Analysis of Counts — 272
  18.1 Introduction — 272
  18.2 The binomial distribution — 272
  18.3 Confidence intervals for a proportion — 275
  18.4 Hypothesis test of a proportion — 277
  18.5 Comparing two proportions — 279
  18.6 The chi-square goodness of fit test — 280
  18.7 $r \times c$ contingency tables — 284
  18.8 $2 \times c$ contingency tables: comparison of several proportions — 286
  18.9 2×2 contingency tables: comparison of two proportions — 287
  18.10 Association of plant species — 289
  18.11 Heterogeneity chi-square — 290

Chapter 19 Some Non-parametric Methods — 293
  19.1 Introduction — 293
  19.2 The Sign test — 294
  19.3 The Wilcoxon single-sample test — 296
  19.4 The Wilcoxon matched pairs test — 297
  19.5 The Mann–Whitney U test — 299
  19.6 The Kruskal–Wallis test — 302
  19.7 Friedman's test — 304

CONTENTS ix

Appendix 1: The normal distribution function 307

Appendix 2: Percentage points of the normal distribution 308

Appendix 3: Percentage points of the t-distribution 309

Appendix 4a: 5 per cent points of the F-distribution 310

Appendix 4b: 2.5 per cent points of the F-distribution 312

Appendix 4c: 1 per cent points of the F-distribution 314

Appendix 4d: 0.1 per cent points of the F-distribution 316

Appendix 5: Percentage points of the sample correlation coefficient (r) when the population correlation coefficient is 0 and n is the number of X, Y pairs 318

Appendix 6: 5 per cent points of the Studentised range, for use in Tukey and SNK tests 319

Appendix 7: Percentage points of the chi-square distribution 321

Appendix 8: Probabilities of S or fewer successes in the binomial distribution with n 'trials' and $p = 0.5$ 322

Appendix 9: Critical values of T in the Wilcoxon signed rank or matched pairs test 323

Appendix 10: Critical values of U in the Mann–Whitney test 324

References 327

Further reading 328

Index 329

Preface

The references at the end of this book confirm that there are many textbooks on statistics for students who are interested in applied biology. Most of these cover the same subject material, although they vary quite widely in styles of presentation. Thus, it is important to answer the question—why has another book been written? It would be pointless to review the same topics on the design and analysis of experiments unless some original features can be detected by the reader in the 19 chapters that make up this book. One claim to originality is that this text is closely linked to the computer outputs from three commonly used statistical packages (Genstat 5—release 4.1, Minitab—release 12.1 and SAS—release 6.12). However, this in itself may not be sufficient to justify the vast amount of time required to produce another text, and provide a satisfactory answer to the above question.

The answer is more closely linked to our concern about the misuse of statistics by many students and their lack of understanding of the basic principles that underlie the statistical techniques they refer to in their dissertations. Because it is now all too easy to carry out inappropriate analyses by computer, the advent of statistical packages has diluted many students' understanding and interest in the basic principles which are the foundation stone of good design. They often look for an analytical method that seems to 'fit' their data. This frequently results in problems of interpretation; the design of experiments and the related data analysis should not be treated as separate components of the experimental process. Design is more important than analysis, because without a good design, the analysis is meaningless. Matching experimental data with a program given in a textbook on computing and statistics after the experiment has been completed is rarely successful and should be avoided.

As external examiners we frequently find computer summaries of data analyses in the appendices of dissertations, although when faced with simple questions the majority of students seem to have very little understanding of the terms and figures given in their printouts. For example, the ANOVA summary usually includes a mean square (MS) column. It is now quite rare for students to give the correct translation of the error (residual) mean square and even more rare for them to know that this is also an estimate of variance. There is always muddle in relation to the constants quoted for simple regression equations, and discussion of the usefulness of slope and intercept is usually missing mainly because the role of the constants and equations is rarely understood.

Thus, the book has mainly been written to demonstrate both the usefulness of statistics (like previous texts), and also to provide a clear explanation of terms, figures and symbols given in computer printouts. We decided not to link these discussions to one statistical package in order to illustrate the diversity of layouts that are used to provide a summary of the same results. For most statistical procedures the three packages have much output in common, but some are better than others for certain options. For the examples for which we give computer outputs, we use the package that we consider illustrates the main points we are trying to make. The book also encourages students to review the underlying principles of many statistical tests before using them in their research work; this point should be noted by supervisors! When interviewing students in relation to their data-handling methods it is sometimes the supervision that can be faulted, especially when students confess that they simply browsed a recommended text in order to find a statistical example which had similarities to their own design and experimental results.

Students reading this book should initially work through the text and exercise examples using a hand calculator. This technique assists understanding and interpretation. When interpretive skills have been achieved, results can then be confirmed by studying the computer outputs provided. Future data can then be immediately analysed using computer packages.

The classical textbooks describe how to do calculations by hand using the *correction factor method*. This method is no longer needed due to the widespread use of computers and hand calculators. However, we do explain it in the chapter on one-way analysis of variance. In subsequent chapters we show how the various sums of squares can be calculated using the standard deviation function provided by most calculators.

Biological statistics (biometry) is relevant to all areas encompassed by the general subject title Applied Plant Biology. It is an essential tool which is used to uncover or discover scientific information contained within raw (new) data. Like applied plant sciences, biometry is a broad-based subject. It is concerned with all aspects of experimentation (design, sampling methods, data analysis, interpretation and discussion) which are relevant to the research objectives. Experiments should always be designed in such a way that questions posed by the objectives have a good chance of being answered. This implies that the method of analysis and statistical tests are determined before the experiment is carried out.

This book introduces students to the important role of biometry in applied plant science research. Because many readers will have little prior knowledge of biological statistics, the first five chapters are used to summarise the basic principles which underpin simple statistical concepts. Although a more advanced treatise is provided in later chapters, the mathematical theory underpinning the techniques described is mainly ignored. Instead, the text provides some description and justification of the most important calculations which are widely used in statistics testing, and a detailed review of computer output which is now commonly presented by students in their dissertations. For each output an interpretation is given, and for many, how most entries can be found from a hand calculator. A novel feature of the book is the inclusion of examples showing how the sums of squares for the various terms in the most popular analysis of variance models are partitioned without using algebra.

Many biology students feel uneasy with statistical calculations and equations, and are rarely concerned with acquiring an understanding of the mathematical theory of statistics. They are mainly interested in learning how to apply a range of statistical tools to design, analyse and then interpret their results. Although this approach to biological statistics is commonplace, it is the authors' view that effort should still be made to understand some of the background calculations associated with many tests used to compare treatment means. This is an important part of comprehending statistical methods; comprehension is still important even in an era when computers usually undertake the time-consuming labour of arithmetic calculations.

When statistical analyses were carried out using hand-operated calculators, the amount of data collected and analysed was mainly controlled by the sluggishness of desk machinery. Because there were no mountains of computer output to file and review, raw data and results were usually pondered and discussed in great detail. Packaged computer programs have revolutionised data-handling techniques. They have removed most of the drudgery from the analysis of large experiments which compare a range of treatments, and eliminated arithmetic errors. In addition, because most programs provide facilities for reviewing raw data, it is now much easier to check that assumptions which underpin many statistical tests are really true before proceeding with an analysis. An initial study using graphs, scatter diagrams and tables is helpful in deciding whether a particular statistical method is really appropriate for the new data being examined.

It could be argued that statistical tables are no longer required to carry out statistical tests when the P-value is given in the computer output. Nevertheless, to understand the P-value, a familiarity with the underlying distribution of the test statistic is required, so we include statistical tables in Appendices 1 to 10 inclusive. They are also required for those readers without a computer or relevant software.

The authors hope that this book will assist students and researchers in crop and plant sciences to explain in simple language the objectives of simple statistical tests, and achieve an understanding of the principles of experimentation. This book can be read at several levels and so will be useful for a wide range of readers. It will be useful for teachers. It will be useful for users of statistical packages who want to interpret the output. It will be useful for researchers who want to know how to design a simple experiment and analyse and present the results. It will also be useful for those who want to know a little of the background theory needed to justify the procedures and how the calculations are performed.

ACKNOWLEDGEMENTS

We wish to thank BASF for permission to reproduce some diagrams and tables. We also thank MINITAB, SAS and Genstat for permission to include computer output. SAS is a registered trade mark of the SAS Institute Inc.

Chapter 1

Basic Principles of Experimentation

1.1 INTRODUCTION

The principles of experimentation can be studied by students who enjoy pure mathematics, and by those who wish to use mathematical principles as tools only for the design and analysis of their experiments. When carrying out research work at an experimental station or university, the applied biologist may be able to discuss the layout and analysis of experimental work with a professional statistician. However, when working in isolated rural areas (especially in developing countries), the experimenter must demonstrate a basic understanding of statistics and also have the confidence to solve design and data-handling problems. Even when professional support is available, it is still essential that the researcher is aware of the wide range of statistical procedures used by applied biologists. More importantly, he or she must have acquired sufficient statistical skills to interpret and discuss experimental results which are analysed using computer packages such as Genstat, Minitab and SAS.

Experimental objectives must be clearly and concisely stated at the outset of an investigational programme. Before starting a field or greenhouse experiment, it is wise to purchase a diary. The first page should be used to give a clear exposition of research objectives. The diary should contain a summary of previous cropping, a description of the treatments, detailed site plans and daily observations. When an experiment is written up, these observations may help to explain results that may at first seem to be anomalous.

1.2 FIELD AND GLASSHOUSE EXPERIMENTS

There are two main systems of experimentation used to explore the effect(s) of experimental treatments on plant development and yield (Figure 1.1). In both, treatments may be applied before, at, as well as after sowing. For example, the objectives of an experiment may be to compare different seed dressings, fertiliser

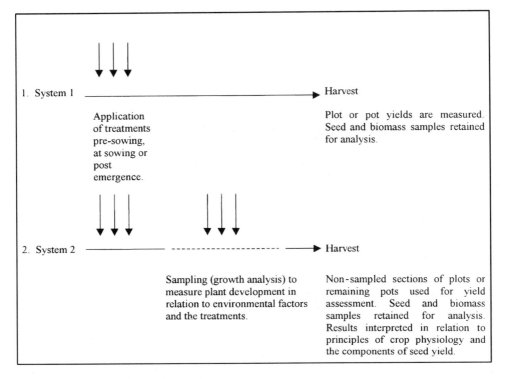

Figure 1.1. Systems of experimentation

placement at drilling, or the post-emergence application of a plant growth regulator at defined growth stages. The first descriptive system shown in Figure 1.1 is widely used by agronomists at arable research centres and farm demonstration sites. However, if the chosen treatments result in a significant increase in yield, the agronomist may be unable to fully explain the field results. It is impossible to provide an in-depth discussion of morphological and physiological factors affected by the treatments unless some additional measurements such as light interception, leaf area index, or crop growth rate are taken during the growing season (System 2).

System 1 is also rather risky because the results of a season's technical work (site mapping, cultivations, sowing, pot and plot maintenance), are solely dependent on data collected at final harvest. It is frustrating when treatment effects which were clearly visible during mid-season are obscured by lodging and seed loss due to thunderstorm damage during the ripening period. Similar end-of-season losses can also occur in glasshouse trials using System 1. In a hydroponic study on the response of wheat to varying concentrations of sulphur, the ears were destroyed by field mice which invaded the glasshouse cubicle just two days before the planned final harvest date. A trial which needed a great deal of maintenance time in relation to monitoring and topping up the different nutrient solutions only provided advice on rodent control! If measurements such as number of tillers per plant and sulphur concentrations in leaves and stems had been taken during vegetative development (System 2), some useful discussions of the effects of sulphur depletion on wheat development may have resulted.

The second system provides useful background information which may be used to interpret and discuss the final results. For example, when regular descriptive samples are taken the effects of treatments on plant morphology and the reproductive components of yield can be determined. If more sophisticated facilities are available, it may also be helpful to carry out physiological measurements such as chlorophyll concentration, light interception and photosynthesis. Subsequently, yield data can be analysed and interpreted in much more detail; additional confidence may then result in the reliability of a new product or variety. If final yield data are lost from System 2, the researcher may still have sufficient information in his or her diary to create a scientific report on vegetative development, flowering, and the early stages of reproductive development. Data collected during the experimental growing season may also be used to look for possible relationships between measurements such as soil temperature, rainfall, emergence and plant development using simple correlation techniques (Chapter 7).

As a disadvantage, the introduction of crop sampling increases experimental costs. Growth analysis is especially tedious and time consuming, and it is essential to ponder original experimental objectives before investing time in measuring components such as leaf area and numbers of seeds in pods or spikelets. The applied plant scientist may enjoy philately as a hobby, but must avoid the temptation to file vast quantities of descriptive data unless the scientific reasons are clearly defined. Computer packages will analyse data and produce means and standard errors. However, they are unable to guide the researcher on interpretative skills and sampling procedures or indicate whether the time invested in measuring a particular parameter is really worthwhile.

1.3 CHOICE OF SITE

The choice of field and the siting of a trial are probably the most important decisions made by the researcher. Incorrect choice of site may mean that the experimental results are difficult to interpret or are even meaningless. There are several common-sense starting points. It is essential to choose a uniform site as trials are designed to detect differences between small plot areas. Thus any variations in soil texture or pH within a site may partly or completely mask treatment effects. Local knowledge is clearly important, and it may be wise to consult historical records of drainage systems and former field boundaries. Previous site management must be researched because many legume and oilseed crops should be sown only once in a 5-year rotation — the principles of crop rotation must be adhered to when the experimental objectives are being defined.

An accessible site makes for ease of sampling, although it is worth remembering that trials situated near parking areas and footpaths may be subjected to damage by trespassers. It is best to avoid compacted headlands and wooded areas as downdraughts from trees can cause severe lodging. In order to minimise edge effects it is usually advisable to surround the experimental site with the crop species being studied. Frequently, plots will be positioned so that they can be sprayed with protective agrochemicals using the tramlines of a surrounding commercial crop.

When undertaking an agronomic research programme, it is valuable to have growth room and glasshouse facilities in addition to field sites. This enables detailed physiological work to be continued during the winter months. In theory, research glasshouses should provide a reasonably uniform environment in which light intensity, temperature and daylength can be controlled by computer technology. In practice, the glasshouse can be a more variable environment to work in than the field. There are frequently wide and sudden fluctuations in temperature between overcast and sunny periods even in the winter months, and variation in light profiles especially when neighbouring research cubicles are using a different daylength regime. In order to minimise these problems, it is important to re-randomise the position of the growing containers on the greenhouse benches from time to time, and surround the experimental unit with guard pots to reduce edge effects. It is essential to monitor pests and diseases—mildew, botrytis, aphids and white fly are common problems in the glasshouse, and while biological systems for insect control can be used, it is wise to have insurance supplies of agrochemicals in store, especially for the control of fungal diseases.

1.4 SOIL TESTING

Large experimental sites are more likely to include variations in soil texture, thus for most arable crop investigations the total experimental area rarely exceeds 1 ha. Larger plots are usually necessary in grassland or grazing studies in order to accommodate fencing, gangways, weighing pens and sufficient replication of the livestock assigned to individual grazing treatments.

Regardless of size it is essential to provide soil analyses before the treatments are applied. The purpose is to assess the adequacy, surplus or deficiency of available nutrients for crop growth. A standard soil analysis package measures soil pH and estimates available concentrations of phosphorus, potassium and magnesium. Some minor elements such as boron and copper can also be measured using soil samples, while others, for example manganese, are usually assessed from plant samples. For nutrient and pH assessment, soil is usually removed from the top 10–20 cm. Although it may be valuable to study nutrient levels in deeper profiles, it is often backbreaking work if traditional soil augers are used. Mechanical soil sampling equipment is available, but this is more expensive than hand-operated augers and rarely available at Experimental Stations in developing countries.

The number of soil samples must be sufficiently large to be representative of the experimental site. Within each site samples should be taken across a W pattern; 6–10 separate samples being taken along each arm of the W. Samples are usually bulked into one bag. The bulked sample comprises around 1 kg of soil which is taken to represent an entire site containing approximately 2000 t soil/ha to a ploughing depth of 20 cm. Clearly, using a small bulked sample is not a precise measure of nutrient or acidity levels and the problem of accuracy is increased by the use of very small soil subsamples for nutrient analysis. Although the researcher may have invested a great deal of time and physical effort in order to obtain a representative soil sample, the amount of soil used by the analytical chemist is often tiny. For example, when determining ammonium nitrate and nitrite nitrogen levels from fresh soil, a subsample of 20 g soil is used. For potassium, magnesium, sodium and manganese only 5 g of dried soil is required when using ammonium acetate extraction techniques. As a result it is important to assess the number of subsamples in relation to the variation between replicates analyses. If the level of variation between three or four replicates is high additional subsamples must be analysed, although for some analyses such as sulphate-sulphur this decision will be expensive. Unfortunately, all soil laboratory procedures are time consuming and costly, but this must not deter the researcher from clearly defining the pH, soil texture and nutrient status of the experimental site prior to drilling.

1.5 SATELLITE MAPPING

The careful control of inputs or precision farming is not new, and many farmers have selectively applied some agrochemicals for many years using their local knowledge of individual fields. For example, patch spraying with graminicides or the application of lime to sections of fields or headlands can often be cost effective.

New technology has been developed which may supersede conventional soil-sampling procedures for assessing the causes of yield variations that occur within most large arable fields. It is based on satellite-controlled navigation systems or GPS (Global Positioning Satellites). GP yield monitoring systems can be fitted to combine harvesters, with up to 500 grain weight checks/ha during harvest. The computer system can create a colour-coded yield map of each field. This shows yield variations across large areas, which would have been difficult to detect from auger samples. It has resulted in the development of management systems for the precise application of nutrients, and as a result manufacturers have now launched computer-controlled fertiliser spreaders.

The success of satellite precision farming systems will ultimately be controlled by their cost effectiveness, and the reliability of microchips linked to machinery which is designed to work under field conditions. For experimentation, it must be remembered that the most common reason for yield variations within a field is soil type. Boundaries between different soil textures were formerly defined by hedgerows, thus it may be easier to consult historical farm maps before using satellite technology. At present this exciting technology is of little value for choosing the position of an experimental site within a field. Background variation is still best defined by conventional soil analysis and historical research, and ultimately controlled by the correct choice of experimental design.

1.6 SAMPLING

A crop sample is a small portion of a population taken for detailed study. It may be a length of row, quadrat area, a number of plants or pots taken at random. Hopefully, it will be representative (large) enough to inform the researcher what he or she needs to know about the whole population of a particular treatment.

Some crop-sampling procedures provide excellent exercises for improving physical fitness! Hand lifting, bagging and labelling samples of potatoes and sugar beet requires stamina on hot summer days. Locating and then removing quadrat areas of oilseed rape from dense, tangled and lodged crops is a challenge of patience and technical skill for the research agronomist. It is important to prepare bags and labels before going into the field, and to check that each plot has been allocated a sample bag (using the field plan) before starting work. Luggage labels made from cardboard are ideal — if felt tip pens are used check that the ink is waterproof. If cold storage facilities are not available, it is best to process samples block by block — assuming a block design has been used (Chapter 10). This minimises problems associated with wilting and loss of dry matter.

As statistical packages are now readily available, it is tempting to carry out data analyses before examining the form of distribution that is associated with the sampled data. This examination is highly recommended. Crop experimentation encompasses both discrete and continuous distributions. Variables such as number of branches, number of pods per plant and number of seeds per pod have discrete distributions, while those such as plant height and dry weight have continuous distributions (Chapters 2 and 3).

Sampling schemes should only be agreed after a careful assessment has been made of plant establishment. They must avoid bias, but at the same time take into account variability in establishment possibly caused by poor drill calibration. The number of plants in individual rows of adjacent plots of winter wheat is shown in Table 1.1.

There are large differences between rows within each plot with rows 5 and 3 having the lowest plant populations. The position of these two rows varied according to the direction of drilling, the problem having mainly arisen due to shallower coulter depth. In this experiment the background variation in establishment was high, and as a result a random sampling system based on a length of one row would have been inappropriate. Instead, a quadrat area encompassing eight rows was randomly removed on each sampling occasion following establishment. It is always important to determine the reliability of the sampling system by examining the level of variation in plant establishment between replicate samples taken from the same treatment.

The morphology of individual crop plants varies widely in an unevenly established crop. Between-plant variation (plot background variation) is especially problematic to sampling procedures in poorly established precision-sown or transplanted crops such as maize and tobacco. Sampling difficulties encountered in many field experiments are clearly illustrated by data obtained from two experimental plots of winter oilseed rape. Quadrat samples (0.33 m^2) were randomly taken when the lowermost terminal raceme buds were yellowing. The individual dry weights of all plants were recorded and the background variation is summarised in Table 1.2.

Table 1.1. The number of winter wheat plants in six plots and individual rows (replicate 1) of 0.5 m prior to the application of husbandry treatments

	Plot Number					
Row	↓ 1	↑ 2	↓ 3	↑ 4	↓ 5	↑ 6
1	45	73	46	73	55	72
2	35	47	31	54	40	41
3	37	**23**	27	**24**	28	**19**
4	28	18	32	29	29	25
5	**14**	52	**10**	57	**7**	45
6	28	70	25	71	23	56
7	37	71	38	79	41	79
8	39	76	54	90	56	84

Notes: Each plot measured 12 × 1.8 m and contained 10 rows—the inner 8 rows were used for sampling. Plants in 0.5 m row lengths were counted.
↑ ↓ = direction of drilling.

Table 1.2. Ranges of plant dry weights (g)

	Quadrat plant number	Mean plant weight (g)	Range of plant weight
Treatment 1 (+128 kg N/ha)	22	7.81	0.36–16.65
	45	5.95	0.30–16.00
	63	4.36	0.31–13.12
	39	5.21	0.39–20.09
	31	6.73	0.57–17.80
	32	8.35	0.76–27.51
	36	5.11	0.49–14.87
	37	6.34	0.24–19.53
Means	38.1	6.23	
Treatment 2 (+236 kg N/ha)	50	7.79	0.60–26.47
	33	7.34	0.48–19.71
	39	6.34	0.83–34.41
	28	9.13	0.73–31.85
	48	9.26	0.55–26.29
	26	6.56	1.35–29.84
	26	8.58	0.40–17.46
	28	12.22	0.74–36.30
Means	34.8	8.40	

For both nitrogen fertiliser treatments the numbers of plants varied widely, and each quadrat contained a number of very small and large specimens. Possible treatment differences may not have been detected if only small plants had been subsampled from each quadrat for growth analysis. It is essential to use a subsampling system which eliminates bias because in practice there is always a tendency to avoid large plants as their subsequent analysis in the crop laboratory is time consuming.

The wide range of plant-to-plant variation shown in Table 1.2 cannot be quantified accurately if only small numbers of plants are studied. Yet subsample sizes in many research papers on rapeseed agronomy and physiology have only consisted of 3–20 plants per treatment. A detailed statistical study of data given in Table 1.2 indicated that for the sample mean plant weight to be within 1 g of the true weight (with 95% confidence), a random sample of approximately 600 plants would be required! As this is time consuming it again raises the question of the value of overcollecting descriptive data unless the accuracy of a sampling system is known. In System 2 it may be better to minimise the time invested in crop description (traditional growth analysis), in order to study in more detail environmental and physiological parameters affecting canopy development and crop yield. The latter approach (crop modelling), has been made more accessible with the availability of field recording equipment which is directly linked to computers.

Although statisticians will always recommend random sampling, this may not be practical in tall, high-density plots of crops such as oilseed rape without causing damage to surrounding areas. In this situation it may be wiser to adopt a 'step-ladder' sampling system in which a quadrat enclosing inner plot rows is first removed from a uniform area at either the top or bottom of each experimental plot. After leaving a discard distance a second quadrat can be removed on a later date using the first sampled area as a working base. This process can then be repeated. An unsampled, undisturbed area of plot must be left for commercial yield assessment. However, the importance of selecting the sampling units in such a way that they shall be as representative as possible of the entire population cannot be overemphasised.

Chapter 2

Basic Statistical Calculations

2.1 INTRODUCTION

When all statistical analyses were carried out using hand-operated calculators, the amount of experimental data collected and analysed was partly controlled by the sluggishness of early desk machinery. Because each analysis may have taken many minutes or even hours to complete, raw data were pondered in detail before calculations were attempted.

Package computer programs have revolutionised attitudes to data-collection and data-handling systems in applied biological research. Sadly, they also seem to have diluted many students' understanding of basic statistics. It is now far too easy to rely uncritically on computer output and to carry out sophisticated analyses which may be inappropriate and lead to misleading conclusions.

Computer technology has greatly improved presentational but not interpretative skills. For example, during oral examinations many students are unable to explain basic statistical terms such as standard error and variance, even when exquisitely presented tables and figures created by computer technology include summaries of statistical tests. Their attitude now seems to be 'don't think, use the computer', and if the output looks good then include it in the dissertation to impress the examiner!

The main objective of this chapter is to provide a definition and clear understanding of some basic statistical terms which are commonly used when analysing data collected from field and glasshouse experiments. For simplicity, only a small number of observations is included in the analyses so that the reader can check the results using a hand calculator.

2.2 MEASUREMENTS AND TYPE OF VARIABLE

The unit on which measurements are made may be a whole plot, a small area of a plot, a single plant, a stem, a leaf, etc. Suppose the experimental unit is an individual plant. For each, measurements may be made on several variables, such as height, weight, leaf area or number of internodes. Variables may be discrete, continuous or categorical.

A **continuous** variable is one that can take any value in a certain range. For instance, plant height is a continuous variable. If one plant has a height of 20 cm and another a height of 21 cm, it is possible to find a third plant with a height of between 20 and 21 cm. For continuous variables, measurements are approximate because they have to be rounded off to a whole number or to a fixed number of decimal places.

A **discrete** variable is one which can only take certain values. An example is the number of seeds in a pod. This number must be an integer such as 0, 1, 2, 3, etc. We cannot have a pod with 2.1 seeds.

A **categorical** variable is formed when data are classified into categories. For example, each plant measured could be classified according to variety. In this case variety is a category, sometimes called a **classification variable**. If the varieties are given names, there is no natural order. If there were three varieties, we could assign the numbers 1, 2 and 3 to them, but it would be meaningless to do any calculations on these numbers. However, it may be meaningful to count the number of plants of each variety. These data may be summarised in a table or a bar chart.

2.3 SAMPLES AND POPULATIONS

One of the main objectives of statistical analysis is to find out as much as possible about a population. Most populations are far too large to be measured. For example, suppose the population under study is a field of wheat. You may want to know the average yield per plant. As resources are not available to measure every wheat plant, a random sample can be taken. The average yield of these plants is an estimate of the mean yield of all plants in the field. The estimate calculated could be 'a long way' from the true value. A statement is needed of how close the sample mean is likely to be to the population mean. For example, it would be helpful to state with 95% confidence that the mean yield per plant lies between 25 and 30 g. The calculation of a **confidence interval** requires some background theory, and in the following discussion a population of field plants is assumed.

N = population size which may be the total number of wheat plants in the field. It is likely to be very large and unknown. For example, a farm crop of wheat in the UK will consist of approximately 250 plants/m^2, or 2.5 million plants per hectare.

μ = the **population mean**. This is rarely known. It may be the mean yield per plant of all the plants in the field. If all the plants were measured, μ would be calculated by adding up all the yields and dividing by N. The formula for μ is

$$\mu = \frac{\Sigma x}{N}$$

x is the symbol used for yield, and Σ is the summation sign. It means add up all the xs (the yields).

As it would be impractical to assess the yield of 2.5 million individual plants, an estimate of μ can be found by taking a sample from the population. To be unbiased

BASIC STATISTICAL CALCULATIONS

and representative of the population, the plants to be chosen for inclusion in the sample must be selected at random. In this way all individuals in the population have an equal chance of being included in the sample.

n = sample size (this may be the number of plants included in the sample)
x_1 = yield of first plant in the sample
x_2 = yield of second plant in the sample
x_n = yield of nth plant in the sample.

Using the sigma notation, $\Sigma x = x_1 + x_2 + \ldots x_n$
$\bar{x}$ is the symbol for **sample mean** and its formula is

$$\bar{x} = \frac{\Sigma x}{n}$$

Example 2.1

The heights of a random sample of 5 plants are 14.8, 15.2, 17.4, 11.6 and 12.5 cm. The mean height is

$$\bar{x} = \frac{14.8 + 15.2 + 17.4 + 11.6 + 12.5}{5} = 14.30 \text{ cm}$$

The mean is a measure of location or central tendency. Another measure of location is the median.

2.3.1 Median

If there are n numbers, the median is the $(n+1)/2$ ranked number. If n is odd, this is the middle number after sorting them in order of magnitude, and if n is even it is the average of the middle two.

The data of the last example after sorting are: 11.6, 12.5, 14.8, 15.2 and 17.4. The median is therefore 14.8. If 13.1 is added, the median is the average of 13.1 and 14.8, namely 13.95.

The median is preferred to the mean when the distribution is very *skew* (non-symmetrical). For instance, if 17.4 is replaced by 37.4 in the original data set, the median is still 14.8, but the mean is 18.3.

The distribution is *positively skewed* when there is a small proportion of unusually high values which normally results in the mean being larger than the median. The distribution is *negatively skewed* when there is a small proportion of unusually low values which normally results in the mean being smaller than the median.

2.3.2 Population Variance

The population variance is denoted by σ^2 and it is the average of the squared deviations from the population mean. It is a measure of the variation in the values and the formula is

$$\sigma^2 = \frac{\Sigma(x - \mu)^2}{N}$$

It cannot be calculated because it is impossible to measure all the x values. It is estimated by calculating the sample variance.

2.3.3 Sample Variance

An unbiased estimator of the population variance is the sample variance, denoted by s^2. Its formula is

$$s^2 = \frac{\Sigma(x - \bar{x})^2}{n - 1}$$

where $\bar{x}$ is the sample mean and n is the number of sample observations.

To calculate s^2, we find the sum of the squares of the deviations from the sample mean and divide by the **degrees of freedom** ($n - 1$). Division by n would give a biased estimate of σ^2. The sample variance is used in hypothesis testing and in calculating confidence intervals (Chapters 4 and 5).

Example 2.2 Sample variance for plant heights in Example 2.1
Table 2.1 shows the original heights, their deviations from the sample mean and the squares of these deviations. Recall that the mean is 14.30 cm.

$$\bar{x} = \frac{\Sigma x}{n} = \frac{71.5}{5} = 14.3 \quad \text{and} \quad s^2 = \frac{\Sigma(x - \bar{x})^2}{n - 1} = \frac{21.20}{4} = 5.30$$

Notice that $\Sigma(x - \bar{x}) = 0$. It is always true that the sum of the deviations from the sample mean add to zero.

2.3.4 Degrees of Freedom

Only $n - 1$ of the deviations are free to vary. Once the sample is taken, the sample mean is fixed. If $n - 1$ of the deviations from the sample mean are calculated, the nth deviation is fixed, as all n deviations must add to zero. As a result, there are $n - 1$ degrees of freedom (df) associated with this estimator of population variance. If the population mean was known and substituted for $\bar{x}$ in the formula for s^2, there would be n df because in this situation where $n - 1$ of the deviations from μ are known, the nth deviation from μ cannot be predicted. The sum of the deviations of the n sample values from the sample mean is zero, but the sum of the deviations of the n sample values from the population mean is not zero.

Exercise 2.1
Use the above method to find the sample variance of the numbers 13.1, 16.4, 19.5, 22.0, 25.5, 18.7. The answer is 18.58.

Table 2.1. Corrected sums of squares for Example 2.2

	x	$(x - \bar{x})$	$(x - \bar{x})^2$
	14.8	0.5	0.25
	15.2	0.9	0.81
	17.4	3.1	9.61
	11.6	−2.7	7.29
	12.5	−1.8	3.24
Total	71.5	0	21.20

2.3.5 Corrected Sum of Squares

A measure of variation which is frequently used in later chapters is the corrected sum of squares. This is the sum of the squares of the deviations from the sample mean: it is denoted by Sxx and its formula is

$$Sxx = \Sigma(x - \bar{x})^2$$

The sample variance can thus be written as $s^2 = Sxx/(n-1)$ and Sxx can be written as $Sxx = (n-1)s^2$.

For Example 2.2, $\qquad Sxx = 21.20$

2.3.6 The Computational Formula for Sxx

When calculating the deviations from the sample mean by hand, rounding-off errors will occur if $\bar{x}$ is not recorded to a sufficient number of decimal places. If a large number of decimal places are used, the calculations become tedious. This problem can be avoided by using the following alternative formula:

$$Sxx = \Sigma x^2 - \frac{(\Sigma x)^2}{n}$$

Using this version, the corrected sum of squares is the uncorrected sum of squares minus (the square of the sum divided by n). $(\Sigma x)^2/n$ is called the correction factor and denoted by CF.

Example 2.3

Now recalculate s^2 for Example 2.2 using the correction factor method. Table 2.2 shows the details.

$$Sxx = \Sigma x^2 - \frac{(\Sigma x)^2}{n} = 1043.65 - \frac{(71.5)^2}{5}$$

$$= 1043.65 - 1022.45$$

$$= 21.20 \text{ as found earlier}$$

Table 2.2. Sums of squares for Example 2.2

	x	x^2
	14.8	219.04
	15.2	231.04
	17.4	302.76
	11.6	134.56
	12.5	156.25
Total	71.5	1043.65

2.3.7 Standard Deviation

The standard deviation is the square root of the variance and is measured in the original units. If the x values are measured in cm, the variance is in cm². As this is a difficult term to work with, the problem is removed by taking the square root. Thus, the standard deviation is in cm.

Population standard deviation $= \sigma$
Sample standard deviation $= s = \sqrt{s^2}$

For Example 2.2, $s = \sqrt{5.30} = 2.302$ cm

For most distributions which are fairly symmetrical, about 95% of the population lies within two standard deviations of the mean.

2.3.8 The Coefficient of Variation (CV)

The CV is the standard deviation expressed as a percentage of the mean. It is independent of the units of measurement.

For the population $\qquad CV = \dfrac{\sigma}{\mu} \times 100\%$

For a sample $\qquad CV = \dfrac{s}{\bar{x}} \times 100\%$

For Example 2.2, $\qquad CV = \dfrac{2.302}{14.3} \times 100\% = 16.10\%$

The concept of coefficient of variation can be better understood by considering the following two data sets:

I: 2.1   3.5   4.7   5.2   6.4
II: 102.1   103.5   104.7   105.2   106.4

They both have the same variation. You should verify that their sample standard deviations are both 1.645. However, the variation within set I is very large in relation

BASIC STATISTICAL CALCULATIONS   15

to its mean of 4.38. This is expressed by a coefficient of variation of 37.56%. Set II is not very variable in relation to its mean of 104.38. The CV is only 1.58%.

The calculations described so far are very important. Practise them until confident that you understand the concepts of mean, variance and standard deviation. In research, you may be dealing with large amounts of data and will use a computer to do the calculations. Practise using a hand calculator with small data sets, to help you understand and interpret computer output.

2.3.9 The Hand calculator

There are many makes and models to choose from. You should obtain a calculator which gives means and standard deviations. Look for a model with SD mode or STATS mode. The buttons labelled σ_{n-1} or s will give the sample standard deviation.

2.3.10 The Weighted Mean

Suppose 3.5, 4.8 and 5.2 are three sample means. What is the overall mean? Is it $(3.5 + 4.8 + 5.2)/3 = 4.5$? The answer is only 4.5 if each mean is based on the same number of values. If the means are based on 4, 5 and 10 values respectively, the overall mean is

$$\bar{x} = \frac{n_1\bar{x}_1 + n_2\bar{x}_2 + n_3\bar{x}_3}{n_1 + n_2 + n_3} = \frac{4 \times 3.5 + 5 \times 4.8 + 10 \times 5.2}{4 + 5 + 10} = \frac{14 + 24 + 52}{19} = \frac{90}{19} = 4.74$$

The values of n_1, n_2 and n_3 are the weights.

2.3.11 The Harmonic Mean

In Chapters 9 and 10 you will learn how to compare several treatment means using a Least Significant Difference test. The calculated LSD value assumes that each mean is based on the same number of values called the number of replications per treatment. A solution used by SAS when the number of replications is unequal is to use the harmonic mean of the number of replications and not the ordinary or arithmetic mean.

The harmonic mean is the reciprocal of the mean of the reciprocals. For example, if the numbers of replications are 3, 4 and 5, the arithmetic mean is 4. The mean of the reciprocals is $(1/3 + 1/4 + 1/5)/3 = (47/60)/3 = 47/180$. Hence the harmonic mean is $180/47 = 3.83$.

Chapter 3

Basic Data Summary

3.1 INTRODUCTION

In Chapter 2 we discussed how to summarise a few sample numbers in the form of basic statistics such as mean and standard deviation. In practice, we often have a large amount of data to summarise. Before submitting data to sophisticated statistical analyses it is advisable to obtain a numerical and graphical summary to check for 'outliers' and to display the distribution.

3.2 FREQUENCY DISTRIBUTIONS (DISCRETE DATA)

In a set of discrete data, many of the values are repeated. The distribution can be summarised in a frequency table and illustrated in a line diagram or bar chart. The calculation of the mean is made easier by multiplying each distinct x-value by its frequency (the number of times it appears) and the formula for the sample mean becomes:

$$\bar{x} = \frac{\Sigma xf}{n} \quad \text{where} \quad n = \Sigma f$$

Example 3.1

The number of tillers were counted on each plant in a random sample of 60 barley plants. Data are summarised in a frequency table (Table 3.1) and illustrated in a line diagram (Figure 3.1).

$x =$ number of tillers per plant
$f =$ number (frequency) of plants with x tillers

Table 3.1 shows that 3 plants each have 1 tiller, 4 plants each have 2 tillers, 8 plants each have 3 tillers, etc. The total number of tillers altogether on the 60 plants is

$$\Sigma xf = 1 \times 3 + 2 \times 4 + 3 \times 8 + 4 \times 16 + 5 \times 13 + 6 \times 9 + 7 \times 5 + 8 \times 2$$
$$= 3 + 8 + 24 + 64 + 65 + 54 + 35 + 16$$
$$= 269$$

BASIC DATA SUMMARY

This is the sum of all the 60 x-values; the total number of tillers. Three of the x-values are equal to 1, four are equal to 2, etc. Hence the sample mean number of tillers per plant is

$$\bar{x} = \frac{\Sigma x f}{\Sigma f} = \frac{269}{60} = 4.48$$

The sample variance and standard deviation can be found by using the formula

$$Sxx = \text{sum of squares} - \frac{\text{square of sum}}{n}$$

The calculations are summarised in Table 3.2.

The sum of the squares of all the 60 numbers is $\Sigma x^2 f = 1369$. This is because the sum of the squares of the three ones is $1^2 \times 3 = 3$, and the sum of the squares of the four twos is $2^2 \times 4 = 16$, etc.

The sum of all the 60 numbers is $\Sigma x f = 269$, and the square of the sum divided by n is $269^2/60 = 1206.02$ where $n = \Sigma f = 60$. Hence

$$Sxx = \Sigma x^2 f - \frac{(\Sigma x f)^2}{n} = 1369 - 1206.02 = 162.98$$

$$s^2 = \frac{Sxx}{n-1} = \frac{162.98}{59} = 2.762$$

The sample standard deviation, s, is the square root of $s^2 = \sqrt{2.762} = 1.662$ tillers per plant.

Table 3.1. Frequency table for Example 3.1

x	1	2	3	4	5	6	7	8	Total
f	3	4	8	16	13	9	5	2	60

Table 3.2. Calculations for Example 3.1

x	1	2	3	4	5	6	7	8	Totals
f	3	4	8	16	13	9	5	2	60
xf	3	8	24	64	65	54	35	16	269
$x^2 f$	3	16	72	256	325	324	245	128	1369

Note: Row 4 is row 1 times row 3.

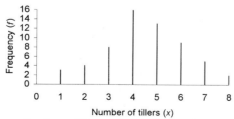

Figure 3.1. Line diagram for data of Example 3.1. *Note*: a line diagram is often presented as a bar chart for greater visual impact

3.2.1 The Mode

The mode is the most frequent number. In this example it is 4.

3.2.2 The Median

In this example, there are 60 numbers. The rank of the median is $(n+1)/2 = 31.5$, hence it is the average of the 30th and 31st ranked numbers. The first 3 numbers are equal to 1. The first 7 numbers are equal to 2 or less. The first 15 numbers are equal to 3 or less. The next 16 numbers are equal to 4. Therefore the 30th and 31st numbers are both equal to 4. Hence, the median $= 4$.

3.2.3 Use of Spreadsheets

You can easily carry out the calculations involved in finding the mean and standard deviation of a frequency distribution by entering the data for x and f in the first two columns of a spreadsheet. You can then instruct the computer to calculate the entries for xf and x^2f in the next two columns. The sums of columns 2, 3 and 4 can be found and entered in the formulae used to find $\bar{x}$ and s^2.

3.3 FREQUENCY DISTRIBUTIONS (CONTINUOUS DATA)

Consider variables such as height and weight which are measured on a continuous scale. No two plants will have exactly the same weight if measurements are made with sufficient accuracy. Important features can be seen if the data are grouped into classes. If weights are recorded to the nearest gram, this is a form of grouping. A plant whose weight is recorded as 80 g has a real weight between 79.5 and 80.5 g, provided the scales are not biased.

Example 3.2

The following data are the yields (g) of 80 small equal-sized plots of barley:

95	70	68	88	79	92	64	83	67	63
56	70	53	78	71	62	42	80	50	68
78	104	62	66	90	86	66	82	83	56
82	90	71	77	93	68	91	98	79	75
92	93	73	79	95	78	77	108	86	87
68	68	49	75	82	61	68	65	56	96
52	61	87	79	64	64	84	63	64	44
87	63	65	64	81	72	62	58	84	67

BASIC DATA SUMMARY   19

The sample mean and standard deviation can be found by the method described for Examples 2.2 or 2.3. You should verify that $\bar{x} = 73.96$ and s is 14.024 by entering these 80 numbers into your calculator in SD or Stats mode.

Notice that several plots have yields of 64 g. This is because a form of grouping has already been applied by recording the weights to the nearest gram. These plots have yields of between 63.5 and 64.5 g.

The distribution of weights can be seen more easily by grouping them into classes. The first step is to find the range of weights. The smallest weight is 42 g and the largest is 108 g. The **range** is $108 - 42 = 66$ g. A sensible grouping is to put all weights which are 40 and under 50 to the nearest gram in the first group. This group includes all plots whose actual weights lie between 39.5 and 49.5 g. All weights which are 50 and under 60 to the nearest gram will be put into the second group, and so on. With this grouping, the class boundaries are 39.5, 49.5, 59.5, etc. and the mid-points of the classes are 44.5, 54.5, 64.5, etc. The class interval is constant and equal to 10. The grouped frequency table is shown in Table 3.3. This table shows that 25 plots have yields between 60 and 69 g inclusive, when measured to the nearest gram. The actual weights lie between 59.5 and 69.5, so the class interval is 10 g.

If presented with a table such as this, you can obtain approximate answers for the sample mean and standard deviation by using the method of Example 3.1, where x is the class mid-point and f is the number of values in the class. Use this method to confirm that $\bar{x} = 73.875$ and $s = 14.083$. These values are quite close to 73.96 and 14.024 found using the individual values. The discrepancy is due to assuming that the mean of the values within each class is the class mid-point.

The grouping into classes is arbitrary. We could have used smaller or larger class intervals, or made the lower boundary of the first class different from 39.5. Also, we could have had classes of unequal width.

3.3.1 The Histogram

The tally column in Table 3.3 gives an idea of the distribution of weights. A clearer representation of the distribution is obtained by drawing a histogram and Figure 3.2 shows a histogram for the data of Example 3.2.

Table 3.3. Grouped frequency table for data of Example 3.2

Class interval (g)	Mid-point, x (g)	Tally	Frequency (f)
39.5–49.5	44.5	111	3
49.5–59.5	54.5	11111 11	7
59.5–69.5	64.5	11111 11111 11111 11111 11111	25
69.5–79.5	74.5	11111 11111 11111 11	17
79.5–89.5	84.5	11111 11111 11111	15
89.5–99.5	94.5	11111 11111 1	11
99.5–109.5	104.5	11	2

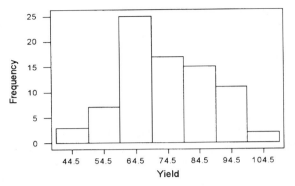

Figure 3.2. Histogram of data of Example 3.2 produced by Minitab

The area of each rectangle is proportional to the frequency. If the classes are of equal width, the heights represent the frequencies. If the top two classes had been combined, the height of the last rectangle would have been $(11 + 2)/2 = 6.5$. The histogram shows that the distribution of weights is not symmetrical. It is slightly positively skewed (skewed to the right); the mean (73.96) is greater than the median (72.5).

Once you have entered your data into a computer, you can easily obtain a histogram displayed on the screen. **Outliers** (unusual data values which may be due to recording errors) are easily identified and should be investigated. A character histogram (Figure 3.3) is often better at showing skewness and outliers.

```
         Histogram of yield    N = 80

         Midpoint         Count
            40              1    *
            45              1    *
            50              3    ***
            55              4    ****
            60              6    ******
            65             14    **************
            70             11    ***********
            75              5    *****
            80             12    ************
            85              9    *********
            90              6    ******
            95              5    *****
           100              1    *
           105              1    *
           110              1    *
```

Figure 3.3. Character histogram of data for Example 3.2

Do not confuse a bar chart with a histogram. The bars of a bar chart represent distinct categories, such as different varieties. The bars do not have to be contiguous. The bars of a histogram represent the values of a continuous variable divided into class intervals and are contiguous (no gaps between them).

3.3.2 Quartiles and Ranges

The **lower quartile** (Q1) is such that 25% of the observations are less than Q1.

The **upper quartile** (Q3) is such that 75% of the observations are less than Q3 and 25% are above.

The second quartile is the median (Q2 = M). The median is such that 50% of the observations are less than M.

The **interquartile range** is the upper quartile minus the lower quartile. IQR = Q3 − Q1.

The range

This is the largest value minus the smallest value and it is very sensitive to outliers.

3.3.3 Other Graphical Methods

A **boxplot** is a graphical device for displaying the quartiles and the range of a distribution. A boxplot for the data of Example 3.2 produced by Minitab is shown in Figure 3.4.

This boxplot shows that the yields vary from 44 to 108 g. The box includes the middle 50% of yields. The lower end of the box is the first quartile (Q1 = 64). The upper end of the box is the third quartile (Q3 = 84). The cross in the box indicates the median (M = 72.5).

Other visual methods of distribution summary are the **dot plot** and the **stem and leaf plot**. Figures 3.5 and 3.6 show Minitab versions of these for Example 3.2.

The stem-and-leaf plot is similar to the character histogram but gives more information. The first column shows cumulative frequency, the second column is the stem and the third column is the leaf which represents the final digit. For this example it shows that the two smallest yields are 42 and 44 g. The next highest yield is 49 g and the highest yield is 108 g. Twenty-three yields are less than or equal to 64 g and 35 yields are less than or equal to 68 g. Thirty-nine yields are greater than or equal to 75 g and 28 yields are greater than or equal to 80 g. Six yields are in the median class from 70 to 73. This

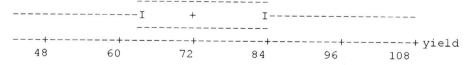

Figure 3.4. Boxplot for data of Example 3.2

Figure 3.5. Dotplot for data of Example 3.2 showing 80 yields (g)

diagram can also show whether any bias has taken place during measurement when recording the final digit. In this example an eight appears as the final digit 13 times whereas a zero only appears six times. Should we regard this as suspicious?

These graphical methods are useful for exploratory data analysis. They enable you to compare several distributions and check for symmetry or lack of it (skewness). They are particularly useful in highlighting any unusual values (outliers) which may or may not be due to recording errors.

3.4 DESCRIPTIVE STATISTICS

While graphical methods give visual summaries of a distribution, descriptive statistics give numerical summaries of central tendency and dispersion and can be used to compare distributions and carry out further analyses.

```
    2      4 24
    3      4 9
    6      5 023
   10      5 6668
   23      6 1122233344444
   35      6 556677888888
  (6)      7 001123
   39      7 55778889999
   28      8 012223344
   19      8 667778
   13      9 0012233
    6      9 5568
    2     10 4
    1     10 8
```

Figure 3.6. Stem-and-leaf plot for data of Example 3.2 showing 80 yields (g)

Example 3.3

If you enter the data of Example 3.2 as a single column in a Minitab worksheet and ask for a display of the descriptive statistics you should obtain the following output:

```
Variable            N        Mean    Median    TrMean    StDev    SE Mean
yield              80       73.96     72.50     73.97    14.02       1.57

Variable  Minimum    Maximum       Q1       Q3
yield       42.00     108.00    64.00    84.00
```

We have already explained most of these quantities. The standard error (SE Mean) is explained in Section 4.8. The trimmed mean (TrMean) is the mean after removing the smallest and largest 5% of values.

Chapter 4

The Normal Distribution, the *t*-Distribution and Confidence Intervals

4.1 INTRODUCTION TO THE NORMAL DISTRIBUTION

Suppose you are interested in the distribution of plant heights in a field; call the total collection of heights the **population** of interest. An idea of plant height distribution can be obtained by taking a large random sample of plants and drawing a histogram and the shape of the histogram will depend on the widths of the classes chosen and their class boundaries. If you were able to measure all plants in the field, the heights could be grouped into a very large number of classes. Each class width would be very small. As a result, the mid-points of the tops of the rectangles could be joined to form a smooth curve.

For many variables such as plant height and yield the smoothed-out histogram is approximately symmetrical about the mean, and about 95% of the population observations lie within 2 standard deviations of the mean. The common occurrence of such distributions has led to the importance of a theoretical curve used to describe them. It is called the **normal distribution** (Figure 4.1).

Data collected during sampling are never perfectly distributed in this way. However, many variables associated with crop growth analyses have distributions close enough to normality for important inferences to be made about them based on information obtained from random sampling and the mathematical properties of the normal distribution.

Some Properties of the Normal Distribution

(1) This curve is symmetrical about its mean and is steepest at one standard deviation from the mean.
(2) About 67% of the population is within one standard deviation of the mean.
(3) 90% of the population is within 1.645 standard deviations of the mean.

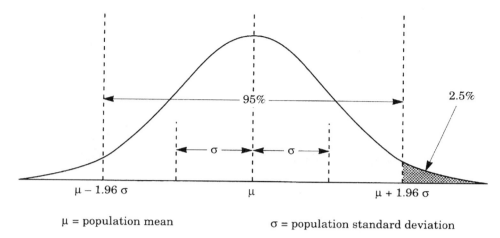

Figure 4.1. The normal distribution curve

(4) 95% of the population is within 1.96 standard deviations of the mean, and hence only 2.5% of the population have values which are greater than 1.96 standard deviations above the mean.
(5) 99% of the population is within 2.576 standard deviations of the mean.
(6) 99.9% of the population is within 3.29 standard deviations of the mean.

Example 4.1
From a large random sample of cereal plants the average height was calculated to be 80 cm and the sample standard deviation was found to be 5 cm. If the population of heights is normally distributed you can say (approximately) that 95% of the plant heights in the population lie between $80 - (1.96 \times 5)$ and $80 + (1.96 \times 5)$ cm, i.e. between 70.2 and 89.8 cm.

Verify that 99% of the heights are estimated to lie between 67.1 and 92.9 cm, and that only 0.05% of the heights are predicted to be greater than 96.45 cm.

4.2 THE STANDARD NORMAL DISTRIBUTION

You may wish to estimate the proportion of the cereal population of Example 4.1 having heights less than a specified value. Taking the area under the normal curve as 1 and the specified height as x, the required proportion is the area under the curve to the left of x. This area can be found using tables of the standard normal distribution which is a normal distribution with a mean of zero and a standard deviation of 1. Thus the x-value needs to be standardised before looking in the tables. This involves finding the number of standard deviations from the mean for our given x value. The

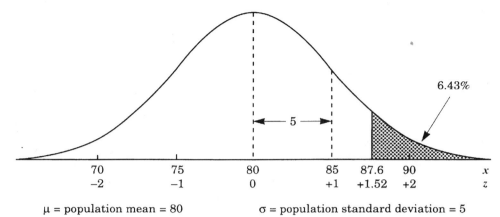

μ = population mean = 80 σ = population standard deviation = 5

Figure 4.2. Normal distribution with mean 80 and standard deviation 5

result is called the z-value of x. To find z first subtract μ from x and divide the result by σ, i.e.

$$z = \frac{x - \mu}{\sigma}$$

Figure 4.2 shows a normal distribution of cereal plant heights with a mean of 80 cm and a standard deviation of 5 cm. The x scale represents the heights in cm, and the z scale represents the standardised heights, i.e. the heights in units of standard deviations from the mean.

Referring to Example 4.1, suppose you wish to estimate the proportion of the population of plants with heights less than 87.6 cm. You first calculate the corresponding z-value using

$$z = \frac{x - \mu}{\sigma} = \frac{87.6 - 80}{5} = 1.52$$

This shows that 87.6 is 1.52 standard deviations above the mean. These values are shown in Figure 4.2.

You now consult the tables of the normal distribution function (Appendix 1). Corresponding to $z = 1.52$ the table value is 0.9357. The conventional notation for this is $\Phi(1.52) = 0.9357$.

Think of the symbol Φ as meaning 'the table value of' or 'the area to the left of'. The table shows that 93.57% of the heights are expected to be less than 87.6 cm. In Figure 4.2, the area under the curve to the left of the vertical line at $x = 87.6$ is 93.57% of the total area. You can also deduce that $100 - 93.57 = 6.43\%$ of plants are expected to have heights greater than 87.6 cm. This is represented by the area under the curve to the right of $x = 87.6$ ($z = 1.52$). In mathematical notation $1 - \Phi(1.52) = 1 - 0.9357 = 0.0643$.

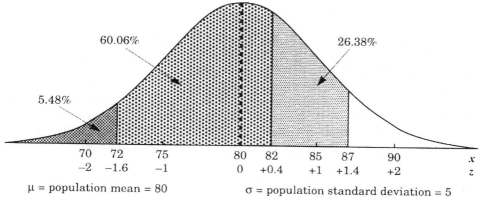

Figure 4.3. Diagram to explain Example 4.2

4.3 FURTHER USE OF THE NORMAL TABLES

Suppose you wish to estimate the proportion of the population with heights between two specified values. Call these values x_1 and x_2. The first step is to find the two corresponding z-values:

$$z_1 = \frac{x_1 - \mu}{\sigma} \quad z_2 = \frac{x_2 - \mu}{\sigma}$$

If these two z-values are both positive or both negative you look in the tables ignoring the signs and subtract the smaller table value from the larger. Hence, the proportion of the population with heights between x_1 and x_2 is the absolute magnitude of $\Phi(z_2) - \Phi(z_1)$.

If the two z-values are of opposite sign; which will happen if one x-value is above the mean and the other is below, then the proportion of the population with heights between x_1 and x_2 is found by adding the two table values and subtracting 1, i.e.

$$\Phi(z_1) + \Phi(z_2) - 1$$

The reasoning behind these rules will become apparent after studying the following example.

Example 4.2
A cereal population has heights which are normally distributed with a mean of 80 cm and a standard deviation of 5 cm. Find the proportion of plants having heights (1) less than 72 cm, (2) between 82 and 87 cm, and (3) between 72 and 82 cm. The answers to these questions are explained with help from Figure 4.3.

(1) For $x = 72$, $z = \dfrac{72 - 80}{5} = -1.60$

This indicates that 72 is 1.6 standard deviations below the mean.

The tables do not refer to negative z-values so we have to make use of the symmetry of the normal curve. The area to the left of -1.6 is the same as the area to the right of $+1.6$. The table value of $+1.6$ is $\Phi(1.60) = 0.9452$ of the total area under the curve, and this is the area to the left of $z = +1.6$. Hence the area to the right of $z = +1.60$ is $1 - 0.9452 = 0.0548$. This is also the area to the left of $z = -1.60$. In mathematical notation $\Phi(-1.60) = 1 - \Phi(+1.60) = 1 - 0.9452 = 0.0548$. **Thus 5.48% of the plants should have heights less than 72 cm.**

(2) For $x = 82$, $z = \dfrac{82 - 80}{5} = +0.40$ and for $x = 87$, $z = \dfrac{87 - 80}{5} = +1.40$

Thus 82 is 0.4 standard deviations above the mean, and 87 is 1.4 standard deviations above the mean.

The area under the curve between $z = +0.4$ and $+1.4$ is the same as the area to the left of $z = +1.4$ minus the area to the left of $z = +0.4$. Hence, the required area is the table value of 1.4 minus the table value of 0.4, namely $0.9192 - 0.6554 = 0.2638$. In mathematical notation the required area is $\Phi(1.40) - \Phi(0.40) = 0.9192 - 0.6554 = 0.2638$. **Thus 26.38% of the plants should have heights within the range 82 to 87 cm.**

(3) For $x = 72$, $z = \dfrac{72 - 80}{5} = -1.60$ and for $x = 82$, $z = \dfrac{82 - 80}{5} = +0.40$

Hence one z-value is positive and one is negative so we are unable to subtract two table values as in (2) above. Referring to Figure 4.3, the area between $z = -1.60$ and $z = +0.4$ is made up of two parts: first, the area between $z = -1.60$ and $z = 0$, and second, the area between $z = 0$ and $z = +0.4$. By symmetry the first area is the same as the area between $z = +1.6$ and $z = 0$ which is the table value of 1.6 minus the table value of 0. The table value of 0 is 0.5 because 50% of the area below the curve is to the left of the centre. Hence the first area is $0.9452 - 0.5$. Similarly, the second area is the table value of 0.4 minus 0.5 which is $0.6554 - 0.5$. On adding these two areas we obtain $0.9452 + 0.6554 - 1 = 0.6006$. The required area is thus obtained by adding the table values for the two z scores ignoring signs and then subtracting 1. In mathematical notation

$$\Phi(+0.40) - \Phi(-1.60) = [\Phi(1.60) - \Phi(0)] + [\Phi(0.40) - \Phi(0)]$$
$$= [\Phi(1.60) - 0.5] + [\Phi(0.40) - 0.5]$$
$$= \Phi(1.60) + \Phi(0.40) - 1$$
$$= 0.9452 + 0.6554 - 1 = 0.6006$$

Thus 60.06% of the plants should have heights within the range 72 to 82 cm.

Remember: If the two z-values have opposite signs add the two table values and subtract 1. If they are of the same sign just subtract the two table values.

4.4 USE OF THE PERCENTAGE POINTS TABLE (APPENDIX 2)

This table gives the z-values corresponding to specified areas in the top tail of the standard normal distribution. For example, if the area in the top tail is 2.5%, the z-value is 1.96. The interpretation of this statement is that in any normally distributed population, 2.5% of the values are greater than 1.96 standard deviations above the mean. By symmetry, if the area in the bottom tail is 2.5%, the corresponding z-value is -1.96, i.e. 2.5% of the values are less than 1.96 standard deviations below the mean. These two statements together imply that 95% of the area lies between $z = -1.96$ and $z = +1.96$, the interpretation being that 95% of the values in any normal distribution are within 1.96 standard deviations of the mean. The above conclusions are illustrated in Figure 4.4.

Using the percentage points table you should verify that 90% of values are within 1.645 standard deviations of the mean, and 99% of values are within 2.576 standard deviations of the mean. These statements and additional relationships are summarised in Table 4.1.

Example 4.3

A plant population of a cereal variety has heights which are normally distributed with a mean of 80 cm and a standard deviation of 5 cm. (a) Above what height are the top 20% of plants? (b) Below what height are the bottom 4% of plants?

(a) From the percentage points table, for $P = 20$, $z = 0.8416$, i.e. the top 20% of heights are 0.8416 or more standard deviations above the mean, hence

$$x = \mu + 0.8416\sigma = 80 + (0.8416 \times 5) = 80 + 4.208 = 84.208$$

In other words 20% of the plants have heights of 84.2 cm or more.

(b) From the percentage points table, for $P = 4$, $z = 1.7507$. Using the symmetry of the normal curve, this implies that 4% of the plants have heights which are 1.7507 or more standard deviations below the mean, hence

$$x = \mu - 1.7507\sigma = 80 - (1.7507 \times 5) = 80 - 8.7535 = 71.2465$$

In other words 4% of the plants have heights of 71.25 cm or less.

4.5 THE NORMAL DISTRIBUTION IN PRACTICE

Most of the statistical procedures studied in later chapters are only valid on the assumption that data are normally distributed. In practice, no real distribution is exactly normal, but for most procedures the deviation from normality has to be fairly marked to seriously invalidate the conclusions. In some cases, transformation of data (Chapter 14) will produce normally distributed values that can be analysed by standard techniques.

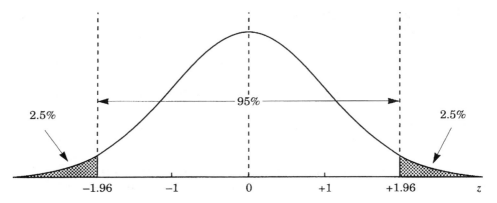

Figure 4.4. The standard normal distribution curve

Table 4.1. Some areas under the standard normal distribution curve

Area in top tail (P)	Area in both tails	Central area	z-value
5%	10%	90%	1.645
2.5%	5%	95%	1.960
1%	2%	98%	2.326
0.5%	1%	99%	2.576
0.1%	0.2%	99.8%	3.090
0.05%	0.1%	99.9%	3.291

If only a small amount of data are available, it is not usually possible to test whether they come from a normally distributed population. Experience should be a guide. If all plants in a large plot are treated exactly the same, their heights will still vary. Most plant heights should be within a limited range and symmetrical about the mean. Thus if a large sample of plants is selected at random and a histogram of their heights made, this should indicate whether it is reasonable to assume a normal distribution. The assumption that a set of data is from a normal distribution can also be investigated by obtaining a **normal probability plot**. Figure 4.5 shows such a plot for the data of Example 3.2 produced by asking for a normality test in Minitab. As the points are close to a straight line and the P-value is greater than 0.05, there is no reason to doubt the assumption of normality.

Many variables measured in plant science are discrete, for example the number of pods on a plant. Suppose the numbers of pods on five sampled rapeseed plants are 10, 15, 48, 90 and 210. It is unlikely that number of pods is a symmetrical variable. Fortunately, a consequence of the **central limit theorem** is that the means of random samples from the same population tend to follow a normal distribution even if the individual values do not. If the distribution of the original values is very skew, the means need to be based on larger samples than if this distribution is symmetrical. For this reason it is better to use the mean number of pods per plant of several

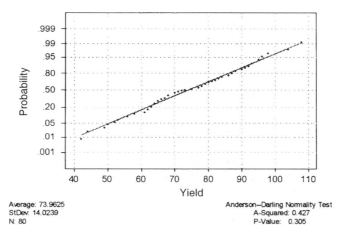

Figure 4.5. Normality test for data of Example 3.2

randomly selected plants as the data value to represent the plot. These means from plots treated alike should be approximately normally distributed. The central limit theorem links many parameters of crop growth analysis with the normal curve, and is used to find confidence intervals for population means.

4.6 INTRODUCTION TO CONFIDENCE INTERVALS

The mean yield of a new cereal variety has been estimated as 8.5 t/ha. Suppose that this overall mean was based on data from ten experimental sites each containing four plots (40 experimental plots). The results provide a useful guide to the variety's future commercial value. Yet the plant breeder is fully aware that the 40 individual plot yields used to obtain the overall mean were not all equal to 8.5 t/ha.

The decision to recommend a particular variety is greatly helped if a confidence interval around the mean (8.5 t/ha) can be given. This is defined by two values, one above and one below the mean. The confidence interval is given at an arbitrarily selected level of probability, for example 95%. This implies that the given interval has a 95% chance of containing the 'true' overall variety mean (the population mean, μ). The narrower this interval, the closer the 8.5 t/ha is likely to be to the 'true' mean.

4.7 ESTIMATION OF THE POPULATION MEAN, μ

You may wish to estimate the mean weight of plants in a plot. You could take one random sample of, say, 20 plants and find their mean weight. However, a different random sample of 20 plants would give a different mean weight, and there is no way of telling which of these two means is closest to the true population mean μ, or how far μ is from the closest value. The best you can do is to make a statement in terms of probabilities, and give confidence limits for μ, that is, give a range within which you are

95% certain the true population mean lies. This procedure relies on some statistical theory about the distribution of the values of the sample mean $\bar{x}$, which would be obtained if a very large number of samples, each of 20 plants, were removed.

4.8 THE SAMPLING DISTRIBUTION OF THE MEAN

Assume that the population of plant weights is normally distributed with a mean μ and a variance of σ^2. If a very large number of random samples (each of n plants) is taken, and for each sample the mean, $\bar{x}$, is found, statistical theory shows that the values of $\bar{x}$ are normally distributed with a mean of μ and a variance of σ^2/n. According to the central limit theorem, this is approximately true even if the population is not normally distributed, and this emphasises the importance of the normal distribution. Figures 4.6 and 4.7 illustrate these points.

The standard deviation of the values of $\bar{x}$ is $\sigma/\sqrt{n}$ provided a large enough number of samples is taken, and the size (n) of each sample is small compared with the population size (N). This standard deviation is usually called the standard error of the mean.

In practice you may only have one sample of n values to estimate σ. As the sample estimate of this is given by s, the standard error is calculated from

$$SE = \frac{s}{\sqrt{n}} \text{ which is the same as } \sqrt{\frac{s^2}{n}}$$

4.9 CONFIDENCE LIMITS FOR μ WHEN σ IS KNOWN

The properties of the normal distribution discussed earlier can now be used to find a confidence interval for μ. It has been shown in Figure 4.1 that for any normal distribution, 95% of the values lie within 1.96 standard deviations of the mean. Therefore, if many random samples (each of n plants) are taken, 95% of values of $\bar{x}$ are within 1.96 units of $\sigma/\sqrt{n}$ from μ. Hence, if only one sample of n plants is taken you can be 95% certain that the true population mean weight is within $1.96 \times (\sigma/\sqrt{n})$ of $\bar{x}$, i.e. the interval $\bar{x} \pm 1.96 \times (\sigma/\sqrt{n})$ has a 95% chance of containing the population mean. It is called a 95% confidence interval for μ. Figure 4.8 can be used to illustrate this statement.

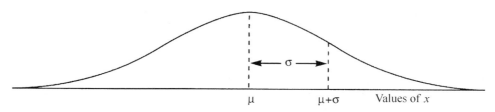

Figure 4.6. Distribution of population values (x)

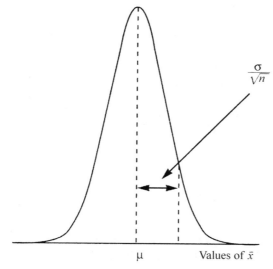

µ = population mean   σ = population standard deviation

Figure 4.7. Distribution of sample means ($\bar{x}$)

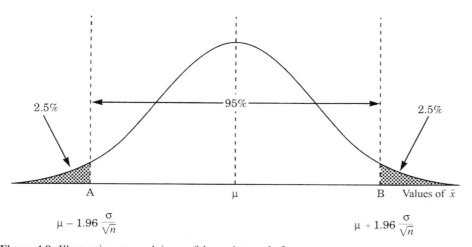

Figure 4.8. Illustration to explain confidence intervals for μ

- The standard deviation of this distribution is $\sigma/\sqrt{n}$.
- 95% of values of $\bar{x}$ are within 1.96 standard deviations of the mean, μ.
- Therefore, the point A is $\mu - 1.96(\sigma/\sqrt{n})$ and the point B is $\mu + 1.96(\sigma/\sqrt{n})$.

Now take just one random sample of n values and calculate the interval $\bar{x} \pm 1.96(\sigma/\sqrt{n})$. If $\bar{x}$ falls between A and B this interval will include μ. The chance of $\bar{x}$ falling between A and B is 95%, so there is a 95% chance that the interval calculated will contain μ. Hence,

$\bar{x} \pm 1.96\sigma/\sqrt{n}$ is called a 95% confidence interval for μ

If this formula is used every time a new sample of size n is taken from the same population (assumed infinite), 95% of such intervals would include μ. When this interval is constructed as a result of taking a single sample, the interpretation is that there is a 95% chance that the interval contains the population mean. There is a 5% chance that the interval does not contain μ.

A 99% confidence interval for μ is $\bar{x} \pm 2.58(\sigma/\sqrt{n})$. The 2.58 is found by looking up $P = 0.5$ in the percentage points table (Appendix 2) of the normal distribution. If the central area is to be 99%, the two tails must add up to 1% and so the top tail has to be 0.5%. Unfortunately, the population standard deviation, σ, is unlikely to be available, and so you have to use the sample standard deviation. As a result of this the confidence intervals have to be modified.

4.10 CONFIDENCE LIMITS FOR μ WHEN σ IS UNKNOWN — USE OF THE t-DISTRIBUTION

In practice, σ is unknown so it is estimated by calculating s, the sample standard deviation (Chapter 2). This estimation creates extra uncertainty about the true value of μ, so the confidence interval has to be widened to take this into account. For 95% intervals, the 1.96 is increased by an amount which depends on the sample size, n. This adjustment will become clearer when Example 4.4 is studied. If the sample size is large, s is likely to be close to σ so no increase is necessary. The number to replace the 1.96 is found from tables of the t-distribution with $(n-1)$ degrees of freedom (Appendix 3). This distribution is similar to the normal distribution but more spread out. The amount of spread is less, the greater the degrees of freedom, so that for large samples the normal distribution can be used. Verify that the figures in the last row of the t-table in Appendix 3 appear in the percentage points table of the normal distribution in Appendix 2.

Example 4.4
For a sample of size $n = 10$, $(n-1) = 9$ so in order to calculate a 95% confidence interval for μ look up the 2.5 percentage point of the t-distribution on 9 degrees of freedom. Look at the row for $(n-1) = 9$ in the table, and the column for the top tail percentage. For a 95% confidence interval the top tail percentage is 2.5, thus the table value is denoted by $t_{(n-1, 2.5\%)}$. You find $t_{(9, 2.5\%)} = 2.262$. Hence the 1.96 is replaced by 2.262, and σ is replaced by s.

In summary

- If σ is known, a 95% confidence interval for μ is given by $\bar{x} \pm 1.96(\sigma/\sqrt{n})$
- If σ is unknown, a 95% confidence interval for μ is now given by $\bar{x} \pm 2.262(s/\sqrt{n})$
- In general, the formula for a 95% confidence interval for μ when σ is unknown is

$$\bar{x} \pm t_{(n-1, 2.5\%)} \times \frac{s}{\sqrt{n}} \quad \text{where} \quad \frac{s}{\sqrt{n}} \quad \text{is called the standard error of the mean}$$

- Note also that $s/\sqrt{n}$ can also be expressed as $\sqrt{s^2/n}$

For a 99% confidence interval the top tail percentage is 0.5%, so the formula becomes

$$\bar{x} \pm t_{(n-1, 0.5\%)} \times \frac{s}{\sqrt{n}}$$

Note: For small samples the use of the *t*-distribution is only valid if it can be assumed that the population is normally distributed, otherwise the above formulae are considered approximations. You should always obtain a graphical summary to see if the assumption of normality is reasonable. For seriously skewed data the quotation of a standard error may be misleading as the mean minus two standard errors could be negative. In this case a transformation may be required (Chapter 14).

Example 4.5
The following numbers represent the heights (cm) of 10 plants taken at random from a plot. Find 95% and 99% confidence intervals for the mean height of all the plants in the plot:

72.3, 78.9, 82.6, 71.8, 86.1, 80.5, 72.0, 91.8, 77.3, 88.2

The sample size n is 10, so there are $(n-1) = 9$ degrees of freedom. From t tables (Appendix 3), $t_{(9, 2.5\%)} = 2.262$ and $t_{(9, 0.5\%)} = 3.250$. The sample mean $\bar{x}$ is found to be 80.15 and the sample variance s^2 is found to be 49.945. The standard error is

$$\frac{s}{\sqrt{n}} = \sqrt{\frac{s^2}{n}} = \sqrt{\frac{49.945}{10}} = 2.235$$

A 95% confidence interval for μ is $\bar{x} \pm t_{(9, 2.5\%)} \times \frac{s}{\sqrt{n}}$

$$= 80.15 \pm (2.262 \times 2.235)$$
$$= 80.15 \pm 5.056$$
$$= (75.09, 85.21)$$

You are thus 95% confident that μ lies between 75.09 cm and 85.21 cm. The lower limit is 75.09 cm and the upper limit is 85.21 cm.

A 99% confidence interval for μ is $\bar{x} \pm t_{(9, 0.5\%)} \times \frac{s}{\sqrt{n}}$

$$= 80.15 \pm (3.250 \times 2.235)$$
$$= 80.15 \pm 7.264$$
$$= (72.89, 87.41)$$

You are thus 99% confident that μ lies between 72.89 cm and 87.41 cm.

Note: To be more confident you must have a wider interval. A larger sample size will tend to give a narrower interval as $s/\sqrt{n}$ will tend to be smaller.

4.11 DETERMINATION OF SAMPLE SIZE

In order to estimate the mean yield per plant with a specified degree of precision, a knowledge of the variation in the plant yields of the population is required. This is not known exactly, but may be guessed using the results of previous experiments on the same crop, or estimated from a pilot study.

Example 4.6
It is required to estimate the mean grain yield per wheat plant to within 0.5 g with 95% confidence. How many plants should a random sample consist of?

Recall that a 95% confidence interval for the population mean μ when σ is known is given by $\bar{x} \pm 1.96(\sigma/\sqrt{n})$. If $\bar{x}$ is to differ from μ by less than 0.5 g with 95% confidence, you require a sample of n plants such that $1.96(\sigma/\sqrt{n})$ is less than 0.5. Although σ is unknown, previous studies may suggest that it is approximately 2 g. If $\sigma = 2$, you require n such that $1.96 \times (2/\sqrt{n})$ is less than 0.5. This implies that $\sqrt{n}$ is greater than $(1.96 \times 2)/0.5 = 7.84$. Hence n must be greater than $7.84^2 = 61.5$. However, n has to be an integer so at least 62 plants should be sampled to give the required degree of precision.

In general, to estimate μ to within δ with 95% confidence you require n at least the next integer greater than $(1.96\sigma/\delta)^2$. The problem is that you need a prior estimate of σ.

4.12 ESTIMATION OF TOTAL CROP YIELD

The total yield of a plot can be estimated by taking a random sample of n individual plants from the plot and finding the mean yield per plant. A 95% confidence interval for the mean yield per plant can then be found using the formula

$$\bar{x} \pm t_{(n-1, 2.5\%)} \times \frac{s}{\sqrt{n}}$$

The lower and upper limits of this interval are multiplied by the total number of plants in the plot to obtain the corresponding confidence interval for the total plant yield of the experimental area.

This method is suitable if the plot is not large and it is easy to count the plants. Precision-sown crops such as maize, dwarf French beans and Brussels sprouts are ideal. If the plot is large, there is likely to be a wider range of variation, and a simple random sample of individual plants may be unrepresentative. Furthermore, the total number of plants in the plot will have to be estimated leading to extra uncertainty.

In the case of densely sown row crops such as cereals and oilseed rape, it would be extremely tedious and time consuming to make an accurate count of total plants in the plot. Sampling a few individual plants is likely to give a very misleading estimate of total plot yield. A much more reliable estimate can be obtained by taking all the plants from several quadrats placed at random in the plot. The total yield for each quadrat is found, and a 95% confidence interval for the mean yield of an area the

NORMAL DISTRIBUTION, t-DISTRIBUTION AND CONFIDENCE INTERVALS

size of the quadrat is calculated. The lower and upper limits of this interval are multiplied by the ratio of the area of the plot to the area of the quadrat in order to obtain 95% confidence limits for the total yield of the plot. The following example shows the details of the calculations.

Example 4.7

The total dry weights (g) of winter oilseed rape plants from eight quadrats (each measuring a third of a square metre) are given below. It is assumed that these quadrats were placed at random within the plot. If the area of the plot was 100 m², find a 95% confidence interval for the total plant dry weight of the plot:

171.8, 267.7, 274.7, 203.2, 208.6, 267.2, 184.1, 234.5

Verify that the sample mean, $\bar{x}$, is 226.5 and the sample standard deviation, s, is 40.35. A 95% confidence interval for the total dry weight per quadrat is

$$226.5 \pm t_{(7, 2.5\%)} \times \frac{40.35}{\sqrt{8}}$$

$$= 226.5 \pm 2.365 \times \frac{40.35}{\sqrt{8}}$$

$$= 226.5 \pm (2.365 \times 14.26)$$

$$= 226.5 \pm 33.7$$

The lower limit is $226.5 - 33.7 = 192.8$ g and the upper limit is $226.5 + 33.7 = 260.2$ g. The area of each quadrat is a third of a square metre, therefore the ratio of the area of the plot to the area of a quadrat is 300. The confidence limits for the total dry weight of the plants in the plot are $300 \times 192.8 = 57\,840$ g and $300 \times 260.2 = 78\,060$ g. Thus, you are 95% confident that the total dry weight of all the plants in the plot lies between 57.8 kg and 78.1 kg.

This wide range is due to the very large plant-to-plant variation. To achieve a greater degree of precision, more quadrats would be needed. To find how many would be required, the method described in Section 4.11 can be used with the unit of measurement being a quadrat rather than an individual plant.

Note: This method of estimating plot yield does not have to involve weighing individual plants — all the plants from the quadrat can be bulked and then weighed. Clearly, a large number of plants must be sampled in order to estimate crop yields when plant-to-plant variation is very high.

Chapter 5

Introduction to Hypothesis Testing

5.1 THE STANDARD NORMAL DISTRIBUTION AND THE *t*-DISTRIBUTION

In Chapter 4 it was stated that

- The sampling distribution of the mean ($\bar{x}$), based on samples of size n, is approximately normal with a mean μ and a standard deviation $\sigma/\sqrt{n}$.
- If you subtract the mean from a value taken from a normal distribution, and divide by the standard deviation, the result is a z-value from a standard normal distribution.

Hence, if $\bar{x}$ is the mean of a random sample,

$$z = \frac{\bar{x} - \mu}{\sigma/\sqrt{n}}$$

is a value from a normal distribution with mean 0 and variance 1. In practice σ is unknown and is estimated from the sample by s. If the value of $(\bar{x} - \mu)/(s/\sqrt{n})$ is calculated each time a new random sample is taken, theory predicts that these values follow a *t*-distribution with $n - 1$ degrees of freedom. Hence,

$$t = \frac{\bar{x} - \mu}{s/\sqrt{n}}$$

The *t*-distribution is very close to the standard normal distribution when n is large, because s is likely to be close to σ for each sample. For small samples, the *t*-distribution is more spread out because the values of s are likely to vary from sample to sample. In Figure 5.1, the dotted curve represents the standard normal distribution, and the full curve represents the *t*-distribution with 4 df.

The striped shading in Figure 5.1 represents the top 2.5% and the bottom 2.5% of the standard normal distribution ($z = 1.96$), while the black shading represents the corresponding percentages of the *t*-distribution with 4 df ($t_{(4, 2.5\%)} = 2.776$). These

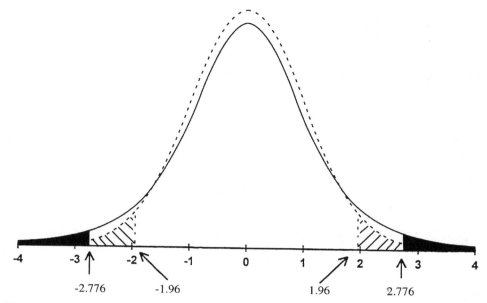

Figure 5.1. The *t*-distribution with 4 df and the standard normal distribution

values can be found from tables of the *t*-distribution (Appendix 3). Look in row 4 (df = 4 if sample size $n = 5$), column 2.5. The value is $t_{(4, 2.5\%)} = 2.776$. Now look in the last row (infinity), corresponding to a very large sample. The value is $t_{(\infty, 2.5\%)} = 1.96$. This means that if

$$t = \frac{\bar{x} - \mu}{s/\sqrt{n}}$$

is calculated each time a random sample of size $n = 5$ is selected from a normal distribution with mean μ, then 2.5% of the values of *t* are expected be greater than 2.776 (5% of the values are expected to lie between -2.776 and $+2.776$). If *n* is very large, 2.5% of the *t*-values are expected to be greater than 1.96, and 5% are expected to lie in the range -1.96 to $+1.96$. In this case it would not matter if the distribution sampled is not normal.

5.2 THE SINGLE-SAMPLE *t*-TEST

This is one of the simplest hypothesis tests, and is used to test whether the mean of a normally distributed population is a specified value. If the population distribution is markedly non-normal you should consider a non-parametric alternative (Chapter 19) or a transformation (Chapter 14). Before giving an example of its use, some background theory is given as an aid to understanding.

The null hypothesis is that the population mean is μ_0. This hypothesis is denoted by H_0 and stated as

$$H_0: \mu = \mu_0$$

An alternative hypothesis could be that μ is not equal to μ_0. This is a two-tailed alternative and stated as

$$H_1: \mu \neq \mu_0$$

To test the null hypothesis, a random sample is taken from the population and the sample mean $\bar{x}$, and the sample variance s^2, are calculated. The value of t is found using the formula

$$t = \frac{\bar{x} - \mu_0}{s/\sqrt{n}}$$

$s/\sqrt{n}$ is known as the **standard error of the mean** (*SE*) and is more easily calculated as $\sqrt{s^2/n}$.

Hence the formula for t can be written as $t = \dfrac{\bar{x} - \mu_0}{SE}$

In general, there is a 95% chance that t will lie between $-t_{(n-1, 2.5\%)}$ and $+t_{(n-1, 2.5\%)}$ if the null hypothesis that $\mu = \mu_0$ is true. If you take a random sample of n values and the resulting t-value is outside this range, reject the null hypothesis H_0 against a two-tailed alternative H_1 at the 5% level of significance. In particular, for a sample size of $n = 10$, the degrees of freedom are 9 and the table value is $t_{(9, 2.5\%)} = 2.262$, so H_0 would be rejected if the calculated t-value is greater than 2.262 in magnitude. This is an example of a **two-tailed test** because prior to collecting the sample data you do not know if $\bar{x}$ will be greater or less than μ_0 and you have to allow for both possibilities.

If the alternative hypothesis is that μ is greater than μ_0, stated as $H_1: \mu > \mu_0$, reject H_0 at the 5% level if the calculated t-value is greater than $t_{(n-1, 5\%)}$. This is because there is a 5% chance that the calculated t-value would exceed the 5% table value if the null hypothesis were true. For a sample of size 10, the 5% table value is $t_{(9, 5\%)} = 1.833$. This is an example of a **one-tailed test**. You should not carry out this one-tailed test unless you have good reason to believe that $\bar{x}$ cannot be less than μ_0.

5.2.1 Summary of the Single-sample *t*-test

(1) Set up the null hypothesis $H_0: \mu = \mu_0$. You believe the population mean to be μ_0.
(2) Before collecting data decide whether the test is to be one-tailed or two-tailed.
For a two-tailed test the alternative hypothesis is $H_1: \mu \neq \mu_0$.
For a one-tailed test the alternative hypothesis is either or $H_1: \mu > \mu_0$ or $H_1: \mu < \mu_0$.
Decide a level of significance (5%, 1% or 0.1%, etc.). See Section 5.5

(3) Collect the data values. They should be a random sample of n values from a normal distribution.
(4) Calculate

$$t = \frac{\bar{x} - \mu_0}{SE} \quad \text{where} \quad SE = \sqrt{\frac{s^2}{n}}$$

(5) Consult the t-table (Appendix 3). If the test is two-tailed at the 5% level, look in row $n - 1$ (there are $n - 1$ degrees of freedom) and column 2.5 (the tables give the top tail percentages). If the test is two-tailed at the 1% level, look in row $n - 1$ and column 0.5. If the test is one-tailed at the 5% level, look in row $n - 1$ and column 5%.
(6) If the magnitude of the calculated t-value is greater than the table value, reject the null hypothesis at the chosen level of significance.

5.2.2 Further Discussion of the t-test

It should be clear that the evidence against the null hypothesis is not solely based on the difference between the sample mean $\bar{x}$, and the mean μ_0 expected if the null hypothesis is true. If the value of $\bar{x} - \mu_0$ is large, this difference may not be significant if there is a large variation in yields and the sample size is small. On the other hand, if $\bar{x} - \mu_0$ is small, this difference may be significant if the variation in the yields is small and the sample size is large. This difference is called the **error in the mean**. It measures the extent to which the sample mean differs from the assumed population mean. This error must be compared to the **standard error** of the mean $s/\sqrt{n}$ in order to carry out a test of the null hypothesis that the population mean is a specified value. Now you should understand the origin of the expression standard error.

Example 5.1
Suppose that the mean yield of a standard (recommended) linseed variety when grown in South-east England is 2.0 t/ha. The yields obtained from a random sample of 6 plots sown with a new variety in the same region were 2.6, 2.1, 2.5, 2.4, 1.9, and 2.3 t/ha. Do these results provide sufficient evidence to conclude that the yield of the new variety is different from the standard?

The null hypothesis is
$$H_o: \mu = 2.0$$
Before data are collected it is unknown whether the yields of the new variety will be larger or smaller than the standard, so a two-tailed test is required. Hence the alternative hypothesis is $H_1: \mu \neq 2.0$.

The null hypothesis is rejected if the average yield of the new variety is sufficiently larger or smaller than 2.0. This implies rejecting H_0 if the calculated t-value is sufficiently different from zero (large and positive or large and negative).

Assuming that the yields come from a normal distribution, it is valid to carry out a single-sample t-test with $\mu_0 = 2.0$. You decide to carry out the test at the 5% level of significance. This implies there is a 5% chance of rejecting the null hypothesis if it is true. The degrees of freedom (df) are $n - 1 = 5$. From tables, $t_{(5, 2.5\%)} = 2.571$. Hence, if the null hypothesis is true, there is a 95% chance that the calculated t-value will lie between -2.571 and $+2.571$ and a 5% chance that it will lie outside this region. If it lies outside, reject the null hypothesis (at the 5% level) that the population mean is 2.0. That is, reject H_0 if the magnitude of the calculated t-value is greater than 2.571.

Now calculate t. The first step is to calculate $\bar{x}$ and s^2 (Chapter 2). Confirm that $\bar{x} = 2.3$ and $s^2 = 0.068$. Now calculate the standard error of the sample mean by dividing s^2 by 6 ($n = 6$). Take the square root and confirm that $SE = 0.1065$.

If your calculator has an SD mode, the value of SE can be found directly. Enter the six numbers and press the n key to check you have entered six numbers. Press the $\bar{x}$ key to confirm the mean is 2.3. Press the σ_{n-1} (or s) key to get s and square the result to get $s^2 = 0.068$. Divide by 6 and take the square root to obtain $SE = 0.1065$ correct to four decimal places. Calculate t as follows:

$$t = \frac{\bar{x} - \mu_0}{SE} = \frac{2.3 - 2.0}{0.1065} = \frac{0.3}{0.1065} = 2.817 \quad \text{on 5 df}$$

As this value is greater than 2.571, you reject H_0, that the population mean is 2.0 at the 5% level of significance.

Figure 5.2 shows the t-distribution for 5 df, the table values (-2.571 and $+2.571$) for carrying out a two-tailed test at the 5% level, and the calculated value of 2.817.

The shaded area is called the **critical region** of the test. The calculated value of 2.817 is in this critical region, so reject H_0 at the 5% level. There is strong evidence that the new variety gives a different yield from the standard variety. The sample mean of 2.3 t/ha is significantly different from 2.0 t/ha. However, H_0 would not be rejected at the 1% level because $t_{(5, 0.5\%)} = 4.032$ and the calculated t-value of 2.817 is less than this.

5.3 THE *P*-VALUE

When a computer is used to carry out a t-test, a P-value is given in the output. This value is the probability of observing a larger t-value (in magnitude) than the one actually obtained, assuming the null hypothesis to be true. In Figure 5.2, P is the sum of the areas to the right of 2.817 and to the left of -2.817. It is less than 0.05 because the area outside the region -2.571 to 2.571 is 5% (0.05). It is greater than 0.01 because the area outside the region -4.032 to $+4.032$ is 1% (0.01). Thus P is between 0.05 and 0.01 as the t-value is between 2.571 and 4.032.

Before computers were widely available it was not possible to calculate P exactly and tables were used to find an approximate value. The result of a t-test would be summarised by a statement such as $P < 0.05$ if the result was significant at the 5% level but not at the 1% level. $P < 0.01$ implies significance at the 1% level but not at the 0.1% level (a highly significant result). $P < 0.001$ implies significance at the 0.1% level (a very highly significant result).

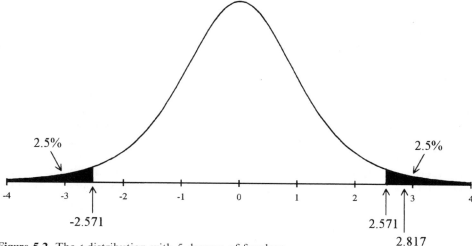

Figure 5.2. The *t*-distribution with 5 degrees of freedom

The Minitab output for this one-sample *t*-test is

```
Test of mu = 2.000 vs mu not = 2.000
Variable    N      Mean    StDev    SE Mean      T         P
yield       6      2.300   0.261    0.106        2.82      0.037
```

It shows that the *P*-value for the two-tailed test is 0.037 and means that the chance of obtaining a *t*-value as far from zero as 2.817 is 0.037 (3.7%) if the null hypothesis is true. This is also the chance of observing a sample mean as far from 2.0 as 2.3 assuming the population from which the sample came is normally distributed with a mean of 2.0 and a variance estimated as 0.068 based on a random sample of six values. As a mean of 2.3 was obtained, we conclude that this is unlikely to have happened by chance if the population mean is 2.0. A more likely explanation is that the population mean is not 2.0.

The smaller the *P*-value, the greater the evidence against H_0. A value of less than 0.05 is generally regarded as 'significant'.

Note: The *P*-value for a one-tailed test is half that for the corresponding two-tailed test. This means that the former could provide a significant result ($P < 0.05$) whereas the latter may not. You should therefore not carry out a one-tailed test for this reason but for reasons decided upon before data collection. If in doubt, always use a two-tailed test.

5.4 TYPE I AND TYPE II ERRORS

A Type I error is committed if the null hypothesis is rejected when it is true. The probability of a Type I error is the level of significance, denoted by α. For example, if a test is carried out at the 5% level, the value of α is 0.05. If you always carry out tests at the 5% level you will reject 5% of the hypotheses you test when they are really true.

The probability of committing a Type I error is reduced to 0.01 by carrying out the test at the 1% level. A larger value of t is now required for significance ($\bar{x}$ has to be further from μ_0). This increases the likelihood of committing a Type II error. A Type II error is committed if the null hypothesis is accepted when it is really false, which is the same as saying that the null hypothesis is not rejected when it should be rejected. The probability of this type of error is denoted by β.

In Example 5.1, suppose you carry out the test at the 1% level ($\alpha = 0.01$). In advance of collecting the data you decide to reject the null hypothesis if the calculated t-value is greater than 4.032. After collecting the data you calculate the t-value and find it to be 2.817. This is less than 4.032 so you do not reject the null hypothesis at the 1% level. By not rejecting the hypothesis you do not assume it is true. You could be committing a Type II error. The true population mean could be 2.1 or 2.2 or 2.4 or many other values and a t-test would not show that 2.3 was significantly different from any of them at the 1% level. Remember that if a null hypothesis is accepted, this does not mean it is true; it just means that the data are consistent with the null hypothesis. The data are also consistent with many other hypotheses.

If the chance of a Type I error is reduced, the chance of a Type II error is increased and vice versa. This means that if you reduce the chance of rejecting a true hypothesis, you increase the chance of accepting a false hypothesis and vice versa. Table 5.1 gives a summary of Type I and Type II errors.

The **power** of a test is the probability of rejecting the null hypothesis when it is false. The power can be increased by having more replications per treatment or by improving the experimental design.

5.5 CHOICE OF LEVEL OF SIGNIFICANCE

What factors influence the decision to carry out a test at the 5%, 1% or 0.1% level of significance? The choice depends on how serious are the consequences of making a wrong decision. Referring to Example 5.1, if the test is carried out at the 5% level, there is a chance of 5% that the new variety will be recommended when it is not really any better than the standard. To reduce the chance of this kind of error (Type I error) to 1%, it would be necessary to carry out the test at the 1% level. If a significant result is obtained at the 1% level, it is much more likely to have resulted from a real difference rather than a chance difference between the new variety and the standard. If it is very important to be sure of an increase in yield before recommending a new variety, the level of significance should be set at the 0.1%. This is a high level of significance. It is then very unlikely that the new variety will be recommended if it is not better than the standard.

Table 5.1. Type I and Type II errors

Hypothesis	Accept	Reject
True	No error	Type I error (probability $= \alpha$)
False	Type II error (probability $= \beta$)	No error

If the new variety is better than the standard, it still may not be recommended because its mean yield may not be large enough to be significantly different from the standard at the 0.1% level. In this case a Type II error will have been committed.

5.6 THE USEFULNESS OF A TEST

The single-sample t-test is of limited use in making comparisons. Although it can be used to test whether the yield of a new variety differs from 2.0 t/ha, it cannot be used to test whether the yield of a new variety gives a different yield from a standard variety unless the latter has been tested on a number of plots in similar conditions. In this case an independent samples t-test (Chapter 6) or a paired samples t-test (Section 5.8) would be used.

It is also important to realise that in practice decisions are seldom made based on the results of a single test. If a test shows a significant result, it is good practice to repeat the analysis with new data to see if the same result is obtained. Remember that if the null hypothesis is really true, there is a 5% chance that a significant result will be found if the test is carried out at the 5% level of significance.

Exercise 5.1
Last year a crop gave a yield of 8 kg/m^2. This year a new fertiliser was applied and the yields from six randomly chosen plots were 8.1, 8.7, 9.2, 7.8, 8.4 and 9.4 kg/m^2. Do the data provide sufficient evidence that the yield has increased this year? Carry out an appropriate test, stating any assumptions you make.

Carry out a single-sample t-test with the null hypothesis that the population mean is 8.0. As you expect the yield to increase, and would not consider using the new fertiliser if the yield decreased, you carry out a one-tailed test. To have a good chance of detecting a positive result, you use the 5% level. Confirm that the mean yield of the six plots is 8.6 with a standard error of 0.254 and the t-value is 2.36 on 5 df. From tables, $t_{(5, 5\%)} = 2.015$. The calculated t-value is greater than the table value, so the result is significant at the 5% level. If you use a computer, you should obtain the P-value of 0.032 which is the probability of obtaining a t-value greater than 2.36 if the null hypothesis is true.

Although there is evidence that the yield has increased, this is not a properly designed experiment. The increase in yield could be due to a range of environmental factors. A valid appraisal of the new fertiliser could be obtained by carrying out an experiment using twelve adjacent plots on a uniform site. The new fertiliser is applied to six of the plots at random, the other six receive the standard fertiliser. The overall mean difference in yield is now more likely to be due to the difference between the fertilisers. An independent samples t-test (Section 6.2) would be used for the analysis. Another design would be to have six pairs of plots with one member of each pair chosen at random to receive the new and the other the standard fertiliser. The pairs would be chosen such that environmental conditions between members of a pair are more alike than those between pairs. In this case a paired samples t-test (Section 5.8) would be used.

5.7 ESTIMATION VERSUS HYPOTHESIS TESTING

Estimation is often of more value than hypothesis testing. The rejection of a null hypothesis may not be of much value if you already know or suspect a big difference. A confidence interval gives an indication of the probable size of the difference. When a sample is taken, it is important to quote the sample mean and standard error so that the reader is in a position to carry out further analyses such as hypothesis testing or the calculation of confidence intervals. Many statisticians recommend that confidence intervals should used in preference to hypothesis testing.

Consider the data of Example 5.1. We have already shown that the mean yield of the new variety (2.3) is significantly different from 2.0. At the 5% level of significance the data are not consistent with the hypothesis that the population mean is 2.0. The value of 2.3 is a point estimate of the true but unknown population mean, μ. The t-test shows that this mean is likely to be greater than 2.0 but does not indicate by how much. A confidence interval estimate (Section 4.10) conveys more information.

The formula for a 95% confidence interval for μ is $\bar{x} \pm t_{(5, 2.5\%)} \times SE$ (Section 4.10). As there are 6 values, the degrees of freedom are 5, and the table value is 2.571. Hence the interval is

$$2.3 \pm (2.571 \times 0.1065) = 2.3 \pm 0.27 = (2.03, 2.57)$$

You are 95% confident that the true value of the mean yield of the new variety lies between 2.03 and 2.57 t/ha. Note that this interval does not include 2.0. Compare these results with the Minitab output:

Variable	N	Mean	StDev	SE Mean	95.0% CI
yield	6	2.300	0.261	0.106	(2.026, 2.574)

Note: A two-tailed test at the 5% level is equivalent to finding a 95% confidence interval. If the hypothesised mean (in this case 2.0) is not in the interval, the null hypothesis is rejected at the 5% level. This interval (in this case 2.03 to 2.57) is the range of possible null hypothesis values that would not be rejected. The sample mean of 2.3 is consistent with the population mean being any value in this range. For instance, if the t-test was carried out with the null hypothesis that the population mean was 2.5, the result would not be significant, as 2.5 is within the confidence interval.

5.8 THE PAIRED SAMPLES t-TEST

This test can be used to compare two treatments when applied to pairs of similar experimental units such as halves of a leaf, two leaves of a plant or two adjacent plots of land. For example, to compare a standard (recommended) variety with a new variety, several pairs of adjacent plots could be used. One member of each pair would be planted with the standard and the other with the new variety. The environmental differences between pairs are likely to be larger than those between members of a pair. As this is taken into account in the analysis, it is more likely that a real between-treatment difference (which may have been missed in the absence of pairing) is detected. If the two varieties give equal population yields, you would expect the mean of the within pair differences to be zero. Under the null hypothesis of equal population means, the differences are expected to come from a normal distribution with a mean of zero, so the paired t-test is a single sample t-test of the differences. It is also a special case of a randomised blocks analysis of variance (Chapter 10) when there are just two treatments. If you cannot make the assumption of normality you may consider a non-parametric alternative (Chapter 19).

Example 5.2

Two varieties of wheat were tested using two similar plots on eight different farms. One plot on each farm was selected at random to receive variety A, and the other variety B. All plots were planted on the same day and managed identically. The yields (kg/plot) and plot differences are shown in Table 5.2. The null hypothesis is that the population mean difference is zero versus the alternative that it is not. A two-tailed single sample t-test at the 5% significance level is performed on the differences, with $\mu_0 = 0$.

The mean difference is $\bar{d} = 12.2/8 = 1.525$. This is the same as the A mean (14.875) minus the B mean (13.350). Confirm that the sample variance of the differences is $s^2 = 2.299$ and hence the standard error is $\sqrt{s^2/n} = \sqrt{2.299/8} = 0.536$. The t-value is

$$t = \frac{\bar{d} - 0}{\sqrt{s^2/n}} = \frac{1.525}{0.536} = 2.845$$

As there are 8 pairs there are 7 degrees of freedom so the table value (Appendix 3) is $t_{(7, 2.5\%)} = 2.365$. The calculated t-value is greater, so the null hypothesis is rejected at the 5% level ($P < 0.05$). The exact P-value is 0.025 (Minitab output). A 95% confidence interval for the mean difference is

$$\bar{d} \pm t_{(7, 2.5\%)} \times SE = 1.525 \pm (2.365 \times 0.536) = 1.525 \pm 1.268 = (0.26, 2.79)$$

You are therefore 95% confident that the population mean yield of variety A exceeds that of variety B by between 0.26 and 2.79 kg/plot. This interval does not include zero which confirms rejection of the null hypothesis at the 5% level. Compare these results with the Minitab output:

```
Paired T for Variety A - Variety B

              N     Mean     StDev    SE Mean
Variety       8    14.88     3.46      1.22
Variety       8    13.35     2.77      0.98
Difference    8     1.525    1.516     0.536

95% CI for mean difference: (0.256, 2.794)
T-Test of mean difference=0 (vs not=0): T-Value=2.84 P-Value=0.025
```

Table 5.2. Yields (kg) from plots grown with two varieties of wheat on eight farms

Farm	1	2	3	4	5	6	7	8	Total
Variety A	17.8	18.5	12.2	19.7	10.8	11.9	15.6	12.5	119.0
Variety B	14.7	15.2	12.9	18.3	10.1	12.2	13.5	9.9	106.8
Difference (d)	3.1	3.3	−0.7	1.4	0.7	−0.3	2.1	2.6	12.2

Note: The result of this test may be open to question if the farms have widely differing soil types. There may be an interaction due to some soil types favouring variety A and others variety B. There is less likely to be interaction if the experiment is carried out on pairs of plots within a single farm, but then the results would be only relevant to environmental conditions similar to those on that farm.

Chapter 6

Comparison of Two Independent Sample Means

6.1 INTRODUCTION

In Example 5.1 we considered the problem of testing whether the mean yield of a new variety was different from a standard by measuring the yields from several plots. It was not clear how the 2.0 t/ha yield of the standard variety was estimated. It may have been based on data from a previous season(s) or a range of sites. Hence the observed difference may not have been due to the introduction of a new variety.

Most experiments are designed to compare treatments and estimate the differences between them. In a simple trial two treatments can be compared to discover if they have different effects on the yield of a particular species. The treatments could be two fertilisers, fungicides, or concentrations of the same fertiliser. In most cases one treatment is a control. For example, to find the effect of spraying a plant growth regulator (PGR), the trial would include sprayed and unsprayed (control) plants. All plots or pots would otherwise be managed identically.

Example 6.1
Suppose a claim is made that a novel PGR gives a 6.2% increase in the yield of oilseed rape when applied (using a single concentration) at early stem extension. This raises several questions:

(1) How reliable is this percentage difference? Are similar yield benefits obtained on all soil types? Is the experimental result repeatable?
(2) How many experimental and control plots or pots were harvested? What method was used to assess yield?
(3) Was the claimed response solely due to the application of water which was used to dilute the concentrated PGR in the spray tank? Is it necessary to compare treated plots with watered and totally untreated dry control plots in this type of trial?

Table 6.1 shows the yields of oilseed rape obtained from 20 plots.

> Returning to question one: how reliable is this percentage difference? For the 10 control plots there was a wide variation in recorded yield (2.23–3.28 t/ha). If by chance the yield of the seventh plot (*) had been 3.23 t/ha instead of 2.23 t/ha the control mean would have been 2.86 t/ha. Similarly, if the eighth (*) growth regulator plot yield (3.35 t/ha) had been recorded as 2.35 t/ha, the treatment mean is reduced to 2.82 t/ha.
>
> **By altering just two values** the conclusion has been changed, for the evidence based on overall means now suggests that the growth regulator reduces yield. Thus great care must be taken when making claims based on small amounts of variable data. **Overall means may be meaningless** if the range of variation on which they are based is not indicated, and it may be impossible to answer the above questions if the experiment is designed incorrectly.
>
> There are many designs, the simplest being the **completely randomised design** (CRD). One of the advantages of this is that it is not necessary to have equal replications per treatment, and when there are just two treatments, the mean yields can be compared using an **independent samples *t*-test**.

Do the results given in Example 6.1 demonstrate beyond all reasonable doubt that the PGR increases the yield of oilseed rape? In order to answer this question we begin by assuming there is no real difference between the two treatments. In statistical language the null hypothesis, H_0, is assumed to be true. The two sets of 10 yields are assumed to be random samples from the same population, or populations with the same mean. However, even if this hypothesis were really true, sampling would result in the sample means being different. How big a difference in means could we reasonably expect under the null hypothesis? This depends on the sample size and the variability of yields in the populations from which our results are

Table 6.1. Yields of oilseed rape from treated and control plots

	Yield (t/ha)	
	Control	+ growth reg.
	2.65	3.02
	3.28	2.21
	2.62	2.29
	2.84	3.21
	2.61	3.30
	2.80	3.13
	2.23*	2.86
	2.45	3.35*
	2.95	2.72
	3.12	3.16
Means	2.76	2.93

Difference between means 0.17
Percentage difference = +6.2

Table 6.2. Yields (t/ha) of oilseed rape from 20 plots for each of two varieties

	Variety 1		Variety 2	
	Control	+ growth reg.	Control	+ growth reg.
	3.5	5.5	3.3	5.0
	4.6	4.1	4.8	3.0
	4.0	4.5	5.7	5.6
	4.3	6.1	2.6	3.0
	4.0	4.6	4.8	7.7
	4.6	5.3	3.8	5.3
	5.0	5.4	5.9	5.0
	3.9	3.9	3.0	4.1
	3.5	4.4	4.3	4.6
	4.8	6.1	4.0	6.6
Mean	4.22	4.99	4.22	4.99
Difference		0.77		0.77
% difference		18.25		18.25

random samples. Table 6.2 illustrates the effect of plot-to-plot variation on the interpretation of treatment differences.

The difference between the sample means is the same (0.77) for both varieties. For variety one, the plot-to-plot variation within treatments is smaller than it is for variety two. This suggests that the observed difference of 0.77 is unlikely to be due to chance for the first variety but it could be for the second. The reader is asked to carry out t-tests in Exercise 6.1 to confirm that for variety one, the means are significantly different, but for variety two they are not. You will conclude that for variety two there is insufficient evidence that the growth regulator is having an effect. Note that this does not mean it has no effect; it just means that we have been unable to show that it does. An experiment with more plots may be more conclusive.

6.2 THE INDEPENDENT SAMPLES t-TEST

This test can be used to analyse data from experiments designed to compare two treatments assuming the following:

- The experimental units or plots are divided into two groups at random; treatment one is applied to the units of group one and treatment two to the units of group two.
- Data from each treatment are from normal distributions with the same variance.
- Data from units receiving the same treatment are independent.

If the two treatment means are found to be significantly different this is likely to be due to the treatments.

The test can also be applied in observational studies where independent random samples are taken from two populations but it will not provide the reason for any significant difference. If the assumptions underlying the *t*-test are seriously in doubt a Mann–Whitney test (Section 19.5) should be considered.

Example 6.2

In order to compare a new variety of wheat with a standard, a CRD is used with six plots receiving the new variety and ten plots receiving the standard. The yields (t/ha) are as follows:

New variety: 2.6 2.1 2.5 2.4 1.9 2.3
Standard variety: 1.7 2.1 2.0 1.8 2.3 1.6 2.0 2.1 2.2 1.9

The null hypothesis is that the two varieties give equal yields and this is stated as $H_0: \mu_1 = \mu_2$. We decide on a two-tailed test **before** collecting the data because we do not know which variety will perform better. Hence the alternative hypothesis is $H_1: \mu_1 \neq \mu_2$.

6.2.1 Using a Computer

The reader may use a computer package to carry out a two sample *t*-test, in which case the data should be entered in two columns; the first should include the treatment (variety) codes and the second the yields. Outputs from Minitab (a), SAS (b) and Genstat (c) for Example 6.2 are shown in Output 6.1.

Each output shows the variety means (2.30 and 1.97), the *t*-value (2.71) and the *P*-value ($P = 0.017$). You conclude that the two means are significantly different because of the small probability (1.7%) that the observed difference (0.33) has arisen by chance. Alternatively you can state that you are 98.3% confident that the two varieties give different yields. The Minitab and Genstat outputs show that you are 95% confident that the population mean yield for the new variety exceeds that for the standard by between 0.068 and 0.592 t/ha.

One of the conditions for this *t*-test to be valid is the assumption of equal population variances. This can be tested using an *F*-test (Section 6.5). The SAS output shows a *P*-value of 0.6296. This is not less than 0.05 so you can assume equal variances.

The *P*-value of the *t*-test is the probability of obtaining a difference in means at least as great as 0.33 if the null hypothesis is true. It is also the probability of obtaining a *t*-value greater than 2.71 in magnitude. Before computers were available the *P*-value could not be calculated very easily and so *t*-tables had to be used. From these you could only conclude whether *P* was less than particular values such as 10%, 5%, 1%, 0.1% etc. by comparing a calculated *t*-value with a table value. A calculated value greater than the 5% table value indicates a *P*-value less than 5% (0.05). In this example the degrees of freedom are $(n_1 + n_2 - 2) = 14$. The *t*-table value to be used in this two-tailed test is $t_{(14, 2.5\%)} = 2.145$ (row 14, column 2.5 of Appendix 3). The calculated *t*-value (2.71) is greater than this so you reject the null hypothesis at the 5% level.

COMPARISON OF TWO INDEPENDENT SAMPLE MEANS

Output 6.1 Computer output from three packages for the *t*-test applied to the data of Example 6.2

(a)
```
Two sample T for Yield

Variety    N    Mean    StDev    SE Mean
1          6    2.300   0.261    0.11
2         10    1.970   0.221    0.070

95% CI for mu (1) − mu (2) : ( 0.07, 0.592)
T-Test mu (1) = mu (2) (vs not =) : T = 2.71  P = 0.017  DF = 14
Both use Pooled StDev = 0.236
```

(b)
```
TTEST PROCEDURE

Variable: YIELD

VARIETY  N       Mean         Std Dev        Std Error      Minimum       Maximum
-----------------------------------------------------------------------------------
1        6    2.30000000    0.26076810     0.10645813    1.90000000    2.60000000
2       10    1.97000000    0.22135944     0.07000000    1.60000000    2.30000000


Variances      T      DF      Prob > |T|
-----------------------------------------
Unequal      2.5901   9.3       0.0285
Equal        2.7056  14.0       0.0171

    For H0: Variances are equal, F' = 1.39  DF = (5,9)  Prob > F' = 0.6296
```

(c)
```
***** Two-sample T-test *****

Sample      Size    Mean     Variance
New          6      2.300    0.06800
Standard    10      1.970    0.04900

*** Test for evidence that the distribution means are different ***

    Test statistic t = 2.71 on 14 df.
    Probability level (under null hypothesis) p = 0.017

95.0 % Confidence Interval for difference in means: ( 0.06840 , 0.5916 )
```

Exercise 6.1
Use a computer to carry out separate *t*-tests for the two varieties in Table 6.2.
- For both varieties the difference in means is 0.77.
- For variety one the *P*-value is 0.02 which indicates that the difference is significant.
- For variety two the *P*-value is 0.20 indicating no significant difference.
- Discuss the results.

6.2.2 Using a Calculator

In order to fully appreciate the computer output, we now show how to carry out the calculations for Example 6.2 by hand (the cookbook approach) and as an aid to understanding we later justify the method by presenting some background theory (Section 6.4).

To test the null hypothesis that the two treatments give the same yields, the difference between the two sample means is calculated and divided by their standard error of difference (*SED*) to give the *t*-value. The formula for *SED* will be discussed later.

$$t = \frac{\bar{x}_1 - \bar{x}_2}{SED}$$

The degrees of freedom (df) for this test are found by adding the degrees of freedom from each sample.

$$df = (n_1 - 1) + (n_2 - 1) = n_1 + n_2 - 2$$

where n_1 is the number of yields in the first sample and n_2 is the number of yields in the second.

The *t*-value is compared with the tabulated value in the *t*-table (Appendix 3). If the test is carried out at the 5% level and is a two-tailed test, this is found in the row corresponding to df and the column headed 2.5. If the calculated *t*-value is greater in magnitude than the table value, the null hypothesis is rejected. To carry out the test at the 1% you look in the column headed 0.05.

For Example 6.2, the degrees of freedom are $(n_1 + n_2 - 2) = 14$, so the *t*-table value for testing at the 5% level is $t_{(14, 2.5\%)} = 2.145$. Hence, if the calculated *t*-value is greater in magnitude than 2.145 we shall reject the null hypothesis at the 5% level. If it is greater than $t_{(14, 0.5\%)} = 2.977$ we shall reject at the 1% level.

You should be familiar with how to calculate the sample means and variances from Chapter 2. The calculations leading to the *t*-value are summarised in Table 6.3. You should check them carefully with your calculator.

- Next calculate the pooled estimate of the assumed common population variance:

$$s_p^2 = \frac{Sxx_1 + Sxx_2}{n_1 + n_2 - 2} = \frac{0.340 + 0.441}{14} = 0.0558$$

Table 6.3. Preliminary calculations leading to the *t*-test of Example 6.2

	New variety	Standard variety
n	6	10
$\bar{x}$	2.30	1.97
s^2	0.068	0.049
Sxx	0.34	0.441

- Then calculate the standard error of the difference between the two means:

$$SED = \sqrt{s_p^2\left(\frac{1}{n_1}+\frac{1}{n_2}\right)} = \sqrt{s_p^2\left(\frac{n_1+n_2}{n_1 \times n_2}\right)} = \sqrt{\frac{0.0558 \times 16}{60}} = 0.122$$

- Finally, calculate the t-value:

$$t = \frac{\bar{x}_1 - \bar{x}_2}{SED} = \frac{2.30 - 1.97}{0.122} = 2.70$$

Now compare this t-value with $t_{(14, 2.5\%)} = 2.145$.

The calculated t-value of 2.70 is greater than 2.145, so the observed difference in means is significant at the 5% level. It is not significant at the 1% level because 2.70 is not greater than $t_{(14, 0.5\%)} = 2.977$. This shows that if the null hypothesis is true, the probability of obtaining a t-value as far from zero as 2.70 is between 5% and 1%. The exact probability is the P-value of 0.017 (1.7%) which is given in Output 6.1. We conclude there is evidence that the two varieties give different yields. Remember, the smaller the P-value, the greater the evidence against the null hypothesis.

6.3 CONFIDENCE INTERVALS

In Example 6.2 the t-test showed evidence of a difference in the population mean yields. The magnitude of the difference can be estimated with a confidence interval. A 95% confidence interval for the true difference $(\mu_1 - \mu_2)$ between the variety means is

$$(\bar{x}_1 - \bar{x}_2) \pm t_{(n_1+n_2-2, 2.5\%)} \times SED$$

For Example 6.2 the t-table value on 14 df is 2.145, so the 95% confidence interval is

$$(2.30 - 1.97) \pm (2.145 \times 0.122) = 0.33 \pm 0.26 = (0.07, 0.59)$$

We are 95% confident that the mean yield of the new variety exceeds that of the standard by between 0.07 and 0.59 t/ha. This is shown in the Minitab and Genstat outputs.

The 95% interval includes all the null hypothesis values of $\mu_1 - \mu_2$ that would not be rejected at the 5% level on a two-tailed test. Hence, the calculation of a confidence interval is equivalent to carrying out a hypothesis test. For this example, the interval does not include zero, so the hypothesis that the difference in population means is zero is rejected at the 5% level.

6.4 THE THEORY BEHIND THE t-TEST

- The yields from the plots receiving treatment one have a sample mean $\bar{x}_1$ and a sample variance s_1^2. It is assumed that they are a random sample from a population having a mean μ_1 and variance σ_1^2.

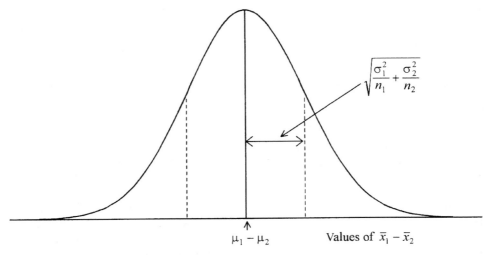

Figure 6.1. Normal distribution of possible values of $\bar{x}_1 - \bar{x}_2$

- The yields from the plots receiving treatment two have a sample mean $\bar{x}_2$ and a sample variance s_2^2. It is assumed that they are a random sample from a population having a mean μ_2 and variance σ_2^2.

Statistical theory shows that the quantity $(\bar{x}_1 - \bar{x}_2)$ is a value from a normal distribution with a mean $(\mu_1 - \mu_2)$ and variance

$$\left(\frac{\sigma_1^2}{n_1} + \frac{\sigma_2^2}{n_2} \right)$$

This is illustrated in Figure 6.1.

This is the distribution that would be predicted if the experiment was repeated indefinitely. After each experiment a different value of $(\bar{x}_1 - \bar{x}_2)$ would be obtained. These values would be approximately normally distributed with the mean and standard deviation shown in Figure 6.1.

The value of $(\bar{x}_1 - \bar{x}_2)$ found from a single experiment is considered to be a single value from this normal distribution. If the mean is subtracted and the result divided by the standard deviation, a value from a standard normal distribution is obtained. This is denoted by z:

$$z = \frac{(\bar{x}_1 - \bar{x}_2) - (\mu_1 - \mu_2)}{\sqrt{\frac{\sigma_1^2}{n_1} + \frac{\sigma_2^2}{n_2}}}$$

In practice, σ_1^2 and σ_2^2 are both unknown. We estimate them by s_1^2 and s_2^2, the sample variances.

The test is simpler if we assume that the two population variances are equal $(\sigma_1^2 = \sigma_2^2 = \sigma^2)$. This is a reasonable assumption if the two treatments are applied to similar material. In this case, the effect of treatment one is to change the yields of each experimental unit or plot by a fixed amount, δ_1. All plots receiving treatment two are expected to have their yields changed by δ_2. If this is true, the variation in yields is not affected by the treatments. This is the **additivity** assumption (Chapter 14).

- If $\sigma_1^2 = \sigma_2^2 = \sigma^2$, then s_1^2 and s_2^2 are both estimates of σ^2.
- If $n_1 = n_2$, a pooled estimate of σ^2 is the average of the two s^2 values.
- If n_1 and n_2 are not equal (Example 6.2) we calculate a weighted average of the two s^2 values. The weights are the respective degrees of freedom, $(n_1 - 1)$ and $(n_2 - 1)$. Hence the pooled estimate of σ^2 is

$$s_p^2 = \frac{(n_1 - 1)s_1^2 + (n_2 - 1)s_2^2}{(n_1 - 1) + (n_2 - 1)}$$

In the formula for z we substitute s_p^2 for both σ_1^2 and σ_2^2. The result is a value from a t-distribution with $(n_1 - 1) + (n_2 - 1) = (n_1 + n_2 - 2)$ degrees of freedom. The formula then becomes

$$t = \frac{(\bar{x}_1 - \bar{x}_2) - (\mu_1 - \mu_2)}{\sqrt{s_p^2 \left(\frac{1}{n_1} + \frac{1}{n_2}\right)}}$$

$\sqrt{s_p^2 \left(\frac{1}{n_1} + \frac{1}{n_2}\right)}$ is called the standard error of the difference between the two sample means (*SED*).

This formula can be used to test hypotheses about the difference between μ_1 and μ_2. When μ_1 and μ_2 represent the yields of two treatments, we usually wish to test the null hypothesis that $\mu_1 = \mu_2$. In this case **the test statistic is**

$$t = \frac{\bar{x}_1 - \bar{x}_2}{SED}$$

6.4.1 Summary of Procedure

- Set up the null hypothesis H_0: $\mu_1 = \mu_2$. You believe the two population means are equal.
- Before collection of data decide whether the test is to be one-tailed or two-tailed. For a two-tailed test the alternative hypothesis is H_1: $\mu_1 \neq \mu_2$.
  For a one-tailed test the alternative hypothesis is either H_1: $\mu_1 > \mu_2$ or H_1: $\mu_1 < \mu_2$.
- Before collecting the data, decide on the level of significance desired (5%, 1% or 0.1%, etc.).
- Collect the data values. They are assumed to be random samples from normal distributions with the same variance but possibly different means. The assumption of equal variances should be checked before proceeding with the t-test by carrying out an F-test (Section 6.5).
- Calculate

$$t = \frac{\bar{x}_1 - \bar{x}_2}{SED}$$

where $SED = \sqrt{s_p^2 \left(\frac{1}{n_1} + \frac{1}{n_2}\right)}$ and $s_p^2 = \frac{(n_1 - 1)s_1^2 + (n_2 - 1)s_2^2}{n_1 + n_2 - 2}$

Note that $(n_1 - 1)s_1^2 = Sxx_1$, the sum of the squares of the deviations of the sample 1 values from the sample 1 mean. Similarly $(n_2 - 1)s_2^2 = Sxx_2$, the sum of the squares of the deviations of the sample 2 values from the sample 2 mean. Hence the formula for s_p^2 can be written

$$s_p^2 = \frac{Sxx_1 + Sxx_2}{n_1 + n_2 - 2}$$

- Consult the *t*-tables. If the test is two-tailed at the 5% level, look in row $(n_1 + n_2 - 2)$, the degrees of freedom row, and column 2.5 (the tables give the top tail percentages). If the test is two-tailed at the 1% level, look in row $(n_1 + n_2 - 2)$ and column 0.5. If the test is one-tailed at the 5% level, look in row $(n_1 + n_2 - 2)$ and column 5%.
- If the magnitude of the calculated *t*-value is greater than the table value, reject the null hypothesis at the level of significance decided upon.

6.5 THE *F*-TEST

One of the conditions for the independent samples *t*-test to be valid is that the two population variances σ_1^2 and σ_2^2 are equal. The calculation of the pooled estimate of variance, s_p^2, should only be carried out if the assumption of equal variances is justified. The *F*-test is a test of this assumption. To test the null hypothesis that $\sigma_1^2 = \sigma_2^2$ divide the larger s^2 value by the smaller s^2 to obtain the variance ratio:

$$VR = \frac{\text{larger } s^2}{\text{smaller } s^2}$$

The test of equality of variances is two-tailed. By calculating the variance ratio as the ratio of the larger variance estimate to the smaller variance estimate, a ratio bigger than one is always obtained. The test then becomes one-tailed as we only reject the null hypothesis for sufficiently large values of the variance ratio.

If you use the 5% *F*-table (Appendix 4a) you are, in effect, carrying out the two-tailed test of equality at the 10% level. To undertake this two-tailed test at the 5% level you need to carry out the one-tailed test at the 2.5% level. To do this consult the 2.5% table of the *F*-distribution (Appendix 4b). Look down the column corresponding to the degrees of freedom for the larger s^2 value and the row corresponding to the degrees of freedom for the smaller s^2 value. If the calculated VR is less than the table value, the two sample variances are not significantly different and the pooled estimate s_p^2 can be used in the *t*-test.

COMPARISON OF TWO INDEPENDENT SAMPLE MEANS 59

Example 6.3
We now carry out the F-test for the data of Example 6.2. We found $s_1^2 = 0.068$ and $s_2^2 = 0.049$:

$$VR = \frac{\text{larger } s^2}{\text{smaller } s^2} = \frac{0.068}{0.049} = 1.39 \text{ (SAS output in Output 6.1)}$$

If the two population variances were equal we would expect this ratio (often called the calculated F) to be 1. To find if 1.39 is significantly greater than 1 we carry out the F-test.

The degrees of freedom for the larger s^2 are $(6 - 1) = 5$, and the degrees of freedom for the smaller s^2 are $(10 - 1) = 9$. The F-table value at the 2.5% level is $F_{(5,9,2.5\%)} = 4.484$. This is found by looking in column 5 and row 9 of Appendix 4b. The calculated value (1.39) is less than the table value (4.484). Therefore the two sample variances are not significantly different and the calculation of s_p^2 is valid. Figure 6.2 showing the F-distribution with 5 and 9 df illustrates the results of this test. The area under the curve to the right of 1.39 is obviously greater than 0.025 which shows that the two-tailed test is not significant at the 5% level. This area is doubled to find the correct P-value for the two-tailed test. Output 6.1b shows it to be 0.6296.

6.6 UNEQUAL SAMPLE VARIANCES

If the F-test shows that s_1^2 and s_2^2 are not significantly different we can proceed with the t-test as described above. If the F-test shows that they *are* significantly different, we should not pool the two variance estimates. Instead we should calculate the t-value using the formula

$$t = \frac{\bar{x}_1 - \bar{x}_2}{\sqrt{\frac{s_1^2}{n_1} + \frac{s_2^2}{n_2}}}$$

Unfortunately, this t-value does not follow the t-distribution exactly.

However, if n_1 and n_2 are both large (at least 30), the approximation is good and the infinity row (corresponding to large samples) in the t-table can be used.

If the samples are small, an approximate test can be carried out using a complicated formula for the degrees of freedom. This formula is

$$df = \frac{(\nu_1 + \nu_2)^2}{\left(\frac{\nu_1^2}{n_1 - 1} + \frac{\nu_2^2}{n_2 - 1}\right)}$$

where $\nu_1 = s_1^2/n_1$ and $\nu_2 = s_2^2/n_2$.

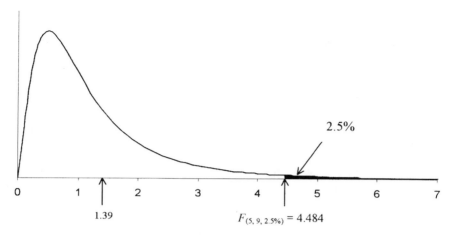

Figure 6.2. The F-distribution with 5 and 9 degrees of freedom

The answer for df is rounded down to the next smallest integer and the ordinary t-table used. If you use the unequal variance option for Example 6.2 in Minitab you will get 9.3 for the df, and this is also shown automatically by SAS (Output 6.1).

6.7 DETERMINATION OF SAMPLE SIZE FOR A GIVEN PRECISION

Suppose we wish to compare two treatments. How many replicates per treatment do we need if a difference in means of δ is to be declared significant at the 5% level? This question is difficult to answer because it depends on σ^2, the variance in yields of plots treated alike. This variation is unlikely to be known before the experiment, but it can be guessed from results of past similar experiments.

A difference in means of δ will be found significant at the 5% level if r, the number of replications per treatment is, such that

$$t \times \sqrt{\frac{2 \times \sigma^2}{r}} < \delta$$

This implies that r must be at least $2t^2\sigma^2/\delta^2$ where t is the t-table value. The table value depends on the number of replicates, but for a reasonable number $t = 2$ is a good first approximation.

Example 6.4

Suppose that σ^2 is assumed to be 40, and you wish to declare a difference in means of 5 to be significant. What is the minimum replication required?

Using 2 as an approximation for t, you require r at least

$$\frac{2 \times 4 \times 40}{25} = 12.8$$

Because r must be a whole number, use 13 as the first approximation. As $t_{(13, 2.5\%)} = 2.160$, the second approximation for r is $(2 \times 2.16^2 \times 40)/25 = 14.93$. Confirm that the third approximation is 14.53, using $t_{(15, 2.5\%)} = 2.131$. Hence you require at least 15 replications per treatment.

This method gives only a rough idea of the number of replicates needed. In theory, the more replicates, the better the precision of the comparison of the treatment means. In practice, the variation in the measured yields may increase with replication because if the plots cover a large area, the plot-to-plot variation is likely to be larger than if a smaller number of plots are used over a small area. Also, with larger experiments the non-sampling errors will tend to increase as more mistakes are made in making measurements. As the experiment increases in size, more workers may be required, and this will increase the variation in the measured yields.

An alternative method of obtaining a value of σ^2 to use in the above formula is to substitute a value of coefficient of variation which is typical of the type of experiment. For example, if the CV is 20%, the standard deviation is 20% of the overall mean. If the overall mean yield for the two treatments is expected to be in the region of 50, then σ should be about 20% of 50, which is 10. A value of 100 can then be used for σ^2 when determining the sample size required.

Exercise 6.2
An experiment was designed to compare two varieties of spring barley. Twenty plots were used, ten being randomly allocated to variety A and ten to variety B. Unfortunately one plot was destroyed. The yields (t/ha) from the remaining plots were as follows:

Variety A	3.6	3.4	4.3	4.8	3.5	4.4	4.8	3.9	4.7	3.4
Variety B	4.9	4.0	4.9	4.1	4.4	5.2	4.9	5.3	4.6	

Find the standard error of the difference between the two means and hence test the null hypothesis that the two varieties give equal yields.

Exercise 6.3
A pot experiment was carried out to compare the effects of two fertiliser treatments F_1 and F_2 on the yield of tomato plants. The results (kg) were:

Treatment	Mean yield/plant	Standard deviation (s)	No. of plots
F_1	5.68	0.77	10
F_2	7.00	0.95	7

Test the hypothesis that the two fertilisers have equal effects on the yields and calculate 95% confidence limits for the difference in mean yield per plant. Assume that each pot consisted of one plant and that plants were allocated to the treatments at random.

Answer to Exercise 6.2
Variety A: $n = 10$, $\bar{x} = 4.08$, $s^2 = 0.344$ Variety B: $n = 9$, $\bar{x} = 4.70$, $s^2 = 0.210$

$F = 1.64$ on $(9, 8)$ df. From F-tables, $F_{(9,8,2.5\%)} = 4.36$ approx. Calculated F is less than the table value so you can assume similar population variances and calculate a pooled estimate, $s_p^2 = 0.2809$. Using this find $SED = 0.2435$ and $t = -2.546$ on 17 df. From tables $t_{(17,2.5\%)} = 2.110$ and $t_{(17,0.5\%)} = 2.898$. The calculated t-value is greater than 2.110, but not greater than 2.898, so conclude that there is a significant difference between the varieties with respect to yield ($P < 0.05$ for a two-tailed test). A difference in means as great as that observed has a less than 5% chance of occurring if the varieties have equal yielding abilities, assuming a completely randomised design. The exact P-value found by computer is 0.021.

Answer to Exercise 6.3
Using the summary of calculations given in the question, $F = 1.52$ on $(6, 9)$ df. From tables, $F_{(6,9,2.5\%)} = 4.32$. The calculated F is less than the table value so the data can be pooled to get $s_p^2 = 0.7167$, $SED = 0.4172$ and $t = -3.164$ on 15 df. From tables $t_{(15,2.5\%)} = 2.131$, $t_{(15,0.5\%)} = 2.947$, and $t_{(15,0.05)} = 4.073$. As 3.164 lies between 2.947 and 4.073 the difference in means of 1.32 is significant at the 1% level on a two-tailed test. The 95% confidence limit for the true difference in mean yields is $-1.32 \pm (2.131 \times 0.4172) = -1.32 \pm 0.89 = (-2.21, -0.43)$. You are 95% confident that the mean yield of F_2 exceeds that of F_1 by between 0.43 and 2.21 kg per plant.

Note: If F_1 was a standard variety and you wished to know if F_2 gives a significantly increased yield you would have carried out a one-tailed test and stated a one-sided confidence interval. For a one-tailed test at the 5%, 1%, and 0.1% levels, the corresponding table values are $t_{(15,5\%)} = 1.753$, $t_{(15,1\%)} = 2.602$ and $t_{(15,0.1\%)} = 3.733$. The calculated t-value of 3.164 in magnitude is thus significant at the 1% level. The lower limit of a 90% confidence interval for the mean difference is $1.32 - (1.753 \times 0.4172) = 0.59$. Thus you are 95% confident that, on average, the mean yield per plant of variety F_2 exceeds that of variety F_1 by at least 0.59 kg.

Chapter 7

Linear Regression and Correlation

7.1 BASIC PRINCIPLES OF SIMPLE LINEAR REGRESSION (SLR)

The applied biologist frequently finds it necessary to consider the simultaneous variation within a number of different parameters. He or she may wish to investigate relationships between yield, environmental data and yield components. For example, such studies can be used to assess whether an increase or decrease in one variable, such as number of pods per plant, can be explained by change in another such as plant population and light intensity.

Study of simultaneous change in two (or more) variables can be carried out using techniques of regression and correlation. A regression equation is a mathematical relationship which can be used to determine the expected value of a **dependent** (or **response**) variable for a given value of a correlated **independent** (or **predictor**) variable. The independent variable is generally denoted by X and the dependent variable by Y. For example, a study of the effect of change in plant population (X) on the number of pods per plant (Y) is referred to as the regression relationship of Y on X. The results of a regression analysis provide information on the average rate of change in number of pods per unit increase or decrease in plant population. Regression equations can be used to provide a succinct summary of large amounts of experimental and observational data.

SLR can be used in the following circumstances:

- To find out whether two variables are connected by a straight-line relationship. If so, we are interested in the fitted equation.
- If the X-values are fixed by the experimenter (for example, different fertiliser concentrations) and the Y-values are measured (for example, yields) then the relationship between the two variables can be studied and predictions made about future Y-values for given X-values.
- Suppose one variable is difficult, and another is easy to measure. We can carry out a study to find out if these variables are connected by a straight-line relationship.

If they are, then future values of one variable can be estimated from measurements on the other. For example, the leaf area of linseed which has small ovate leaves is more difficult to measure than stem length. If the equation connecting these variables is known, it can be used to estimate leaf area from measurements of stem length.

The methods described in this chapter are illustrated with two data sets. We realise that most readers will be using a computer, and will just want to interpret the output. We cater for the needs of these readers with Example 7.1. However, a full understanding will only come after a consideration of background theory and the assumptions which make the analyses valid. For the data of Example 7.2, we take a step-by-step approach showing how the calculations are made and also give justification and interpretation of the statistical tests carried out. At this stage the reader will be in a strong position to understand the detailed computer output (Output 7.2), and use it selectively for presentation and further discussion.

Example 7.1
Data (Table 7.1) were collected on 34 linseed plants and the results entered in two columns of a computer spreadsheet. For each plant, the stem length was typed in column one and the corresponding leaf area in column two. A scatter diagram was produced with the fitted regression line superimposed (Figure 7.1); Minitab was used to carry out a statistical analysis (Output 7.1).

7.1.1 Interpretation of Computer Output

A full understanding of the output can only be obtained after studying background theory and carrying out the calculations on a simple example (Example 7.2). However, as most students and researchers only consult their computer output, a brief descriptive summary of Output 7.1 is now given.

Table 7.1. Stem length (cm) and leaf area (cm^2) for 34 linseed plants

Length	Area	Length	Area	Length	Area	Length	Area
22	31	50	68	67	108	82	116
23	36	51	80	68	96	84	120
25	35	52	76	70	98	87	126
27	36	55	80	75	102	88	125
30	50	59	78	78	96	89	132
30	49	60	96	79	100	92	138
35	52	60	100	80	104	105	139
38	56	62	86	81	110	107	142
45	55	65	106				

LINEAR REGRESSION AND CORRELATION

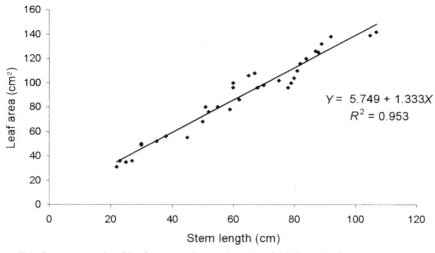

Figure 7.1. Scatter graph of leaf area and stem length of 34 linseed plants

Output 7.1 Regression analysis of linseed data (Minitab output)

```
The regression equation is
LeafArea = 5.75 + 1.33 StemLength

Predictor     Coef       StDev          T           P
Constant      5.749      3.479          1.65        0.108
StemLeng      1.33264    0.05215        25.56       0.000

S = 7.189   R-Sq = 95.3%   R-Sq(adj) = 95.2%

Analysis of Variance

Source           DF      SS         MS          F           P
Regression       1       33750      33750       653.07      0.000
Residual Error   32      1654       52
Total            33      35404

Unusual Observations
Obs         StemLeng    LeafArea    Fit      StDev Fit    Residual    St Resid
16          60          100.00      85.71    1.24         14.29       2.02R

R denotes an observation with a large standardized residual
```

Note: StDev given in this output is an estimate of the standard error.

In Example 7.1 we assume that X (stem length) and Y (leaf area) are connected by the equation

$$Y = \alpha + \beta X$$

where the intercept α is the expected value of Y when X is zero, and the slope β is the expected increase in Y if X is increased by one unit (1 cm).

The equation of the line of best fit has been estimated as $Y = 5.749 + 1.333X$ and superimposed on the graph. As stem length is easy to measure, we can use this equation to estimate a leaf area from a stem measurement. For example, a plant with a stem length of 80 cm is expected to have a leaf area of $5.749 + 1.333 \times 80 = 112.4$ cm^2.

We conclude that when stem length increases by 1 cm, leaf area is expected to increase by 1.333 cm^2. The standard error of this slope is 0.052. On dividing 1.333 by 0.052 we obtain 25.56 as the t-value for testing whether the slope is significantly different from zero. The corresponding P-value is zero to three decimal places; hence there is practically no chance of these data being obtained if the true slope is zero. We are almost 100% confident that the true slope is not zero.

The intercept is 5.749. However, this must be interpreted with caution. A plant with zero stem length cannot have any leaf area! In Example 7.1 there were no plants with stem lengths less than 20 cm, so the form of the relationship may not be linear for plants with stem lengths less than this.

The R^2 value is 95.3%. This indicates that 95.3% of the variation in the leaf area values can be explained by the linear relationship with stem length (see Section 7.6.2 for an explanation of the adjusted R-square of 95.2%). The analysis of variance table shows that the total variation in the leaf area values is 35 404 (their corrected sum of squares) and 33 750 of this is due to the regression. By dividing 33 750 by 35 404 we obtain R^2. The residual, or unexplained, variation is 1654. This is the sum of the squares of the deviations of the observed leaf areas from those predicted by the equation (the fitted values). The F-value is 653.07 which is expected to be about 1.0 if there is no linear regression. The corresponding P-value is 0.000. The high F-value and small P-value indicate a strong linear relationship.

Note 1: As the square root of the F-value is the t-value of 25.56 and the corresponding P-values are the same, the F-test is equivalent to the t-test.

Note 2: For the hypothesis tests to be valid we assume that for each X-value, the corresponding Y-value is a random value from a normal distribution with an unknown variance. This variance is assumed to be the same for all the X-values which are assumed to be either fixed in advance or measured without error.

7.2 EXPERIMENTAL VERSUS OBSERVATIONAL STUDIES

In many experiments, the X-values are under the control of the experimenter. For instance, if several plots are given different amounts of nitrogen fertiliser to discover the effect on yield, then yield is the Y (**dependent** or **response**) variable and nitrogen level is the X (**independent** or **predictor**) variable.

In observational studies it may not be immediately clear which should be the Y and which the X variable. Observations are made on 'cases'. A tree could be a 'case'.

For each tree one could measure its height and the diameter of its trunk at 1.5 m above ground level. Does height depend on diameter or does diameter depend on height? As diameter is much easier to measure, we could use any relationship between these two variables in order to estimate the heights of trees from measurements of their diameter. In this situation we would carry out a regression of height (Y) on diameter (X). When several variables are measured on each 'case', none of them may be considered a response variable. A correlation matrix containing the correlation coefficients between all possible pairs of variables may be a good summary of any possible relationships and can be used to stimulate further research.

7.3 THE CORRELATION COEFFICIENT

This is a measure of the extent to which two variables are linearly related. The correlation coefficient, denoted by r, must lie between -1 and $+1$. Its value does not depend on the units of measurement or on which is the dependent and which is the independent variable, so it is suitable for summarising relationships in observational studies. A regression analysis should be carried out instead when it is clear which is the response variable. Section 7.7 shows how to calculate r and Section 7.9 how to test whether it is significantly different from zero.

A correlation coefficient of $+1$ indicates a perfect positive linear relationship between two variables (Figure 7.2(a)). A correlation coefficient of -1 indicates a perfect negative linear relationship (Figure 7.2(b)). A correlation coefficient of zero may result when there is no relationship (Figure 7.2(c)) or when the relationship is markedly non-linear (Figure 7.2(d)).

Note: A strong linear relationship between two variables, indicated by a correlation coefficient near $+1$ or -1, does not necessarily imply causation. A third unmeasured variable could affect both X and Y.

A correlation coefficient has limited value as a descriptive statistic. It is a single number which summarises the extent to which there is a straight-line relationship. Its size does not tell you whether it is significantly different from zero. A value of 0.8 is not significant if based on only 6 pairs, whereas a value of 0.2 based on 100 pairs is (Section 7.9). Even a non-significant value should be treated with caution as there may be a relationship similar to Figure 7.2(d). There is no substitute for a plot of the data.

7.4 THE LEAST SQUARES REGRESSION LINE AND ITS ESTIMATION

The method almost universally used to calculate estimates of α and β when given a set of n (X, Y) pairs is the method of least squares. To understand this, refer to Figure 7.3.

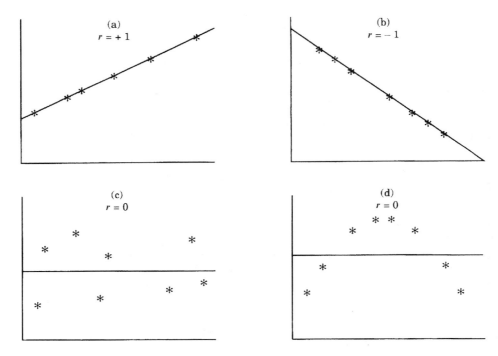

Figure 7.2. Graphs showing correlation coefficient of −1, +1 and 0

The line fitted through the points is called the **least squares** regression line of Y on X. This is drawn so that the sum of the squares of the deviations (the dotted vertical lines) is a minimum. These deviations are called the residuals. For any other line drawn through these data this sum of squares would be larger. Assuming an underlying linear relationship with slope β and intercept α the least squares estimators $\hat{\alpha}$ and $\hat{\beta}$ are calculated using the formulae

$$\hat{\beta} = \frac{Sxy}{Sxx} \quad \text{and} \quad \hat{\alpha} = \bar{Y} - \hat{\beta}\bar{X}$$

- Sxy = corrected sum of products = $\Sigma(X - \bar{X})(Y - \bar{Y}) = \Sigma XY - \frac{(\Sigma X)(\Sigma Y)}{n}$
- Sxx = corrected sum of squares of the X-values = $\Sigma(X - \bar{X})^2 = \Sigma X^2 - \frac{(\Sigma X)^2}{n}$

Thus the equation of the fitted line is $\hat{Y} = \hat{\alpha} + \hat{\beta}X$.

$\hat{Y}$ is the fitted (or expected Y-value) for a given X-value. To draw the fitted line, substitute two extreme values of X into the equation in order to find the two corresponding fitted Y-values, then join up the two points.

Note: In the equation of the fitted line, ($\hat{Y} = \hat{\alpha} + \hat{\beta}X$), if $\bar{X}$ (the mean of the X-values) is inserted we obtain

LINEAR REGRESSION AND CORRELATION

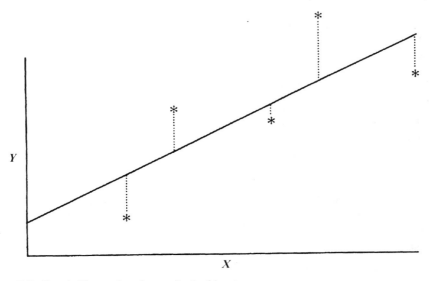

Figure 7.3. Graph illustrating the method of least squares

$$\hat{Y} = \hat{\alpha} + \hat{\beta}X$$
$$= (\bar{Y} - \hat{\beta}\bar{X}) + \hat{\beta}\bar{X}$$
$$= \bar{Y}$$

Thus the fitted line must pass through the point $(\bar{X}, \bar{Y})$.

The adequacy of the model can be investigated by looking at the residuals. For the ith point the residual is $e_i = Y_i - \hat{Y}_i$ where $\hat{Y}_i = \hat{\alpha} + \hat{\beta}X_i$.

The sum of the squares of these residuals, the residual sum of squares (*ResidSS*), is smaller than would be obtained using any other line through the data.

In Example 7.2 we illustrate the method of least squares showing how the calculations can be carried out using a hand calculator or computer spreadsheet.

Example 7.2
Yields (Y) of wheat receiving different amounts (X) of nitrogen fertiliser:

X (kg N/ha)	0	25	50	75	100	125
Y (t/ha)	3.70	4.45	4.75	5.20	5.15	4.95

The preliminary calculations leading to the equation of the fitted line are summarised in Table 7.2.
Check the following calculations:
$n = 6$ $\Sigma X = 375$ $\Sigma Y = 28.20$
$\bar{X} = 62.5$ $\bar{Y} = 4.70$ $Sxx = 10\,937.50$ $Sxy = 110.0$

$$\hat{\beta} = \frac{Sxy}{Sxx} = \frac{110.00}{10\,937.50} = 0.010057$$

$$\hat{\alpha} = \bar{Y} - \hat{\beta}\bar{X} = 4.70 - 0.010057 \times 62.5 = 4.0714$$

So the fitted equation is $\bar{Y} = 4.0714 + 0.010057X$.

This implies that for every 1 kg/ha increase in nitrogen applied, the yield is expected to increase by about 0.01 t/ha. Figure 7.4 shows the fitted line superimposed on the data points. To draw it note that when $X = 0$, $\hat{Y} = 4.07$ and when $X = 125$, $\hat{Y} = 4.0714 + 0.010057 \times 125 = 5.328$. Plot the points (0, 4.07), (125, 5.33) and join them with a straight line.

The fitted equation can be used for predicting future Y-values from given X-values, but this is not recommended for X-values well outside the range used in estimating the equation. On looking at the points in Figure 7.4 it appears that yields start to decrease after $X = 75$. This suggests that a quadratic curve would be a better fit than a straight line.

Note: Sxx can be obtained directly from your calculator by squaring the sample standard deviation of the X-values and multiplying by $n - 1$ (Section 2.3.5). Also, some calculators allow the values of $\hat{\alpha}$ and $\hat{\beta}$ to be obtained directly.

Table 7.2. Summary of calculations needed to estimate the regression equation

	X	Y	$(X - \bar{X})$	$(Y - \bar{Y})$	$(X - \bar{X})(Y - \bar{Y})$	$(X - \bar{X})^2$
	0	3.70	−62.50	−1.00	62.500	3 906.25
	25	4.45	−37.50	−0.25	9.375	1 406.25
	50	4.75	−12.50	0.05	−0.625	156.25
	75	5.20	12.50	0.50	6.250	156.25
	100	5.15	37.50	0.45	16.875	1 406.25
	125	4.95	62.50	0.25	15.625	3 906.25
Total	375	28.20	0.00	0.00	110.000	10 937.50

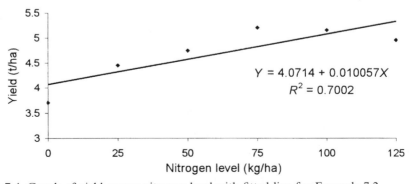

Figure 7.4. Graph of yield versus nitrogen level with fitted line for Example 7.2

7.5 CALCULATION OF RESIDUALS

For each observed (X, Y) pair the residual is the observed Y-value minus the fitted Y-value. In a scatter diagram with the fitted line superimposed, each residual is the height of the point minus the height of the line:

$$\text{Residual} = Y - \hat{Y}$$

Examination of the residuals is important as it allows us to investigate the adequacy of the model. For the least squares regression line, the sum of the squares of the residuals, the **Residual Sum of Squares** ($ResidSS$) is smaller than it would be for any other line through these data.

To calculate the residuals substitute the observed X-values, one at a time, into the fitted equation to give the corresponding fitted Y-values. The residuals are then found by subtracting these fitted values from the observed Y-values. For data of Example 7.2 the fitted Y-value corresponding to $X = 0$ is

$$\hat{Y} = 4.0714 + 0.010057 \times 0 = 4.0714$$

So the corresponding residual is $Y - \hat{Y} = 3.70 - 4.0714 = -0.3714$.

As can be seen from Figure 7.4, the first point is about 0.37 **below** the fitted line. The second fitted Y-value is $4.0714 + 0.010057 \times 25 = 4.3229$ and the corresponding residual is $4.45 - 4.323 = 0.1271$.

Thus the second point is about 0.13 above the line. The other residuals are found in a similar way. The results are summarised in Table 7.3.

Note that the sum of the residuals is zero (which is always true for a least squares regression line) and the sum of squares of these residuals is $ResidSS = 0.4737$. Output 7.2 also shows the standardised residuals. They are independent of the units of measurement and are calculated by dividing the residuals by their estimated standard error. Standardised residuals outside the range -2 to $+2$ are considered to be large and give evidence that the model is not a good fit.

There is a pattern in these residuals because the first one is negative (point below line), the next four are positive (points above line), and the last one is negative (point below line). This suggests a curve would be a better fit than a straight line (Figure 7.4).

Table 7.3. Fitted values and residuals for data of Example 7.2

X	Y	Fitted Y	Residual
0	3.70	4.0714	−0.3714
25	4.45	4.3229	0.1271
50	4.75	4.5743	0.1757
75	5.20	4.8257	0.3743
100	5.15	5.0771	0.0729
125	4.95	5.3286	−0.3786

7.6 THE GOODNESS OF FIT

The residuals can be used to give a measure of how well the fitted line explains the observed data. If the points lie **exactly** on the fitted line (as in Figure 7.2(a)) then the residuals are all zero and $ResidSS = 0$. In this case the variation in the Y-values about their mean $(Syy = \Sigma(Y - \bar{Y})^2)$ is completely explained by the linear regression model. Thus the smallest value $ResidSS$ can take is zero and this occurs when all the points lie exactly on a straight line. The largest value that $ResidSS$ can take is Syy. To understand this, refer to Figure 7.5 which shows the fitted line and a typical observed point P. The horizontal dashed line is drawn through the mean of the Y-values so, for the point P, the vertical distance PR is $(Y - \bar{Y})$. PQ is the residual and QR is $(\hat{Y} - \bar{Y})$, the fitted Y-value minus the Y mean. Thus

$$(Y - \bar{Y}) = \text{Residual} + (\hat{Y} - \bar{Y})$$

This shows that the difference between each Y-value and the Y mean is made up of the residual and the difference between the fitted Y-value and the Y mean.

It can also be proved algebraically that, taking all the points together, the sum of the squares of the deviations of the Y-values about their mean (Syy) is made up of the Residual Sum of Squares $(ResidSS)$ and the sum of the squares of the deviations of the fitted Y-values from the Y mean. This latter sum of squares is called the Regression Sum of Squares which we denote by $RegSS$. Hence

$$\Sigma(Y - \bar{Y})^2 = \Sigma(Y - \hat{Y})^2 + \Sigma(\hat{Y} - \bar{Y})^2 \quad \text{that is}$$
$$Syy = ResidSS + RegSS$$

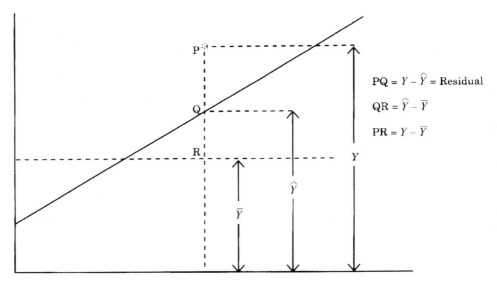

Figure 7.5. Diagram showing the relationship between the fitted values, residuals and the Y mean

LINEAR REGRESSION AND CORRELATION   73

This relationship can be shown to be true for data of Example 7.2 by making the relevant calculations and summarising them in Table 7.4. The $\hat{Y}$-values are in the third column of Table 7.3 and the residuals are in the last column. On rounding off to three decimal places $0.474 + 1.106 = 1.580$ so the relationship is verified in this case.

Note 1: *ResidSS* can also be found by subtraction using the formula $ResidSS = Syy - RegSS$, where $Syy = \Sigma(Y - \bar{Y})^2$ and $RegSS = \Sigma(\hat{Y} - \bar{Y})^2$.

Note 2: *RegSS* can also be found using the formula $RegSS = \hat{\beta}Sxy$.

For Example 7.2, $RegSS = 0.010057 \times 110.00 = 1.1063$ as obtained in Table 7.4.

Returning to our original investigation of the goodness of fit, it is now obvious that *ResidSS* must lie between 0 (when $RegSS = Syy$) and Syy (when $RegSS = 0$). When the fitted line is horizontal all the fitted Y-values ($\hat{Y}$) are equal to the Y mean ($\bar{Y}$) and $RegSS = 0$. A horizontal fitted line implies no linear regression—on average, Y neither increases nor decreases with X. This does not necessarily mean the points are scattered at random; they could form a pattern similar to Figure 7.2(d).

7.6.1 The R-square Value (R^2)

By dividing *RegSS* by Syy we can obtain a measure of the goodness of fit of the points to a straight line. When the fit is perfect $RegSS = Syy$ so this ratio is equal to 1. When there is no linear regression, the ratio is equal to zero. This ratio is usually denoted by R^2:

$$R^2 = \frac{RegSS}{Syy}$$

If we express R^2 as a percentage we can interpret it as the percentage of the variation in the Y-values accounted for by the linear regression model. For Example 7.2, $R^2 = 1.106/1.580 = 0.700$. Hence, we conclude that 70% of the variation in the yield values is accounted for by the linear regression model. This is shown in Output 7.2.

Table 7.4. Table showing calculation of $Syy(1.5800)$, $ResidSS(0.4737)$ and $RegSS(1.1062)$

X	Y	$(Y - \bar{Y})^2$	Residual^2	$(\hat{Y} - \bar{Y})^2$
0	3.70	1.0000	0.1379	0.3951
25	4.45	0.0625	0.0162	0.1422
50	4.75	0.0025	0.0309	0.0158
75	5.20	0.2500	0.1401	0.0158
100	5.15	0.2025	0.0053	0.1422
125	4.95	0.0625	0.1433	0.3951
Total		1.5800	0.4737	1.1062

7.6.2 The Adjusted R-square

The R-square value is adequate as a measure of the goodness of fit for simple linear regression when the estimates of α and β are found by the method of least squares and the assumptions of normality and homogeneity of variance (Section 7.8) are valid. When more parameters are added to the model as in polynomial regression (Chapter 8) or more explanatory variables are added as in multiple regression, R^2 is bound to increase. The adjusted R-square gives a measure of goodness of fit after allowing for the extra terms in the model, so it can be used to compare models using the same X- and Y-values. Its value can be found from the ANOVA table using the formula

$$\text{Adjusted R-square} = \frac{\text{Total mean square} - \text{Residual mean square}}{\text{Total mean square}}$$

For Example 7.2, the total mean square is $1.5800/5 = 0.3160$, and the residual mean square is 0.1184. Hence the adjusted R-square is $(0.3160 - 0.1184)/0.3160 = 0.625$. This is shown in Output 7.2 as 62.5%.

Note: Genstat calls the adjusted R-square the **percentage variance accounted for**.

7.7 CALCULATION OF THE CORRELATION COEFFICIENT

Although R^2 is a measure of the extent to which data lie on a straight line, it is always positive. A correlation coefficient must lie between -1 and $+1$. It is calculated as the square root of R^2 and denoted by r and is given a positive sign if the slope of the fitted line is positive. If the fitted line has a negative slope then the correlation coefficient is negative. For Example 7.2

$$r = \sqrt{0.700} = 0.837$$

An alternative formula for the correlation coefficient is usually given in textbooks and it is found as follows. Replace $RegSS$ by $\hat{\beta} \times Sxy$ in the expression for R^2 and then replace $\hat{\beta}$ by Sxy/Sxx. Then

$$R^2 = \frac{RegSS}{Syy} = \frac{\hat{\beta} \times Sxy}{Syy} = \frac{Sxy \times Sxy}{Sxx \times Syy} = \frac{(Sxy)^2}{Sxx \times Syy}$$

On taking the square root,

$$r = \frac{Sxy}{\sqrt{Sxx \times Syy}}$$

For data of Example 7.2,

$$r = \frac{110.00}{\sqrt{10\,937.5 \times 1.580}} = \frac{110.00}{131.458} = 0.837 \text{ as obtained above}$$

Note: The last formula for r is symmetric in X and Y which indicates that the same answer will be obtained if the roles of X and Y are reversed. It can be shown that the product of the two regression coefficients (slopes) is r^2. When the X-values are fixed by the experimenter it is best not to quote a correlation coefficient as this would imply that X is a random variable.

Exercise 7.1

Use a computer or hand calculator to find the regression equation of stem length on leaf area for the data of Example 7.1. In other words assume the response variable is $Y =$ stem length, and the predictor is $X =$ leaf area. Confirm the fitted equation is $Y = -1.199 + 0.7153X$. From Output 7.1 the fitted equation for the regression of leaf area on stem length is $Y = 5.749 + 1.3326X$ and the R^2 value is 0.953. Now multiply the two regression coefficients to obtain $r^2 = 0.7153 \times 1.3326 = 0.9532$. The square root is the correlation coefficient, $r = 0.976$.

7.8 ASSUMPTIONS, HYPOTHESIS TESTS AND CONFIDENCE INTERVALS FOR SIMPLE LINEAR REGRESSION

So far, our discussion and analysis of the data for Example 7.2 has been descriptive. We have estimated the intercept and slope, and now show how to test reliability and goodness of fit. It is important to realise that these procedures are only valid if certain assumptions are true. We assume that for each observed X-value, the corresponding observed Y-value is just one value taken at random from a normal distribution with mean $\alpha + \beta X$ and variance σ^2. This implies that as X increases, the means of these normal distributions lie on a straight line and each normal distribution has the same variance σ^2 (Figure 7.6).

Suppose the underlying relationship was truly linear, then for a given X, if we were able to make a very large number of independent observations of Y, the mean of all these Y-values would be $\alpha + \beta X$, and would lie on the dotted line of Figure 7.6. A problem associated with many studies in plant science is that, for a given X, we only observe one Y-value which could be a long way from the expected value. In mathematical notation the assumption is that for a given X-value our observed Y-value is given by the equation

$$Y = \alpha + \beta X + \epsilon$$

and that ϵ comes from a normal distribution with mean 0 and variance σ^2. The values of ϵ from successive observations are assumed to be independent. They can never be found but can be estimated, after fitting the line, by the residuals $Y - \hat{Y}$. Consequently an analysis of these residuals can be used to test the assumptions of the model. A plot of the residuals against the fitted values should show a random scatter of points about a horizontal line through the origin. Any distinct pattern in this plot should be taken as evidence that the basic assumptions of the linear regression model

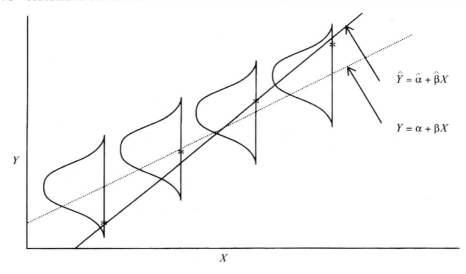

Figure 7.6. Diagram illustrating the assumptions of the linear regression model

are not correct and further analysis should be carried out. In the case where the X variable represents time there is likely to be correlation between the residuals as successive observations tend not to be independent. If the assumptions are seriously in doubt, you may consider the rank correlation coefficient. This is obtained by ranking the two sets of values independently and using the ranks to calculate r.

In Example 7.2 assume the yield of 4.45 t/ha was obtained from a single plot on which 25 kg N/ha had been applied. If a neighbouring plot had been used the yield may well have differed considerably from 4.45 t/ha due to plot-to-plot variation in soil conditions. The plot to which 50 kg N/ha was applied gave a yield of 4.75 t/ha. Perhaps this increase in yield (0.30 t/ha) was due to factors other than an increase in nitrogen level.

Suppose we have four X-values as in Figure 7.6. By chance, the first Y-value is a long way below the mean; the second is a little below its mean; the third is a little above its mean, and the fourth is a long way above its mean. This misleading situation may easily arise by chance in experiments, and so the least squares regression line (the full line) has a much greater slope than the population line shown dotted in Figure 7.6. In practice we have no idea of the true underlying relationship, as the true values of α and β are unknown. Only the experimental data points are available. The best we can do in this situation is to find estimates of α and β by the method of least squares as described earlier and then give confidence intervals for them. First we have to estimate σ^2.

7.8.1 Estimation of σ^2

Assuming that the underlying relationship is linear, and that for each observed X the corresponding Y-value comes from a normal distribution with variance σ^2, the 'best' estimate of σ^2 is

LINEAR REGRESSION AND CORRELATION

$$\hat{\sigma}^2 = \frac{ResidSS}{n-2} = RMS$$

where *ResidSS* is the residual sum of squares and $n-2$ is called the **residual degrees of freedom** and *RMS* is called the **residual mean square**. In effect we have 'lost' two degrees of freedom because we have estimated α and β.

7.8.2 Testing the Significance of the Linear Regression

To test whether there is significant linear regression we set up the null hypothesis that $\beta = 0$. i.e. $H_0: \beta = 0$. If this hypothesis is true we would expect $\hat{\beta}$, and hence *RegSS*, to be zero.

If H_0 is true, the regression mean square ($RegMS = RegSS/1$) is also an estimator of σ^2. If H_0 is not true, *RegMS* estimates $\sigma^2 + \beta^2 Sxx$.

To test H_0 we find the Variance Ratio $VR = RegMS/RMS$ on $(1, n-2)$ df and compare it with *F*-tables (Appendices 4a, 4c and 4d). The calculations leading to VR can be summarised in an Analysis of Variance (ANOVA) table (Table 7.5).

In order to understand the importance of the above discussion, data from Example 7.2 are now re-examined.

Example 7.3 Analysis of variance for the data of Example 7.2
First we present the data again and summarise the results obtained so far.

Nitrogen level (X)	0	25	50	75	100	125
Yield (Y)	3.70	4.45	4.75	5.20	5.15	4.95

$n = 6$ $\bar{X} = 62.5$ $\bar{Y} = 4.70$

$Sxx = 10\,937.5$ $Sxy = 110.00$ $Syy = 1.5800$

$\hat{\alpha} = 4.0714$ $\hat{\beta} = 0.010057$

$RegSS = \hat{\beta} \times Sxy = 1.1063$ $ResidSS = Syy - RegSS = 0.4737$

The Residual Mean Square (*RMS*) is found from

$$RMS = \frac{ResidSS}{n-2} = \frac{0.4737}{4} = 0.1184$$

The Regression Mean Square $= \dfrac{RegSS}{1} = \dfrac{1.1063}{1} = 1.1063$

because the Regression Sum of Squares has one degree of freedom.
The variance ratio is $RegMS/RMS = 1.1063/0.1184 = 9.34$.

Table 7.5. ANOVA table for simple linear regression

Source	DF	SS	MS	VR	
Regression	1	RegSS	RegMS	RegMS/RMS	on $(1, n-2)$ df
Residual	$n-2$	ResidSS	RMS		
Total	$n-1$	Syy			

Note: Syy is also known as the Total Sum of Squares ($TotalSS$).

The results are summarised in the ANOVA table of Output 7.2. Note that the variance ratio (9.34) is denoted by F in this output.

If the null hypothesis of no linear regression were true we would expect the variance ratio to be about 1.00. The greater the value of VR for a fixed number of points, the stronger the evidence against the null hypothesis. If the true slope is zero, the chance of obtaining a VR at least as great as 9.34 is shown in the output to be 0.038 (3.8%). This is called the P-value for testing the significance of the slope. A value less than 0.05 is generally regarded as significant. The greater the value of VR (or F) the smaller the P-value.

7.8.3 Use of the F-tables

If you do not have access to a computer you will need to consult the F-tables (Appendix 4a) to find out if the P-value is less than 0.05. Because the regression sum of squares has one degree of freedom and the residual sum of squares has four degrees of freedom you need to look in column one, row four, of the 5% F-table. The table value is $F_{(1,4,5\%)} = 7.709$. This means that the chance of VR being greater than 7.709 is 5%, so the chance of it being at least as great as 9.34 must be less than 5%; that is, the P-value is less than 0.05. **The result is significant at the 5% level because the calculated value (9.34) is greater than the table value (7.71).**

To find if the P-value is less than 0.01 you need to look in column one, row four, of the 1% F-table (Appendix 4c). The value is $F_{(1,4,1\%)} = 21.20$. As 9.34 is not greater than 21.20, P is not less than 0.01. Because the calculated value (9.34) is between these values we conclude that the linear regression is significant at the 5% level but not at the 1% level. This means that if the true slope is zero, the chance of observing a VR as large as 9.34 is less than 5% but greater than 1%. Output 7.2 shows that this chance is 0.038 (3.8%). We therefore conclude that the true slope is different from zero. Our best estimate of the true slope is 0.010057 which is also given in this output.

7.8.4 Testing the Hypothesis that $\beta = \beta_0$

Often we may know that there is an underlying linear relationship. In such cases we may wish to test whether the true slope is a particular value and/or give confidence intervals for the true slope.

With the same assumptions as before, it can be shown that $\hat{\beta}$ comes from a normal distribution with mean β and variance σ^2/Sxx. The meaning of this statement can be

Output 7.2 Regression analysis of nitrogen data (Minitab output)

```
The regression equation is
yield = 4.07 + 0.0101 nitrogen

Predictor          Coef      StDev        T         P
Constant         4.0714     0.2491    16.35     0.000
nitrogen       0.010057   0.003291     3.06     0.038

S = 0.3441   R-Sq = 70.0%   R-Sq(adj) = 62.5%

Analysis of Variance

Source            DF       SS        MS        F         P
Regression         1    1.1063    1.1063     9.34     0.038
Residual Error     4    0.4737    0.1184
Total              5    1.5800

Obs  nitrogen   yield     Fit    StDev Fit   Residual    St Resid
1         0    3.700    4.071      0.249     -0.371       -1.56
2        25    4.450    4.323      0.187      0.127        0.44
3        50    4.750    4.574      0.146      0.176        0.56
4        75    5.200    4.826      0.146      0.374        1.20
5       100    5.150    5.077      0.187      0.073        0.25
6       125    4.950    5.329      0.249     -0.379       -1.59
```

more clearly understood by referring to Figure 7.6. The fitted line has a slope $\hat{\beta}$ and is based on the four observed points. If the study was repeated for the same X-values it is likely that these four points would be situated differently, and so the fitted line would be in a changed position, and have a different slope, i.e. a changed value of $\hat{\beta}$. If the study was repeated indefinitely, the assumption is that all the calculated values of $\hat{\beta}$ would have a normal distribution with a mean equal to the 'true' unknown value of β and variance σ^2/Sxx. In practice we have only one line, so β is estimated by $\hat{\beta}$ and σ^2 by RMS.

To test the null hypothesis $H_0: \beta = \beta_0$ calculate t as follows and refer to t-tables:

$$t = \frac{\hat{\beta} - \beta_0}{\sqrt{\frac{RMS}{Sxx}}} \quad \text{on } (n-2) \text{ df}$$

In particular, to test whether $\beta = 0$ find

$$t = \frac{\hat{\beta}}{\sqrt{\frac{RMS}{Sxx}}}$$

$\sqrt{\frac{RMS}{Sxx}}$ is called the Standard Error of $\hat{\beta}$

For Example 7.2 the SE of $\hat{\beta}$ is given by

$$SE \text{ of } \hat{\beta} = \sqrt{\frac{0.1184}{10\,937.5}} = 0.00329$$

This is shown as the StDev value for nitrogen in Output 7.2.
For testing the hypothesis, $H_0: \beta = 0$, calculate

$$t = \frac{\hat{\beta}}{SE \text{ of } \hat{\beta}} = \frac{0.010057}{0.00329} = 3.056 \text{ on 4 df}$$

This is shown rounded to 3.06 in Output 7.2.
This calculated t-value (3.056) is significant at the 5% level because it is greater than the t-table value of $t_{(4, 2.5\%)} = 2.776$. The corresponding P-value must therefore be less than 0.05. Output 7.2 shows that it is 0.038.

Note: $t^2 = VR$ when testing the hypothesis that $\beta = 0$, so the F-test and the t-test are equivalent. For this example $3.056^2 = 9.34$.

7.8.5 Confidence Intervals for β

A 95% confidence interval for β is $\hat{\beta} \pm t_{(n-2, 2.5\%)} \times SE \text{ of } \hat{\beta}$. For our example this is:

$$0.010057 \pm (2.776 \times 0.00329) = 0.010057 \pm 0.009133 = (0.000924, 0.01919)$$

7.8.6 Use of Fitted Line for Prediction Purposes

If $X = X_0$ the predicted Y-value is

$$\hat{Y} = \hat{\alpha} + \hat{\beta} X_0$$

Here we are predicting the population mean of the possible Y-values which could be observed when $X = X_0$. It can be shown that the standard error of this predicted mean is

$$SE \text{ of a predicted mean} = \sqrt{RMS \left(\frac{1}{n} + \frac{(X_0 - \bar{X})^2}{Sxx} \right)}$$

In our example, if $X = 120$ the predicted Y mean is

$$\hat{Y} = 4.0714 + 0.010057 \times 120 = 5.2782$$

This mean has a SE of

$$\sqrt{0.1184 \left(\frac{1}{6} + \frac{(120 - 62.5)^2}{10\,937.5} \right)} = \sqrt{0.1184 \times 0.4690} = 0.2356$$

Hence, if nitrogen is applied at the rate of 120 kg/ha the yield, averaged over a large number of plots, is expected to be about 5.28 t/ha with a SE of about 0.24 t/ha. A 95% confidence interval for this predicted mean yield is

$$5.278 \pm t_{(n-2, 2.5\%)} \times 0.236 = 5.278 \pm 0.655 = (4.62, 5.93) \text{ t/ha}$$

In Output 7.2, the column headed StDev Fit shows the standard error of the predicted mean for the X-values of 0, 25, 50, 75, 100, 125. It can be inferred from the above equation that this standard error and hence the width of the confidence interval is smallest when X_0 is equal to the mean of the X-values. Figure 7.7 shows 95% confidence bands which illustrate how the width of this interval varies with X_0.

To predict a single new Y-value from a given X-value, we obtain the same estimate as above but the standard error becomes:

$$SE \text{ of a predicted } Y\text{-value} = \sqrt{RMS\left(1 + \frac{1}{n} + \frac{(X_0 - \bar{X})^2}{Sxx}\right)}$$

You should verify that if $X_0 = 120$ this SE is 0.417 and the corresponding 95% prediction interval is 4.12 to 6.44 t/ha. This is a much wider interval than the one for a predicted mean, and represents the **uncertainty of predicting the yield from a single new plot based on observations taken from several plots**. Prediction bands can be added to the fitted line plot but we do not show them.

It can be deduced from these standard error formulae that precision of estimation can be increased by increasing the number and spread of X-values used to estimate the equation, although in practice this may greatly increase experimental costs. Also the relationship may not be linear over a wide X range. Precision is best when attempting to estimate Y-values for X-values near the X mean.

7.8.7 Standard Error of the Intercept

If $X_0 = 0$ in the formula for the SE of the predicted Y mean we obtain

$$SE \text{ of intercept} = \sqrt{RMS\left(\frac{1}{n} + \frac{\bar{X}^2}{Sxx}\right)}$$

For our example:

$$SE \text{ of } \hat{\alpha} = \sqrt{0.1184\left(0.1667 + \frac{62.5^2}{10\,937.5}\right)} = 0.249$$

This is called StDev for the constant in Output 7.2.

To test whether the true intercept could be zero we find

$$t = \frac{\hat{\alpha}}{SE \text{ of } \hat{\alpha}}$$

For this example:

$$t = \frac{4.0714}{0.249} = 16.35 \text{ on 4 df}$$

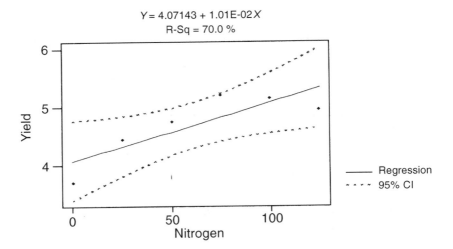

Figure 7.7. Fitted line with 95% confidence bands for Example 7.2

Output 7.2 shows that the P-value for this t-value is 0.000. This means there is no chance of obtaining a t-value this high if the true intercept were zero. Hence we conclude that α could not be zero. This makes sense as a crop yield would still be achieved from residual soil fertility in the absence of inorganic fertiliser.

Note: In some studies the intercept is found to be significantly negative when it is not possible to have a meaningful negative Y-value. This is usually because the straight-line equation was fitted to data using X-values much greater than zero. For small X-values the relationship might have been much different. In this situation the intercept has no physical meaning; it is only important for its use in predicting Y-values over the range of observed X-values.

Caution: Always view a scatterplot of your data before interpreting computer output. In Figure 7.8 there appears to be no relationship between X and Y. The presence of the one unusual observation (outlier) has a major effect on the output. Without the outlier $R^2 = 1.8\%$, the fitted equation is $Y = 2.36 - 0.17X$ and the P-value is 0.675. With the outlier R^2 changes to 83.1%, the fitted equation is $Y = 3.61 + 1.80X$ and P is 0.000. This illustrates the danger of blindly accepting computer output. The presence of the outlier may lead us to the conclusion that there is a very good fit to a straight line when a plot of the data clearly shows this is not the case. However, many packages flag unusual values. The output would include a message that the point (10, 16) has a large influence or **leverage**. This indicates that omission of this point from the analysis would make a large difference to the fitted slope.

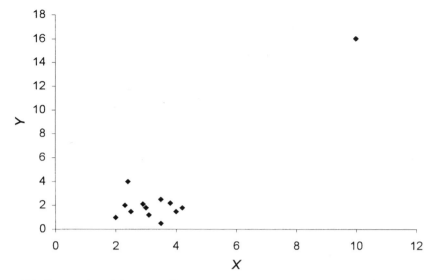

Figure 7.8. Scatterplot with one outlier

7.8.8 Testing the Assumptions

The assumptions which validate significance tests in simple linear regression are stated at the beginning of Section 7.8. The assumption of normality can be checked by producing a histogram and a normal probability plot of the residuals. If the residuals are normally distributed the histogram should look like a normal distribution and the probability plot should be a straight line. The assessment can be done by eye because only serious departures from normality invalidate the tests. The assumption of homogeneity of variance can be checked by plotting the residuals against the X-values or the fitted values. You can make this assumption if the variation in the residuals about zero remains fairly constant as the X or fitted values increase. Figure 7.9 shows the residual diagnostic plots produced by Minitab for Example 7.1. If the assumptions are seriously violated you could consider a transformation of the Y-values.

7.9 TESTING THE SIGNIFICANCE OF A CORRELATION COEFFICIENT

If the assumptions for the regression of Y on X (Section 7.8) are also valid for the regression of X on Y then a t-test of the null hypothesis that the population correlation coefficient (ρ) is zero is valid. The formula for t is $t = r\sqrt{n-2}/\sqrt{1-r^2}$ on $(n-2)$ df where n is the number of (X, Y) pairs.

For the data of Example 7.3, $r = 0.837$, so

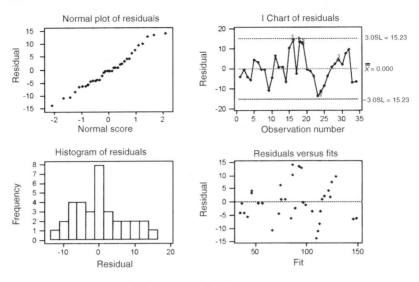

Figure 7.9. Residual diagnostics for Example 7.1

$$t = \frac{0.837 \times \sqrt{4}}{\sqrt{1 - 0.837^2}} = \frac{1.674}{0.5477} = 3.056 \text{ on 4 df}$$

From tables, $t_{(4, 2.5\%)} = 2.776$ so you conclude that r is significantly different from zero.

Note: This test is equivalent to that for testing whether the population regression coefficient (slope) is different from zero (Section 7.8.4).

An alternative to calculating t is to use the table in Appendix 5 which shows critical values corresponding to combinations of n and the level of significance. For example, to carry out a two-tailed test at the 5% level with 6 pairs, the table value (found in row 6, column 2.5) is 0.811. Any calculated correlation coefficient greater than this in magnitude is significant when based on 6 pairs. The table also shows that any value of r greater than 0.197 in magnitude is significant if based on 100 pairs.

We have discussed simple linear regression in some depth. This subject is dealt with in greater depth in more advanced texts such as Zar (1998) and Draper and Smith (1998). There are many other types of curve that can be fitted. We discuss some of them in Chapter 8.

Exercise 7.2
A study was undertaken to find out if tree diameter measurements 1.5 m above ground level can be used to predict heights for a certain species. The following measurements were recorded on 12 trees selected at random:

Diameter X (cm)	9.8	7.2	24.0	7.8	2.5	40.2	15.5	64.4	31.5	10.8	3.0	8.3
Height Y (m)	10.9	9.8	20.5	11.0	5.5	20.4	17.5	26.8	25.6	12.3	7.9	13.6

(a) Find and interpret R^2 and the equation of the least squares regression line of height on diameter and interpret parameter estimates.
(b) Find the fitted values and residuals, draw the fitted line superimposed on the scatter diagram of the observed points and discuss whether a straight-line relationship is appropriate.
(c) Complete the analysis of variance table and test the significance of the straight-line relationship.
(d) Find standard errors for $\hat{\alpha}$ and $\hat{\beta}$ and use them to find 95% confidence intervals for the true intercept and slope. If the intercept is significantly different from zero, explain this result.
(e) A new tree is measured and its diameter found to be 10 cm. Use the equation found in (a) to predict its height and give a 95% prediction interval for this height.
(f) Investigate the assumptions underlying the regression model by making a plot of the residuals against the fitted values.

Answers to Exercise 7.2

(a) The R^2 value is 0.79. This implies that 79% of the variation in height is explained by the linear regression model. The correlation coefficient is 0.89, the square root of 0.79.

The fitted equation is $Y = 8.905 + 0.333X$ (height = 8.905 + 0.333 diameter). From this you conclude that for every cm increase in diameter you expect height to increase by 0.333 m. A tree of zero diameter is expected to have a height of 8.9 m. This does not make sense! The most plausible explanation is that very young trees were not included and for them the rate of increase of height with diameter would have been much greater than 0.333. As a tree ages the growth rate slows down. It is always dangerous to use the results of a regression analysis to make inferences outside the range of X-values used to find the equation. Note also that the two largest trees do not appear to fit the model and have a great influence on the fitted slope. Find the effect of removing these two trees from the model.

(b) The equation can be used to find the fitted Y-values and residuals as shown in the following table in which the trees have been sorted by diameter.

Diameter, X	2.5	3	7.2	7.8	8.3	9.8	10.8	15.5	24	31.5	40.2	64.4
Height, Y	5.5	7.9	9.8	11	13.6	10.9	12.3	17.5	20.5	25.6	20.4	26.8
Fitted Y	9.74	9.90	11.30	11.50	11.67	12.17	12.50	14.07	16.90	19.40	22.29	30.36
Residual	−4.24	−2.00	−1.50	−0.50	1.93	−1.27	−0.20	3.43	3.60	6.20	−1.89	−3.56

A fitted line plot (not shown) indicates that the straight-line model appears satisfactory for small to medium-sized trees, but the relationship does not appear to be linear over the whole range. Try fitting log(Y) versus log(X). You should then obtain $R^2 = 0.93$ indicating a much better fit than the straight-line model, for which R^2 is 0.79. This is equivalent to fitting the model $Y = \alpha X^\beta$.

(c) From the ANOVA table (not shown) $RegSS = 417.428$, $ResidSS = 109.322$, $TotalSS = 526.750$. The calculated variance ratio is 38.18 on 1 and 10 degrees of freedom. This is much greater than $F_{(1,10,0.1\%)} = 21.04$ so you conclude there is a strong linear relationship (computer output gives $P = 0$ to three decimal places). This shows the fitted slope is significantly different from zero. However, some other curve might give a better fit.

(d) SE of $\hat{\alpha} = 1.390$ and 95% CI is (5.81, 12.00). This CI does not include zero so the intercept is significantly different from zero. A possible explanation is given in the answer to (a).

SE of $\hat{\beta} = 0.0539$ and 95% CI is (0.213, 0.453)

(e) Predicted height = 12.23 with SE of 3.47 and a 95% PI = (4.50, 19.96).

(f) A histogram and a normal plot of the residuals suggests the normality assumption may be violated. A plot of the residuals versus the fitted values, given in the table above, show that the shortest trees have negative residuals, the medium-sized trees have mainly positive residuals and the two largest trees have negative residuals. It is difficult to come to any definite conclusions with such a small number of points but there is a suggestion that the residuals may depend on the fitted values. In this case some other model might fit better. We discuss this example further in Chapter 8.

Chapter 8

Curve Fitting

8.1 INTRODUCTION

In Chapter 7 we discussed fitting a straight line to data. While a high value for R^2 represents a good fit, it is possible that another curve may have a significantly higher value of R^2. This may be obvious when you plot the data. In many cases a straight-line relationship is only a reasonable model over a limited range of X-values. For example, the growth of a plant is slow following germination, then accelerates for a limited time and eventually slows during the period of seed development. A straight line may be a good fit in the middle period. In another example, the relationship between the yield of a crop and added fertiliser may be quadratic. This is expected on theoretical grounds as yield increases as fertiliser is added until a maximum yield is reached. Adding more fertiliser after this point will eventually cause a reduction of yield.

In this chapter we review a range of curves that can be fitted but do not give much detail of the calculations due to their complexity. We mainly concentrate on the concepts and interpreting computer output. It should always be borne in mind that whatever curve is fitted, it is only valid over a limited range of X-values. For example, an exponential growth curve may be a good fit to plant growth data, but it could not possibly be valid for much larger X-values as growth cannot accelerate for ever.

The type of curve that should be fitted is often suggested by theoretical considerations. However, when presented with data, the researcher often attempts to fit many curves to find the best one. Having found the best, there is no guarantee that this curve will be appropriate if the study is repeated. You should always try to justify the use of a particular curve on theoretical grounds.

8.2 POLYNOMIAL FITTING

After fitting a straight line (a polynomial of order 1), you could try fitting a quadratic equation (a polynomial of order 2). This would be sensible in the case of an experiment to find out how yield depends on level of fertiliser, because if you fit a quadratic equation you will be able to estimate the optimum level of fertiliser which produces maximum yield.

You could also try fitting a cubic equation (order = 3) or a quartic (order = 4). Polynomials of order 2 and 3 are illustrated in Figure 8.1.

It is doubtful whether fitting a polynomial of order greater than 3 can be justified on theoretical grounds. If a plot of the data suggests a curve, any polynomial will be a good fit over a limited range of the X-values. As you fit higher-order polynomials, you will find the R^2 value increases until it becomes equal to one, corresponding to a perfect fit. The residual degrees of freedom reduce by one every time you increase the order of the polynomial by one. For a straight-line fit, the residual degrees of freedom are $n - 2$, for a quadratic they are $n - 3$ and for a cubic they are $n - 4$. Hence if you only have 4 points, you will get a perfect fit to a cubic equation and will not be able to fit a quartic equation.

You should not attempt to fit a polynomial of order k unless you have at least $k + 2$ points. Ideally you need several more than this to get reliable estimates of parameters. When comparing polynomials of different orders you should not compare the goodness of fits using R^2 as this is bound to increase as k increases. Instead you should compare the adjusted R^2 values. The adjusted R^2 takes into account the number of parameters to be estimated in the model (Section 7.6.2).

The arithmetic involved in fitting relationships other than straight lines is extremely tedious. However, once you have understood the concepts involved in

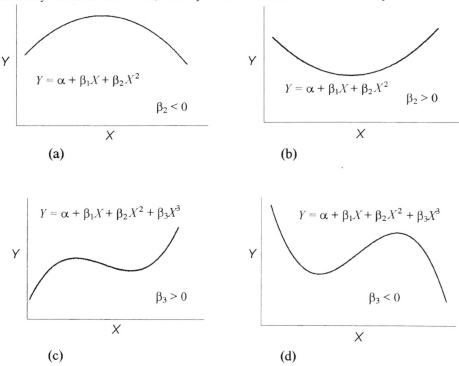

Figure 8.1. Polynomials of degree two (quadratic, (a) and (b)) and three (cubic, (c) and (d))

8.3 QUADRATIC REGRESSION

We now show how to fit a quadratic curve to a set of points and test whether the fit is significantly better than a straight-line fit with reference to Example 7.2.

A glance at Figure 7.4 suggests that a curvilinear relationship exists between yield of wheat and nitrogen fertiliser. Yields appear to increase up to a maximum and then decrease as more nitrogen is added. A quadratic or parabolic curve is often fitted to such data. The mathematical equation is

$$Y = \alpha + \beta_1 X + \beta_2 X^2$$

Estimates of α, β_1, and β_2 are found using the observed X- and Y-values. The formulae for these estimates are complicated and so the experimenter will usually obtain them from a computer output. For each observed X-value the fitted Y-value is obtained using the equation

$$\hat{Y} = \hat{\alpha} + \hat{\beta}_1 X + \hat{\beta}_2 X^2$$

and the corresponding residual is $(Y - \hat{Y})$, the observed Y minus the fitted Y. The residual sum of squares is given by

$$ResidSS = \Sigma(Y - \hat{Y})^2$$

and the regression sum of squares by

$$RegSS = Syy - ResidSS$$

A computer printout gives standard errors and t-values for $\hat{\alpha}$, $\hat{\beta}_1$ and $\hat{\beta}_2$. In each case the t-value is the estimate divided by its standard error and the residual degrees of freedom are $(n - 3)$ as 3 parameters have been estimated. An ANOVA table is also presented — this is similar to the one produced for simple linear regression. The total sum of squares is Syy, the sum of squares of the deviations of the Y-values from their mean. The VR is used to test whether the quadratic model adequately explains the variation in the Y-values.

It can be shown mathematically that the regression sum of squares for fitting a quadratic model is always larger than the corresponding regression sum of squares for fitting a straight line. This implies the R^2 value for quadratic regression will also be bigger than for linear regression. To test whether a parabola is a **significantly** better fit than a straight line find the t-value for $\hat{\beta}_2$ and compare with t-tables on $(n - 3)$ degrees of freedom. See Example 8.1 for details.

Note: The residual degrees of freedom $(n - 3)$, must be at least 1 which implies there must be at least four different X-values. In practice many more are recommended to give several degrees of freedom for residual.

An alternative method of testing if a parabola is a significantly better fit than a straight line is to

(a) Fit a straight line and find the corresponding regression sum of squares $RegSS(Lin)$.
(b) Fit a parabola and find from the computer output the regression sum of squares, $RegSS(Quad)$ and the corresponding residual mean square, $RMS(Quad)$.
(c) Find the variance ratio

$$VR = \frac{RegSS(Quad) - RegSS(Lin)}{RMS(Quad)}$$

(d) Compare VR with F-tables on $(1, n - 3)$ degrees of freedom.

Note: The calculated VR is the square of the t-value for the $\hat{\beta}_2$ term in the computer output so in practice we compare this t-value with t-tables. If our t-value is greater than $t_{(n-3, 2.5\%)}$ we conclude at the 5% level that a parabola is a better fit than a straight line.

Example 8.1

Use the data of Example 7.2 to test whether a parabola is a significantly better fit than a straight line. The data were as follows:

X	0	25	50	75	100	125
Y	3.70	4.45	4.75	5.20	5.15	4.95

The X-values are amounts of nitrogen fertiliser in kg/ha, and the Y-values are yields in t/ha. Minitab was used to fit a quadratic regression model. The Y-values were entered in the first column and the corresponding X-values in the second. The calculator function was then used to enter the values of X^2 in the third column. These columns were labelled yield, nitrogen and nit2 respectively. The results are shown in Output 8.1.

8.3.1 Interpretation of Computer Output

The VR (F) in the ANOVA table is 99.58 (for the straight line it was only 9.34) and the P-value is 0.002. It is less than 0.05, showing that there is significant quadratic regression. It is much less than 0.038 which was obtained from the straight-line fit. If you compare this output with Output 7.2 you will see that the regression sum of squares has increased from 1.1063 (linear) to 1.5566 (quadratic). The difference is $1.5566 - 1.1063 = 0.4503$. This is the extra sum of squares due to fitting a quadratic curve compared to a straight line. These figures are shown in the sequential sum of squares (Seq SS) column. The regression sum of squares due to fitting the straight-line model is 1.1063. This is increased by 0.4503 when the quadratic model is fitted.

Output 8.1 Quadratic regression analysis of data of Example 8.1 (Minitab output)

```
The regression equation is
yield = 3.71 + 0.0320 nitrogen − 0.000176 nit2
```

Predictor	Coef	StDev	T	P
Constant	3.70536	0.08012	46.25	0.000
nitrogen	0.032021	0.003015	10.62	0.002
nit2	−0.00017571	0.00002315	−7.59	0.005

S = 0.08841   R-Sq = 98.5%   R-Sq(adj) = 97.5%

Analysis of Variance

Source	DF	SS	MS	F	P
Regression	2	1.55655	0.77828	99.58	0.002
Residual Error	3	0.02345	0.00782		
Total	5	1.58000			

Source	DF	Seq SS
nitrogen	1	1.10629
nit2	1	0.45027

Obs	nitrogen	yield	Fit	StDev Fit	Residual	St Resid
1	0	3.7000	3.7054	0.0801	−0.0054	−0.14
2	25	4.4500	4.3961	0.0490	0.0539	0.73
3	50	4.7500	4.8671	0.0539	−0.1171	−1.67
4	75	5.2000	5.1186	0.0539	0.0814	1.16
5	100	5.1500	5.1504	0.0490	−0.0004	−0.00
6	125	4.9500	4.9625	0.0801	−0.0125	−0.33

The R^2 value is the regression sum of squares divided by the total sum of squares. This has increased from 70% (linear) to $1.55655/1.5800 = 0.985$, or 98.5% (quadratic). This shows that 98.5% of the variation in the yields is explained by the quadratic regression model.

The adjusted R^2 value is 97.5%. This adjusts the R^2 value to take account of the loss of an extra degree of freedom when comparing with the 70% for the linear model. The fitted quadratic equation is $\hat{Y} = 3.7054 + 0.03202X - 0.0001757X^2$.

The estimate of the curvature coefficient is -0.0001757 with a standard error of 0.0000231. The corresponding t-value is $-0.0001757/0.0000231 = -7.59$. This high t-value and corresponding small P-value (0.005) show that the curvature is significantly different from zero. In other words, the quadratic model is a significantly better fit than the linear model. The negative value indicates that the curvature is convex (with a maximum value) rather than concave (with a minimum).

The estimate of β_1 (the slope when X is zero) is 0.03202 with a standard error of 0.003015. This is significantly different from zero ($P = 0.002$). From Output 7.2 the average slope is estimated as 0.010057.

The estimate of α (the expected value of Y when X is zero) is 3.7054 with a standard error of 0.0801. This is significantly different from zero ($P = 0.000$).

The residual degrees of freedom for the F-test and the t-tests are $n - 3 = 3$ (number of points − number of parameters estimated).

The X (nitrogen) values, Y-values (observed yields), the fitted values and the residuals are shown at the end of Output 8.1. The fitted values are obtained by substituting the X-values in the fitted equation. The fitted values are the heights of the fitted curve at the corresponding X-values.

The standardised residuals (St Resid) are independent of the units of measurement. Standardised residuals outside the range -2 to $+2$ are considered to be large and suggest that the model may not be a good fit. There is no pattern in these residuals. This together with the very high R^2 value shows that the quadratic model is a very good explanation of the yield values. You should check that the quadratic model assumptions are valid by obtaining a residual diagnostics plot similar to that obtained for the linear model (Figure 7.9).

On comparison with Output 7.2 it is seen that these residuals are smaller than the corresponding residuals for the straight-line fit. Their sum of squares is 0.0234 which is less than 0.4737, the residual sum of squares for the linear fit. Figure 8.2 shows the fitted curve superimposed on the scatter diagram of the observed points.

8.3.2 Quadratic Response of Yield to Nitrogen Level

It can be shown that for a quadratic curve, Y is a maximum when

$$X = -\frac{\beta_1}{2\beta_2}$$

For this example maximum yield is expected when

$$X = -\frac{0.03202}{2(-0.0001757)} = 91.12 \text{ kgN/ha}$$

and the corresponding Y is 5.16 t/ha.

Refer to the ANOVA table in Output 8.1. Note that the regression degrees of freedom are 2 because both β_1 and β_2 had to be estimated. On comparison with the corresponding ANOVA table in Output 7.2, the regression sum of squares has increased by $(1.5566 - 1.1063) = 0.4503$. On dividing this increase by the residual

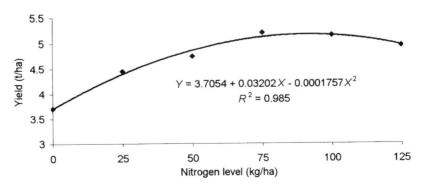

Figure 8.2. Quadratic response of yield to level of nitrogen for Example 8.1

mean square (0.00782) from Output 8.1 a variance ratio of 57.6 on (1, 3) degrees of freedom is obtained. This variance ratio is much higher than the 5% F-table value of 10.13 on 1 and 3 degrees of freedom, so we conclude that a quadratic curve is a significantly better fit than a straight line. Note that the VR of 57.6 is the square of the t-value of -7.59, and this implies that the F-test is equivalent to the t-test carried out for nit2 in Output 8.1.

8.4 OTHER TYPES OF CURVE

Among other types of curve of interest are exponential, power, logarithmic, asymptotic and logistic. The first three types can be fitted using the least squares methods of Chapter 7, because they can be made linear after a logarithmic transformation of the Y- or X-values or both. However, the assumptions of normality and homogeneity of variance may not be valid when applied to the transformed data. The powerful computing facilities now available allow direct curve-fitting methods to be used, making transformations unnecessary.

The asymptotic and logistic curves are intrinsically **non-linear** (they cannot be made linear by a transformation) and more complex methods of fitting have to be used. Computer packages normally require you to supply initial estimates for the parameters. For more details on these and other non-linear curves see Mead, Curnow and Hasted (1993).

The empirical approach to curve fitting is to try many different curves and use the one that fits best. The good fit may be due to chance, may not be repeated with new data and may not make sense biologically. The theoretical approach is to select a type of curve based on biological principles.

8.4.1 Exponential Curves

The equation of an exponential curve is $Y = \alpha e^{\beta X}$ where e is the base of natural logarithms (approximately 2.718) and α is the intercept on the Y axis. If α and β are positive the model is 'exponential growth' (Figure 8.3(a)), the slope increasing with X. It may be reasonable to expect a curve of this type to apply to plant heights measured at daily intervals in the period soon after germination.

If α is positive and β is negative the model is 'exponential decay' (Figure 8.3(b)), the slope decreasing with X. This type of curve may fit data collected on the decay of a chemical from the soil over time.

The equation is equivalent to $\ln(Y) = \ln(\alpha) + \beta X$. Hence, to test the goodness of fit, the methods of Chapter 7 can be used to fit a straight line to a plot of the natural logarithms of the Y-values versus the X-values. The slope of this fitted line is the estimate of β. The intercept is the estimate of $\ln(\alpha)$ so the estimate of α is e raised to the power of this intercept.

Note: The exponential equation is often written as $Y = \alpha \beta^X$, where β has taken the place of e^β. Also we use the notation ln to denote natural logarithms but $\log_e$ (log to the base e) is also in common use.

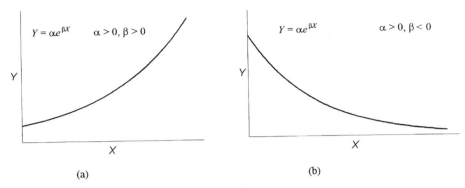

Figure 8.3. Exponential growth (a) and decay (b) curves

8.4.2 The Power Curve

The equation is $Y = \alpha X^{\beta}$ where α is the value of Y when $X = 1$. The exact form depends on the magnitudes and signs of α and β; Figure 8.4 shows various possibilities.

Assuming α is positive and β lies between 0 and 1, the slope is positive but decreases with X, and if β is greater than 1, the slope is positive and increases with X. If $\beta = 1$ the curve becomes a straight line through the origin with slope α. When β is positive the slope is positive and when β is negative the slope is negative. On taking natural logarithms, $\ln(Y) = \ln(\alpha) + \beta \ln(X)$. Hence, α and β can be estimated by fitting a straight line to a plot of $\ln(Y)$ versus $\ln(X)$ using the methods of Chapter 7. The estimate of β is the slope and the estimate of α is e raised to the power of the intercept.

Example 8.2

In Exercise 7.2 you were asked to fit a straight line to data on the heights and diameters of 12 trees where height was the response variable. The R^2 value was 0.79 and a plot of the data suggests that the rate of increase of height with diameter decreases as diameter increases. A better fit should therefore be obtained from a power curve. Table 8.1 shows the log transformations of the diameters and heights.

You should confirm that the fitted equation is $\ln(Y) = 1.436 + 0.471 \ln(X)$ and $R^2 = 0.926$. This compares with 0.79 without the transformation. A fitted line plot can be obtained using the method described in Section 7.4 and is shown in Figure 8.5(a). The estimate of $\ln(\alpha)$ is 1.436 so the estimate of α is $e^{1.436} = 4.204$, and the equation of the fitted curve is $Y = 4.204 X^{0.471}$. Figure 8.5(b) shows a plot of the raw data with the fitted curve. You should check that the assumptions of normality and homogeneity of variance are valid if any significance tests are performed. It would appear from Figure 8.5(b) that the residuals increase with diameter and this together with Figure 8.5(a) suggests that the log transformed data should be used in any tests.

To obtain the fitted curve superimposed on a plot of the data in the original scale, first obtain the fitted values in the log scale. Each fitted value in the original scale is e to the power of the corresponding fitted value in the log scale. Alternatively, substitute the X-values in the equation $Y = 4.204 X^{0.471}$. Using these fitted values the residual sum of squares from the fitted curve is found to be 57.78, which is much less than the 109.32 obtained for the straight-line fit.

A note on R^2: The R^2 value in Example 8.2 is 0.926 as calculated from the transformed data. Another version can be found using the fitted values from the curve. For simple linear regression R^2 is found by dividing the regression sum of squares ($RegSS$) by the total sum of squares (Syy), where $RegSS$ is the sum of the squares of the deviations of the fitted values from the Y mean (Section 7.6). If you

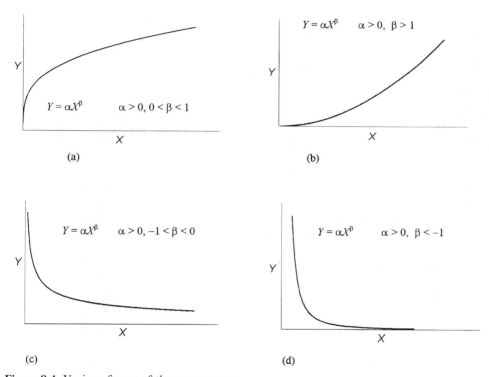

Figure 8.4. Various forms of the power curve

Table 8.1. Data on tree heights (m) and diameter (cm) with log transformations

Diameter, X	2.5	3	7.2	7.8	8.3	9.8	10.8	15.5	24	31.5	40.2	64.4
Height, Y	5.5	7.9	9.8	11	13.6	10.9	12.3	17.5	20.5	25.6	20.4	26.8
ln(X)	0.916	1.099	1.974	2.054	2.116	2.282	2.380	2.741	3.178	3.450	3.694	4.165
ln(Y)	1.705	2.067	2.282	2.398	2.610	2.389	2.510	2.862	3.020	3.243	3.016	3.288

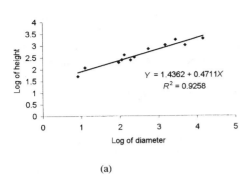
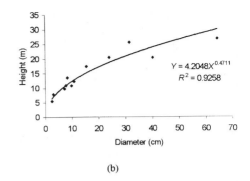

(a)   (b)

Figure 8.5. Data of Example 8.2 (a) log transformed with fitted line (b) raw data with fitted curve

attempt to calculate R^2 for the fitted curve of Example 8.2 by this method you will obtain $RegSS = 552.02$ and $Syy = 526.75$, leading to $R^2 = 1.048$. Clearly this does not make sense as it is greater than one. The explanation is that while the method of least squares was valid for the log transformed data it does not apply to the back transformed estimates of the parameters and the equation $RegSS + ResidSS = Syy$ is no longer true if $RegSS$ is defined as $\Sigma(\hat{Y} - \bar{Y})^2$. This problem is common when curves are fitted using methods other than least squares. In these cases the definition of regression sum of squares as $Syy - ResidSS$ makes sense. Using this definition, $RegSS$ and R^2 for Example 8.2 are 468.97 and 0.890 respectively.

SAS (Output 8.2) uses a direct method of fitting this curve (avoiding transformations) and gives the equation as $Y = 4.8782X^{0.4225}$ and $ResidSS = 51.9084$. SAS gives the regression sum of squares as 3229.1116, being the uncorrected sum of squares of the Y-values (3281.0200) minus $ResidSS$. Using our definition, $RegSS = 526.75 - 51.91 = 474.84$ and $R^2 = 0.901$.

Adjusted R^2: Due to the difficulty of interpreting $RegSS$ in non-linear curve fitting, the adjusted R-square (Section 7.6.2) should be used. It is calculated directly from the total sum of squares (Syy) and the residual sum of squares ($ResidSS$). For this example the total mean square is $526.75/11 = 47.8864$ and the residual mean square is 5.1908. Hence the adjusted R-square is $(47.8864 - 5.1908)/47.8864 = 0.892$.

8.4.3 The Logarithmic Curve

The equation is $Y = \ln(\alpha X^\beta)$ where $\ln(\alpha)$ is the value of Y when $X = 1$. The exact form depends on the magnitudes and signs of α and β; Figure 8.6 shows two possibilities. If data conform to this equation a straight line should be a good fit to a plot of Y versus $\ln(X)$ because the equation is equivalent to $Y = \ln(\alpha) + \beta \ln(X)$.

This type of equation is often found useful in describing the response to a stimulus. To produce equal increases in the response you need to multiply the stimulus by a constant. Thus a graph of response versus the log of the stimulus should be a straight line.

CURVE FITTING 97

Output 8.2 Part of the SAS output for fitting a power curve to the data of Example 8.2

Non-Linear Least Squares Summary Statistics   Dependent Variable HEIGHT

Source	DF	Sum of Squares	Mean Square
Regression	2	3229.1115823	1614.5557911
Residual	10	51.9084177	5.1908418
Uncorrected Total	12	3281.0200000	
(Corrected Total)	11	526.7500000	

Parameter	Estimate	Asymptotic Std. Error	Asymptotic 95 % Confidence Interval	
			Lower	Upper
A	4.878251280	0.77138291315	3.1594922317	6.5970103292
B	0.422491308	0.04766443437	0.3162876622	0.5286949543

8.4.4 The Asymptotic Growth Curve

The equation is $Y = \alpha - \beta\delta^X$. Figure 8.7(a) shows this type of curve for α and β both positive, α greater than β and δ between 0 and 1. When X is very large Y approaches but does not quite reach α, so α is called the asymptote. The intercept on the Y axis is $\alpha - \beta$. The data of Example 8.2 may be expected to follow this curve because as diameter increases with age, height increases at a diminishing rate. Using the standard curves regression option, Genstat (Output 8.3) estimates the fitted equation as $Y = 26.76 - 23.68\,(0.9475)^X$ and the residual sum of squares as 36.91. This is less than the 51.91 obtained by fitting a power curve using SAS and indicates a better fit. The adjusted R-square is $(47.886 - 4.101)/47.886 = 0.914$. This is shown as the percentage variance accounted for (91.4%) in the output.

8.4.5 The Logistic Growth Curve

This type of S-shaped or sigmoid curve can be used to model growth which starts slowly, increases to a maximum rate and then gradually slows approaching zero. The equation is $Y = \alpha/(1 + \beta\delta^X)$. Figure 8.7(b) shows this type of curve for α and β both

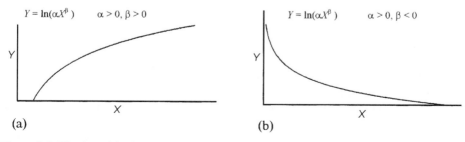

(a)   (b)

Figure 8.6. The logarithmic curve for $\beta > 0$ (a) and $\beta < 0$ (b)

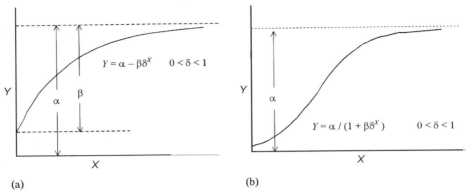

Figure 8.7. Asymptotic (a) and logistic (b) growth curves

positive and δ between 0 and 1. As X increases, Y approaches the asymptotic value of α. The intercept on the Y axis is $\alpha/(1+\beta)$.

Note: The equation may be written in the form

$$Y = \frac{\alpha}{1 + e^{b-cX}}$$

where e^b replaces β and e^{-c} replaces δ.

Example 8.3
Data obtained from sampling an experimental crop of autumn sown winter barley are given in Table 8.2. Several curves may be fitted to these results if the main objective is to practise curve-fitting techniques.

- Increase in total dry matter over the period 100–292 days from drilling can be described by a logistic growth curve ($Y = 13.54/(1 + e^{19.70-0.0864X})$). Figure 8.8 shows that this relationship may not be appropriate because biomass declined after 281 days when the crop was senescing. Also the fit is not good below 204 days. For 100 to 214 days an exponential curve ($Y = 0.0203 e^{0.0233X}$, $R^2 = 0.99$) is a good fit.
- Increase in total dry matter can also be described by a quadratic curve ($Y = 246.42 + 1.9023X - 0.0035X^2$, $R^2 = 0.980$) if data from 220 to 292 days from drilling are considered. In addition, a linear function ($Y = 58.652 + 0.2864X$, $R^2 = 0.985$) could be fitted if only data obtained from the period of maximum growth (214–246 days after drilling) are used. If daily samples had been taken in order to compare weekly dry matter production rates, the separate weekly slopes are likely to vary throughout the growing season.
- Although reproductive development is probably best described by a logistic curve, this function is usually avoided because it is difficult to obtain dry matter data when the reproductive parts are very small. An asymptotic curve (Figure 8.7(a)) is especially appropriate for studying 1000 grain weight data when the date of maximum weight (mass maturity) is required.

Output 8.3 Genstat output for fitting an asymptotic curve to the data of Example 8.2

```
***** Nonlinear regression analysis *****

Response variate:  Height
    Explanatory:   Diameter
    Fitted Curve:  A + B*R**X
    Constraints:   R<1

*** Summary of analysis ***

              d.f.    s.s.      m.s.      v.r.    F pr.
Regression     2     489.84    244.922    59.73   <0.001
Residual       9      36.91      4.101
Total         11     526.75     47.886

Percentage variance accounted for 91.4
Standard error of observations is estimated to be 2.03

*** Estimates of parameters ***

         estimate       s.e.
R         0.9475      0.0149
B       -23.68        2.20
A        26.76        2.32
```

In practice, it is important to avoid the temptation of testing every mathematical function on your experimental data. Although curve fitting is a stimulating academic exercise, the applied biologist must remember that it is not a main experimental objective. A curve for biomass accumulation obscures the fact that the growth rates of different plant parts have a relationship one to another. Increase in total dry weight is influenced by reproductive transition which occurs at early growth stages in many annual crops. In Example 8.3, 2.1 t/ha of the total biomass had accumulated in the reproductive parts on 23 May (241 days after drilling). The researcher must decide whether a sigmoid function (which is frequently used to describe overall plant development from germination to senescence) is of value if the main research objective is to study the physiological factors which influence anthesis and seed development. For a more detailed account of plant growth curve analysis see Hunt (1982).

Table 8.2. Vegetative development of autumn-sown winter barley

Days from drilling, X	100	130	150	198	204	214	220	229	234	241
Total dry matter (t/ha), Y	0.2	0.4	0.8	1.9	2.2	3.2	3.9	6.5	8.5	10.1

Days from drilling	246	252	257	266	273	281	287	290	292
Total dry matter (t/ha)	12.2	12	12.4	13.2	14.5	13.3	13.1	13	12.8

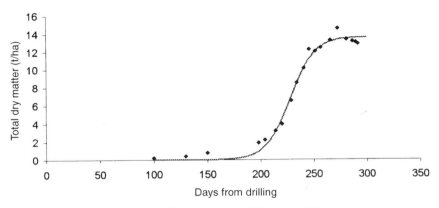

Figure 8.8. A logistic growth curve fitted to data of Example 8.3

It is common to find different types of curves and associated equations in student dissertations because they are easily obtained using computers. Sadly, they are rarely referred to, or used in experimental discussions, and sometimes not even understood! It is important to remember that curve-fitting procedures are only tools that are used to create a scientific foundation which underpin the interpretation and discussion of your experimental data.

8.5 MULTIPLE LINEAR REGRESSION

A possible reason for lack of fit to a straight line or other curve is that other variables not measured are responsible for the variation in the Y-values. For instance, the yield of a crop depends on variables such as temperature, rainfall, sunshine and many more besides level of nitrogen fertiliser. A multiple linear regression model can be set up to attempt to explain the yield variations based on these other explanatory variables. This is a very complex subject which we do not attempt to explain in detail in this book. The interested reader may consult Snedecor and Cochran (1989) and Draper and Smith (1998).

Most computer packages allow you to carry out a multiple regression analysis in a similar way to simple linear regression. You would enter the values of Y, the response variable in one column of the spreadsheet and the values of the explanatory (predictor) variables in other columns. The equation to be fitted is

$$Y = \alpha + \beta_1 X_1 + \beta_2 X_2 + \ldots + \beta_k X_k$$

where k is the number of explanatory values. The computer will provide estimates of α and the βs. These are used to find the fitted Y-values and calculate the residuals and hence the residual sum of squares. The output should include, for each explanatory variable, the estimate of β with its corresponding standard error, t-value and P-value. For example, the estimate of β_1 is the amount Y is expected to increase when X_1 increases by one unit, when the values of all the other explanatory variables are held constant. If the corresponding *P*-value is greater than 0.05 this is evidence that this variable may be dropped from the model without significantly affecting the fit. If two of the explanatory variables are highly correlated with each other, only one will be needed. You should ask for the correlation matrix. The output will show the correlation coefficients for all possible pairs of variables so that those which are significant may be identified. A measure of the goodness of fit is the adjusted R-square (Section 7.6.2) which can be used to compare models with different numbers of explanatory variables. The ordinary R^2 always increases as more variables are added. You should try to end up with as small a model as possible so long as the adjusted R-square is reasonably high.

Chapter 9

The Completely Randomised Design

9.1 INTRODUCTION

The completely randomised design (CRD) is frequently used to compare treatments when environmental conditions are fairly uniform. The principles of replication and randomisation apply. Each treatment is applied at random to several experimental units (plots, pots, individual plants, etc.). With only one unit (replicate) per treatment there would be no estimate of experimental error (variation in the responses of units treated alike) and no way of telling whether observed differences were due to treatments or other uncontrollable factors. Randomisation ensures there is no subjectivity in the allocation of treatments to units and that, on average, the effects of other factors are expected to cancel out when the treatment means are compared.

The independent samples t-test can be used to analyse the results of a completely randomised design with two treatments (Chapter 6). When there are more than two treatments **one-way analysis of variance** is used (Section 9.3). This compares the between-treatment variation with the within-treatment variation and assesses whether the differences in the means are due to chance or treatment effects. This method of analysis can also be used in observational studies to test whether several population means are equal after collecting results from independent random samples.

9.1.1 Advantages of the CRD

- The number of replications need not be the same for each treatment.
- The analysis is straightforward and not unduly complicated by unequal replication or missing data.
- The degrees of freedom for residual (error) are maximised. No other design with the same number of treatments and units provides greater residual degrees of freedom.

9.1.2 Disadvantage

The main disadvantage of this design is low precision if the experimental units receiving the same treatment are not uniform. In this case, real treatment differences are difficult to detect. Precision can theoretically be improved by increasing the number of replications per treatment, although in practice this option may result in some complications. For example, to accommodate a larger number of replications per treatment in a field experiment, a bigger site may be required. In this case, precision may actually decrease due to increased variation between plots receiving the same treatment. Also, as the size of the experiment increases, more workers may be needed. This may lead to extra variation as more errors may be made. Alternatively, it may be possible to increase the precision of treatment comparisons by using the principle of blocking (Chapter 10).

9.1.3 Notation

N = total number of plots (experimental units) in the experiment
t = number of treatments in the experiment
n_1 = the number of plots receiving treatment 1
n_2 = the number of plots receiving treatment 2
.
.
.
n_t = the number of plots receiving treatment t

The total number of plots is $n_1 + n_2 + \ldots + n_t = N$. This is expressed as $\Sigma n = N$.
In the case of equal replication, $n_1 = n_2 = \ldots n_t = r$, and $N = r \times t$ where r is the number of replications per treatment.

9.2 DESIGN CONSTRUCTION

Suppose you wish to construct a field experiment with four treatments, A, B, C and D ($t = 4$), each replicated five times ($r = 5$). You require 20 plots ($N = r \times t = 5 \times 4 = 20$). On a plan of the experimental site the plots are labelled 1 to 20 as follows:

1	2	3	4	5
6	7	8	9	10
11	12	13	14	15
16	17	18	19	20

A valid randomisation procedure is to take 20 slips of paper and number them from 1 to 20. Place them in a small container and thoroughly mix them. Draw out the slips one at a time. Assign treatment A to the plots corresponding to the numbers on the first five slips drawn. Assign treatment B to the plots whose numbers are on the second five slips drawn. Assign treatment C to the plots whose numbers are on the third five slips drawn. Assign treatment D to the remaining plots. Alternative methods of randomisation include using random numbers from tables, a calculator or a computer. You could also use a computer generated design.

9.2.1 Assignment of Treatments to Plots using a Calculator

Most calculators have a key for random number generation which is usually marked RAN#. A random number between 0 and 1 should appear on the screen. Use the following procedure to assign the treatments to the plots:

- Obtain 20 different random numbers. Write these numbers in a column in the order in which they were generated (column one in Table 9.1).
- Rank the numbers from smallest to largest, giving the smallest number rank 1 and the largest number rank 20 (column two in Table 9.1). These ranks correspond to the plot numbers.
- Assign treatment A to the plots having the ranks of the first five random numbers. Assign treatment B to the plots having the ranks of the second five random numbers. Assign treatment C to the plots having the ranks of the third five random numbers. Assign treatment D to the remaining five plots. The complete assignment is shown in Table 9.1 for a particular set of 20 random numbers generated in the order in which they appear in the first column. The field plan with the treatments assigned to the plots is shown in Table 9.2.

You will notice that treatment D has been assigned to plots in the first two rows of the site, while treatment B has been assigned mainly to the next two rows. This could lead to bias if the plots in the first two rows are more fertile than the plots in the next two rows. If you do not like this arrangement, you could carry out another randomisation, although a statistician would argue that while a particular

Table 9.1. Assignment of treatments to plots using random numbers

Random number	Rank (plot)	Treatment	Random number	Rank (plot)	Treatment
0.689	16	A	0.507	14	C
0.007	1	A	0.132	5	C
0.771	18	A	0.794	19	C
0.435	11	A	0.491	13	C
0.266	9	A	0.142	6	C
0.486	12	B	0.257	8	D
0.049	3	B	0.037	2	D
0.886	20	B	0.109	4	D
0.742	17	B	0.209	7	D
0.636	15	B	0.375	10	D

Table 9.2. Field plan showing assignment of treatment to plots

1 A	2 D	3 B	4 D	5 C
6 C	7 D	8 D	9 A	10 D
11 A	12 B	13 C	14 C	15 B
16 A	17 B	18 A	19 C	20 B

arrangement may look biased in one direction, the next experiment could be biased in another direction. This problem should be less likely to occur if the number of replications per treatment is increased, and when results and conclusions are based on several experiments. If you suspect a systematic variation in fertility across the site, blocking should be considered (Chapter 10).

9.3 PRELIMINARY ANALYSIS

The analysis of the CRD is called **analysis of variance** and is illustrated by Example 9.1. As most readers will use a computer, we present a Minitab output and a brief interpretation. Then, for the readers who wish to acquire a deeper understanding, the calculations are carried out step by step supported by a little background theory. Do not worry if you initially find the explanations difficult to follow. Later, a summary is given, and then you will see how easy it is to apply these steps to other examples.

> **Example 9.1**
> An experiment was carried out to compare four wheat varieties (A, B, C and D). A completely randomised design was used, each variety being assigned to five plots. The field plan of Table 9.2 was used. Table 9.3 shows the field plan together with the plot yields (kg/plot).

Table 9.3. Field plan and yields (kg/plot) of three wheat varieties (A, B and C) from 20 plots

A 22.2	D 23.9	B 24.1	D 21.7	C 25.9
C 18.4	D 24.8	D 28.2	A 17.3	D 26.4
A 21.2	B 30.3	C 23.2	C 21.9	B 27.4
A 25.2	B 26.4	A 16.1	C 22.6	B 34.8

Output 9.1 Analysis of variance of wheat yield data (Minitab output)

```
Analysis of Variance for Yield
Source     DF        SS        MS        F         P
Variety     3      188.2     62.7      5.69     0.008
Error      16      176.4     11.0
Total      19      364.6
                                    Individual 95% CIs For Mean
                                        Based on Pooled StDev
Level   N     Mean     StDev   -------+---------+---------+---------
A       5    20.400    3.709   (-------*-------)
B       5    28.600    4.119                             (-------*------)
C       5    22.400    2.701        (-------*-------)
D       5    25.000    2.467                   (-------*------)
                                -------+---------+---------+---------
Pooled StDev = 3.320              20.0        24.0      28.0
```

9.3.1 Using a Computer

Enter the data in a spreadsheet with the variety codes in the first column and the yields in the second column as in the first two columns of Table 9.5. Output 9.1 shows the results using one-way ANOVA in Minitab. The following is a brief explanation. A more detailed explanation is given in Section 9.5.2.

The SS column shows that the total variation in the yields is 364.6. This is the sum of the squares of the deviations of the yields from the overall mean. Of this, 188.2 is explained by variety differences and 176.4 is unexplained. The two figures in the MS column would be the same if there were no variety differences. The first is divided by the second to obtain the F-value of 5.69. Under the null hypothesis of no variety differences in yield, the F-value is expected to be 1.0. The P-value of 0.008 shows that 5.69 is significantly greater than 1.0. A P-value less than 0.05 is generally regarded as evidence of treatment (variety) differences. The greater the F-value, the smaller the P-value and the stronger the evidence of treatment differences.

The StDev column shows the individual sample standard deviations for each group of five yields. If these are not too different they can be pooled to obtain a pooled standard deviation of 3.320 (calculated as the square root of the Error MS). This value can be used to find individual confidence intervals for each mean. The output shows that there is strong evidence of a difference between varieties A and B but not between C and D. There are other methods of making individual comparisons (Chapter 13).

Individual standard errors can be calculated for each mean by dividing the StDev values by the square root of the number of replications (5). Figure 9.1 shows the variety means with their standard errors (mean $\pm$ 1SE).

Note: The conclusions of an analysis of variance are based on certain assumptions, chief of which is that the individual sample variances (obtained by squaring the individual standard deviations) are similar. In Chapter 14 we show how the assumptions can be checked and what to do if they are not valid.

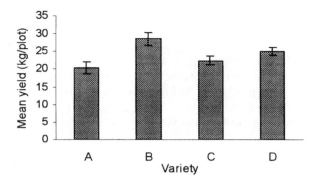

Figure 9.1. Graph showing variety mean yields and individual standard errors

9.3.2 The Step-by-step Calculations

The first step in carrying out the analysis of Example 9.1 by hand is to form a table (Table 9.4) showing the yields, the variety means, and the corresponding sample variances, standard errors and Sxx values. If unsure how to calculate $\bar{x}$, s^2, standard error and Sxx using your calculator in SD mode, consult Chapter 2 and study the instructions.

- Remember that Sxx represents the sum of the squares of the deviations from the sample mean and is equal to $(n-1)s^2$ where n is the number of replications in the sample.

You can get an indication of whether the mean yields differ significantly by comparing the individual s^2 values with the overall s^2 value (based on all 20 yields ignoring treatments). Verify that the overall value is 19.189. It is much bigger than the average of the individual values (11.025) and this difference is probably due to treatment differences.

Table 9.4. Wheat yields (kg/plot) for four varieties with summary statistics

Yields	Variety A	B	C	D
	22.2	24.1	25.9	23.9
	17.3	30.3	18.4	21.7
	21.2	27.4	23.2	24.8
	25.2	26.4	21.9	28.2
	16.1	34.8	22.6	26.4
Total	102.0	143.0	112.0	125.0
Mean, $\bar{x}$	20.4	28.6	22.4	25.0
Variance, s^2	13.755	16.965	7.295	6.085
Standard error	1.659	1.842	1.208	1.103
Sxx	55.02	67.86	29.18	24.34

The grand total $(G) = 482.0$ and the grand mean $(GM) = 24.10$

We now discuss whether there is evidence of treatment differences. As a first step, the means together with their standard errors are presented in Figure 9.1. This gives a simple graphical representation of these data in the form of a bar chart with error bars. The height of each bar represents the mean and the error bars represent one standard error above the mean and one below. For each mean, the standard error gives an indication of its reliability. At first sight, variety B appears to be superior to variety A.

We want to know if the differences between these mean yields are real or due to chance. One possibility worth considering is to calculate individual 95% confidence intervals and substitute them in place of the standard error bars. If there is no overlap this would indicate a significant difference between the means. However, an overlap does not necessarily indicate non-significance at the 5% level. Redraw the graph by multiplying the length of the standard error bars by $t_{(4, 2.5\%)} = 2.776$ to give the 95% confidence interval bars. You will find they all overlap. We show later that the means for varieties A and B are significantly different.

Also look at the overall variation in the yields. This is the sum of the squares of the deviations of the yields from the grand mean. It is often called the **Total Sum of Squares** and we denote this by *TotalSS*. To calculate it we obtain the overall s^2 value and multiply by $N - 1$. For our example $TotalSS = 19.189 \times 19 = 364.6$ (Output 9.1, Table 9.6). We shall see in the next section that this variation is made up of the variation between and within treatments.

9.4 THE ONE-WAY ANALYSIS OF VARIANCE MODEL

Assume that each plot yield is made up of an overall mean, a treatment effect and an unexplained part, called the residual or error. The validity of significance tests depends on further assumptions that the residuals come from a normally distributed population with variance σ^2.

Fit the model: yield = overall mean + treatment effect + residual
Write this as   yield = fitted value + residual
Hence   fitted value = overall mean + treatment effect
And   residual = yield − fitted value

The overall mean is estimated by the grand mean (GM). The treatment effect is estimated as $TE =$ (treatment mean − grand mean) $= (TM - GM)$. The fitted value is estimated as $F = GM + TE = GM + (TM - GM) = TM$. Thus the fitted value is the treatment mean, and this is the yield predicted by the model.

The residual is the difference between the observed yield (Y) and the fitted yield (F). It is estimated as $R = Y - F = Y - TM$. For each yield the deviation from the grand mean can be written as follows:

(yield − grand mean) = (treatment mean − grand mean) + (yield − treatment mean)

That is
$$(Y - GM) = (TM - GM) + (Y - TM)$$

We have calculated each of these three deviations in Table 9.5 and entered their sums of squares in the last row.

The values of $(Y - GM)$ are the deviations in the yields from the grand mean. The sum of the squares of these deviations represents the total variation of the yields. It is called the **Total Sum of Squares** ($TotalSS$). For this example $TotalSS = 364.6$.

The value of $(TM - GM)$ is an estimate of the treatment effect. The sum of the squares of these differences for each yield represents the between-treatment variation which is called the **Treatment Sum of Squares**. We denote it by $TreatSS$. For this example $TreatSS = 188.2$.

The value of $(Y - TM)$ is called the **residual** or error. The residuals represent the variation in the yields of plots treated alike. Their sum of squares represents the within-treatment variation which is called the **Residual Sum of Squares**. However, you will often see it called the **Error Sum of Squares**. We denote it by $ResidSS$. For this example $ResidSS = 176.4$.

You can obtain $ResidSS$ by finding the Sxx values for each treatment separately and adding the results. It represents the variation within the treatments. In Table 9.4, the separate Sxx values are 55.02, 67.86, 29.18 and 24.34. These add to 176.4 which is the $ResidSS$ value.

Table 9.5. Table showing the model terms for Example 9.1

Treatment (variety)	Yield Y	GM	TM	Y−GM	TM−GM TE	Y−TM R	Y−R F
A	22.2	24.1	20.4	−1.9	−3.7	1.8	20.4
A	17.3	24.1	20.4	−6.8	−3.7	−3.1	20.4
A	21.2	24.1	20.4	−2.9	−3.7	0.8	20.4
A	25.2	24.1	20.4	1.1	−3.7	4.8	20.4
A	16.1	24.1	20.4	−8	−3.7	−4.3	20.4
B	24.1	24.1	28.6	0	4.5	−4.5	28.6
B	30.3	24.1	28.6	6.2	4.5	1.7	28.6
B	27.4	24.1	28.6	3.3	4.5	−1.2	28.6
B	26.4	24.1	28.6	2.3	4.5	−2.2	28.6
B	34.8	24.1	28.6	10.7	4.5	6.2	28.6
C	25.9	24.1	22.4	1.8	−1.7	3.5	22.4
C	18.4	24.1	22.4	−5.7	−1.7	−4	22.4
C	23.2	24.1	22.4	−0.9	−1.7	0.8	22.4
C	21.9	24.1	22.4	−2.2	−1.7	−0.5	22.4
C	22.6	24.1	22.4	−1.5	−1.7	0.2	22.4
D	23.9	24.1	25.0	−0.2	0.9	−1.1	25.0
D	21.7	24.1	25.0	−2.4	0.9	−3.3	25.0
D	24.8	24.1	25.0	0.7	0.9	−0.2	25.0
D	28.2	24.1	25.0	4.1	0.9	3.2	25.0
D	26.4	24.1	25.0	2.3	0.9	1.4	25.0
Sum	482.0	482.0		0	0	0	
Sum of squares				364.6	188.2	176.4	

9.4.1 Partitioning the Variance

We have seen that

$$(Y - GM) = (TM - GM) + (Y - TM)$$

It can be shown that not only is this result true for each yield but it is also true for the sums of squares of these deviations:

$$\Sigma(Y - GM)^2 = \Sigma(TM - GM)^2 + \Sigma(Y - TM)^2$$

That is,

$$TotalSS = TreatSS + ResidSS$$

The interpretation is that the total variation in the yields is made up of the between-treatment variation and the within treatment variation. In our example, $TotalSS = 364.6$ and $TreatSS = 188.2$. Expressing $TreatSS$ as a percentage of $TotalSS$ we conclude that 52% of the total variation in the yields can be explained by the variety differences. This is further evidence that there are treatment differences. However, a formal test is provided by the analysis of variance.

9.5 ANALYSIS OF VARIANCE

We now explain the formal test of the null hypothesis that all treatments have the same effect on the yield. If the four (in general, t) varieties (in general, treatments) in Table 9.4 have the same yield, the four sample means are all estimates of a common population mean. For the test to be valid you must assume that that the observed yields for each treatment are independent random samples from normally distributed populations each having the same variance, σ^2. To test whether the mean yields differ significantly, the between-treatment variation is compared with the within-treatment variation. If the former is large compared to the latter, suspect that this is due to real treatment differences.

The null hypothesis implies that differences in the mean yields are due to chance. The calculations leading to the test of this hypothesis are arranged in an **analysis of variance** (ANOVA) table which is described in Section 9.5.2. First some background theory is presented.

9.5.1 Theory

Consider the yields of variety A. These five yields are assumed to be a random sample of five yields taken from the population of all possible yields that could be obtained by growing large numbers of plots under the same conditions. The population from which the five yields is obtained is therefore a hypothetical population. It does not exist in reality. The other three sets of yields are also assumed to be random samples from hypothetical populations.

If these populations have the same variance, σ^2, called the **experimental error variance**, then the individual variance estimates, s_1^2, s_2^2, s_3^2 and s_4^2 can be pooled to give

THE COMPLETELY RANDOMISED DESIGN

a more precise estimate of experimental error. For equally replicated treatments, this estimate is simply the average of the individual s^2 values. In the case of unequal replication, the pooled estimate of σ^2, called s_p^2 is a weighted average of the individual s^2 values. The weights are the individual degrees of freedom, the values of $n - 1$. The formula for s_p^2 is

$$s_p^2 = \frac{(n_1 - 1)s_1^2 + (n_2 - 1)s_2^2 + (n_3 - 1)s_3^2 + (n_4 - 1)s_4^2}{(n_1 - 1) + (n_2 - 1) + (n_3 - 1) + (n_4 - 1)}$$

This is a generalisation of the formula used in the independent samples t-test (Chapter 6). This formula can obviously be generalised to more than four treatments. You will recall (Section 2.3.5) that for any sample of n-values, $Sxx = (n - 1)s^2$. Hence the formula above can also be written as

$$s_p^2 = \frac{Sxx_1 + Sxx_2 + Sxx_3 + Sxx_4}{n_1 + n_2 + n_3 + n_4 - 4}$$

As $n_1 + n_2 + n_3 + n_4 = N$ and $t = 4$, the formula can be written even more concisely as

$$s_p^2 = \frac{\sum Sxx}{N - t}$$

The Sxx values are given in Table 9.4, and by addition $\Sigma Sxx = 176.40$. This is the **Residual Sum of Squares** (*ResidSS*) explained in Section 9.4:

$N = 20$ and $t = 4$, so $N - t = 16$. Hence $s_p^2 = \dfrac{176.40}{16} = 11.025$

s_p^2 is a pooled estimate of the error variance, σ^2. It is also called the **residual mean square** (*RMS*), though many authors call it the **error mean square**. It has $N - t$ **residual degrees of freedom** which can be found by adding up the degrees of freedom for each treatment separately; $(4 + 4 + 4 + 4 = 16)$ in Example 9.1.

Under the null hypothesis that all the treatments have the same effect, statistical theory shows that the sample variance of the treatment means is an estimate of σ^2/r (assuming each treatment is replicated r times; $r = 5$ in Example 9.1). See Section 4.8 for a discussion of the sampling distribution of the mean.

The sample variance of the treatment means for this example is 12.5467. On multiplying by $r = 5$ you obtain 62.733 as our estimate of σ^2 under the null hypothesis. This estimate is called the **Treatment Mean Square** which is denoted by *TMS*. In general it can be calculated by finding the **Treatment Sum of Squares** (*TreatSS*) and then dividing by $(t - 1)$, the number of treatments minus 1.

$(t - 1)$ is called the **treatment degrees of freedom**. Therefore the formula for *TMS* becomes

$$TMS = \frac{TreatSS}{t - 1}$$

TMS is an estimate of σ^2 under the null hypothesis of no treatment differences.

If there are treatment differences the variation in the treatment means will tend to be larger than expected under the null hypothesis and *TMS* will be an estimate of something larger than σ^2. Whether or not the null hypothesis is true, the within-treatment variation summarised by the Residual Mean Square (*RMS*) is an estimate of σ^2. If the null hypothesis of equal treatment effects is true, then similar values for the Treatment Mean Square (*TMS*) and the Residual Mean Square (*RMS*) should be obtained. On the other hand, if the treatments have different effects on yield, *TMS* should be larger than *RMS*. To test the null hypothesis, the ratio of these two variance estimates is calculated. It is called the **Variance Ratio** and denoted by *VR*:

$$VR = \frac{TMS}{RMS}.$$

For Example 9.1

$$VR = \frac{62.733}{11.025} = 5.69$$

A value of *VR* much larger than 1 provides evidence of different treatment effects. A formal one-tailed test involves comparing our value of *VR* with tabulated values of the F-distribution. If testing at the 5% level of significance you look for the value of $F_{(t-1, N-t, 5\%)}$ in the 5% F-table (Appendix 4a). In our example $(t-1)$, the treatment degrees of freedom are 3 and $(N - t)$, the residual degrees of freedom, are $20 - 4 = 16$. So to carry out the F-test at the 5% level for this example you would look in the 5% F-table for the entry in column 3 and row 16. This table value is $F_{(3, 16, 5\%)} = 3.24$. If the calculated *VR* is greater than 3.24 you would conclude there is evidence that the treatments have different effects on yield. In other words the treatment (variety) means are significantly different at the 5% level.

9.5.2 The Analysis of Variance Table

The calculations are summarised in Table 9.6 (the analysis of variance table) and Output 9.1.

Column 1 shows the sources of variation. The total variation in the data (364.6) can be partitioned into two sources. The first source is variation due to treatment differences (188.2) and the second source is due to variation between plots treated alike (176.4).

Column 2 is the degrees of freedom (DF or df):

The total degrees of freedom is $Totaldf = N - 1 = 20 - 1 = 19$.
The treatment degrees of freedom are $Treatdf = t - 1 = 4 - 1 = 3$.
The residual degrees of freedom can be found by subtraction $(19 - 3 = 16)$. It can also be found using $Rdf = N - t = 20 - 4 = 16$.

Column 3 gives sums of squares. Alternative methods of calculating the entries in this column are discussed in Section 9.5.6.

Table 9.6. Analysis of variance of wheat yield data (kg/plot)

Source	DF	SS	MS	VR
Treatment (variety)	3	188.200	62.733	5.69
Residual (error)	16	176.400	11.025	
Total	19	364.600		

The total sum of squares is the corrected sum of squares (the Sxx value) of all $N = 20$ yields. It represents the total variation in the yields and we denote it by $TotalSS$. The treatment sum of squares ($TreatSS$) represents the variation in the yields due to treatment differences, and the residual sum of squares ($ResidSS$) represents the variation in the yields of plots treated alike. The latter can be calculated by finding the corrected sum of squares for each treatment and adding the results. Add the numbers in the last row of Table 9.4 to find $ResidSS = 55.02 + 67.86 + 29.18 + 24.34 = 176.40$. Alternatively, you could use the information already given in Section 9.4.1; $TotalSS = TreatSS + ResidSS$, and hence calculate $ResidSS$ by subtraction:

$$ResidSS = TotalSS - TreatSS = 364.6 - 188.2 = 176.4$$

Column 4 gives the Mean Square (MS) values.

The treatment mean square (TMS) is an estimate of the experimental error variance, σ^2, if the treatments all have the same effect. It is calculated by dividing the treatment sum of squares by the treatment degrees of freedom. The residual (error) mean square (RMS) is an estimate of the experimental error variance whether or not the treatments have equal effects. It is the s_p^2 discussed earlier, and is calculated by dividing the residual sum of squares by the residual degrees of freedom. For this example

$$TMS = \frac{TreatSS}{Treatdf} = \frac{188.2}{3} = 62.733 \quad \text{and} \quad RMS = \frac{ResidSS}{Rdf} = \frac{176.400}{16} = 11.025$$

Note: The entries in the Mean Square (MS) column are found by dividing the entries in the Sums of Squares (SS) column by the corresponding entries in the Degrees of Freedom (DF) column.

Column 5 gives the variance ratio (VR).

The variance ratio for treatments is the ratio of the two variance estimates, TMS and RMS. This is denoted by F in many computer outputs because it is used to carry out the F-test of the null hypothesis that the treatments all have the same effect on yield. The *VR* is calculated by dividing the treatment mean square by the residual mean square. For Example 9.1

$$VR = \frac{TMS}{RMS} = \frac{62.733}{11.025} = 5.69$$

9.5.3 The *F*-test

The *VR* is a value from an *F*-distribution if the null hypothesis of equal treatment effects is true. Under the null hypothesis of equal treatment effects, *VR* is expected to be equal to approximately 1.0 (because *TMS* and *RMS* are then both estimates of σ^2), and has only a 5% chance of being greater than $F_{(Treatdf, Rdf, 5\%)}$, the *F*-table value at the 5% point with *Treatdf* and *Rdf*. If there are real treatment differences, *TMS* estimates something larger than σ^2 as the variation in the treatment means is then larger than can be attributed to chance, and so *VR* could be much larger than 1.0.

If *VR* is larger than $F_{(Treatdf, Rdf, 5\%)}$ we reject the null hypothesis at the 5% level and conclude there are probably real treatment differences. This is a one-tailed test. For our example $VR = 5.69$. From *F*-tables $F_{(Treatdf, Rdf, 5\%)} = F_{(3, 16, 5\%)} = 3.24$.

Look in column 3 and row 16 of the 5% *F*-table (Appendix 4a). Our value of *VR* is greater than the table value of 3.24, so we reject the null hypothesis of equal treatment effects at the 5% level. Thus there is sufficient evidence to conclude that there are real treatment differences. In the context of this example you conclude that the varieties probably have different yields when grown under similar conditions.

9.5.4 The *P*-value

Analysis of variance tables produced by Minitab and SAS have a *P* column to the right of the *F*-column (Output 9.1). In Genstat, *P* is replaced by *Fprob* and *F* is replaced by *VR*. *P* gives the probability of getting a *VR* value greater than that actually obtained, if the null hypothesis of equal treatment effects is true. This probability is called the *P*-value of the *F*-test. If it is less than 0.05 we conclude there are **significant** treatment differences. The larger the calculated *VR*, the smaller the *P*-value and the greater the evidence against the null hypothesis. In advance of data collection, you would expect the *TMS* and *RMS* to be equal and hence the *VR* to be close to 1 if the null hypothesis of equal treatment effects were true. You would expect *P* to be approximately 0.5:

- If *P* is less than 0.01 you conclude the differences in the treatment means are **highly significant**.
- If *P* is less than 0.001 you conclude the differences are **very highly significant**.

Before computers were introduced, it was difficult to calculate *P* exactly. It was only possible to say, after consulting the *F*-tables, that *P* was less than 5% ($P < 0.05$), was less than 1% ($P < 0.01$), or was less than 0.1% ($P < 0.001$). You will often see asterisks representing the significance of results in published work. One star (*) represents significance at the 5% level, but not at the 1% level. Two stars (**) represents significance at the 1% level, but not at the 0.1% level. Three stars (***) represents significance at the 0.1% level.

For Example 9.1:

- The *P*-value is less than 0.05 as our *VR* (5.69) is greater than the 5% table value of 3.24.
- The corresponding 1% table value is 5.29, so *P* is less than 0.01.
- The 0.1% table value is 9.01. The value of *VR* is not greater than this, so *P* is not less than 0.001.
- All you can conclude by looking up tables is that *P* lies between 0.01 and 0.001 and state that the treatment means are significantly different ($P < 0.01$). Output 9.1 shows that *P* is actually 0.008.

Figure 9.2 shows the calculated *VR* in relation to the values of the *F*-distribution with 3 and 16 degrees of freedom. The area to the right of 3.24 is 5%, the area to the right of 5.29 is 1% and the area to the right of 9.01 is 0.1%. Hence, the area to the right of the *VR* (5.69) is between 1% and 0.1% (*P* between 0.01 and 0.001) which means that our value of 5.69 has a less than a 1% chance of occurring if there are no treatment differences in the population. Hence conclude that this high value of *VR* is strong evidence that the treatments (varieties) have different effects.

9.5.5 The Coefficient of Variation

Although the residual mean square gives an estimate of the residual variation, σ^2, its size depends on the units of measurement. If yields are measured in t/ha, the square root of the residual mean square is also measured in t/ha. The coefficient of variation (CV%) is a measure of the residual variation independent of the units of measurement. It is found by dividing the square root of the residual mean square by the grand

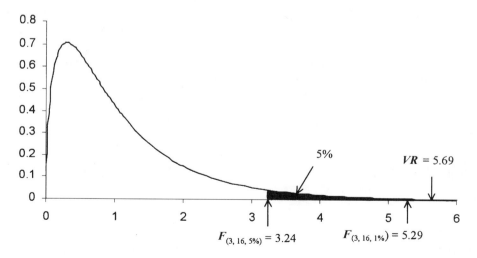

Figure 9.2. The *F*-distribution with 3 and 16 degrees of freedom

mean and multiplying by 100 to express the result as a percentage. For this example it is 13.8%:

$$CV = \frac{\sqrt{RMS}}{GM} \times 100 = \frac{\sqrt{11.025}}{24.1} \times 100 = 13.8\%$$

The coefficient of variation also allows comparisons to be made between similar experiments. For example, field experiments on oilseed rape have been found to have very high coefficients of variation, up to 100% or more! This means high replication must be used in order to detect any real treatment differences. Experiments on cereals generally have smaller coefficients of variation, within the range 10–20%. Thus, treatment differences can be detected with a smaller number of plots per treatment. A knowledge of the likely value of a CV% is of help in planning the number of replications required for an experiment. If the CV% obtained in a trial is well outside the range expected, this indicates that the experiment may not have been properly conducted.

9.5.6 Alternative Methods for Calculating the Sums of Squares

In this section we discuss two alternative ways of calculating the sums of squares in the SS column of the analysis of variance table.

Method 1: The correction factor (CF) method

Before computers were widely available and before pocket calculators with SD mode were obtainable, all calculations had to be carried out by hand. These calculations were done using recipes which minimised the build-up of rounding-off errors. Most standard textbooks on statistical analysis describe these methods and use them mainly because they were written before the advent of calculators with SD mode. Because you may need to consult classic textbooks, it is important you are aware of the calculations using this method. Furthermore, the slowness of the calculating process gives you more time to interpret the results and should lead to a better understanding of the principles of analysis of variance.

Using the data of Example 9.1 (Table 9.4):

Step 1. Find the grand total (G). This is found by adding up all the N numbers:

$$G = \Sigma x = 22.2 + 17.3 + \ldots + 28.2 + 26.4 = 482.0$$

Step 2. Find the correction factor (CF). This is obtained by squaring the grand total and dividing by N:

$$CF = \frac{G^2}{N} = \frac{482.0^2}{20} = 11\,616.20$$

Step 3. Find the uncorrected total sum of squares. This is the sum of the squares of all the N numbers:

THE COMPLETELY RANDOMISED DESIGN   117

$$\Sigma x^2 = 22.2^2 + 17.3^2 + \ldots + 28.2^2 + 26.4^2 = 11\,980.80$$

Step 4. Find the corrected total sum of squares. This is the sum of the squares of the deviations of the N data values from the overall or grand mean. We denote it by *TotalSS*:

$$TotalSS = \Sigma x^2 - CF = 11\,980.80 - 11\,616.20 = 364.60$$

This is the value that should be entered in the Total row and SS column of the analysis of variance table (Table 9.6).

Step 5. Find the treatment sum of squares (*TreatSS*). This is the sum of the squares of the deviations of the treatment means from the overall mean with each squared deviation weighted by the number of replications for the corresponding treatment. To find *TreatSS* by the correction factor method proceed as follows:

First find the individual treatment totals. They are 102.0, 143.0, 112.0 and 125.0. Square each treatment total and divide by the number of replications in that treatment. Add these results for each treatment and finally subtract the correction factor (*CF*):

$$TreatSS = \Sigma \left(\frac{T^2}{r} \right) - CF = \left(\frac{102.0^2}{5} + \frac{143.0^2}{5} + \frac{112.0^2}{5} + \frac{125.0^2}{5} \right) - CF$$
$$= 11\,804.40 - 11\,616.2 = 118.20$$

Step 6. Find the residual sum of squares by subtraction:

$$ResidSS = TotalSS - TreatSS = 364.60 - 188.20 = 176.40$$

Method 2: Using a hand calculator

Use your calculator in SD or STATS mode. Make sure you know how to obtain the standard deviation. The relevant key is either marked σ_{n-1} or s. In what follows SD means sample standard deviation.

Step 1. Enter all N numbers. Press the n key to make sure you have entered all the numbers. Obtain the SD, square the result and multiply by $N-1$ to obtain *TotalSS*. This is the corrected sum of squares of all N values (*TotalSS* = the *Sxx* value for all N values).

Step 2. For each treatment, use your calculator in SD mode to find the treatment total, the treatment mean and the *Sxx* value. Remember $Sxx = (SD)^2 \times (n-1)$. Add the *Sxx* values to obtain the residual sum of squares (*ResidSS* = ΣSxx).

Step 3. Find the treatment sum of squares by subtraction. *TreatSS* = *TotalSS* − *ResidSS*

Note: If the treatments are equally replicated, you can find *TreatSS* directly as follows. Clear the results of previous calculations. Enter the treatment totals found in step 2. Obtain their SD, square the result, and multiply by $t-1$ (the treatment degrees of freedom). This gives the corrected sum of squares of the treatment totals. Divide this result by r (the number of replications per treatment) to get *TreatSS*. The instructions are summarised as:

$$TreatSS: (SD)^2 \times (t-1) \div r =$$

Hence, when the treatments are equally replicated, it is quicker to find *TreatSS* directly and calculate *ResidSS* by subtraction:

$$ResidSS = TotalSS - TreatSS$$

Once the SS column has been completed, the rest of the ANOVA table can be filled in easily.

Method 3: Using the model

For details see Table 9.5. This is **the most instructive method** for calculating the sums of squares as it gives an insight into the meaning of what you are calculating. It is very tedious to carry out in practice. However, the process can be made automatic using a suitable computer spreadsheet.

Summary of the analysis of variance

The ANOVA table of Output 9.1 shows (in our example) that there are significant differences between the four treatment (variety) means because the P-value is 0.008 which is less than 0.05.

9.6 AFTER ANOVA

Note: If the F-test shows no significant difference between the treatment means, further tests should not normally be carried out. However, it is acceptable to make planned comparisons, provided these are independent and number no more than $t - 1$, the treatment degrees of freedom.

The F-test in the analysis of variance table is used to test whether there are any significant differences between the treatment means. It does not tell you where these differences lie, and for Example 9.1 does not provide a method for comparing individual varieties. What you do next depends on your experimental objectives.

The possibilities are:

(1) Carry out tests of individual comparisons that were planned in advance. These may include pairwise comparisons or other contrasts where groups of treatments are compared.
(2) Calculate confidence intervals for effects of interest. You could calculate (i) confidence intervals for particular treatment means, (ii) confidence intervals for the difference between treatment means.
(3) Fit a regression line or curve to the treatment means if the treatments represent increasing or decreasing levels of a factor.
(4) In variety trials, it may be valid to carry out tests which compare all possible pairs or treatment means.

THE COMPLETELY RANDOMISED DESIGN

We describe some of these possibilities in this section and also in Chapter 13. They all use the pooled residual mean square (RMS) which can be found from the analysis of variance table. This is the s_p^2 discussed earlier.

9.6.1 t-Tests

In Example 9.1, you could carry out an independent samples t-test on the data for varieties A and B, ignoring data on varieties C and D. However, the precision of the comparison is increased by using the data from all the treatments because a more precise estimate of experimental error is obtained if the within-treatment variations (s^2 values) are not markedly different. Assuming the yields from plots treated alike are random samples from normal distributions with the same variance σ^2, these s^2 values are assumed to be estimates of σ^2. By pooling these estimates you obtain the s_p^2 value which is the residual mean square (RMS) in the analysis of variance table. You can use this pooled estimate to carry out t-tests which have been planned in advance. The formula for the t-test is (Chapter 6)

$$t = \frac{\bar{x}_1 - \bar{x}_2}{SED}$$

where

$$SED = \sqrt{s_p^2 \left(\frac{1}{n_1} + \frac{1}{n_2} \right)}$$

the standard error of the difference between two means.

The s_p^2 value should be obtained by pooling the data from all the treatments (not just the two being compared) provided the conditions for pooling are valid. This pooled value is the RMS value from the ANOVA table. To test whether the mean yields of variety A and B differ significantly we carry out the t-test as follows.
For this example

$$n_1 = n_2 = r = 5$$

Hence

$$\frac{1}{n_1} + \frac{1}{n_2} = \frac{1}{r} + \frac{1}{r} = \frac{2}{r}$$

and therefore, replacing s_p^2 with RMS

$$SED = \sqrt{\frac{2 \times RMS}{r}} = \sqrt{\frac{2 \times 11.025}{5}} = 2.10$$

Because you are using a pooled estimate of error variance, the pooled degrees of freedom for the t-test are found by adding the degrees of freedom for each sample. This is called the residual degrees of freedom (Rdf). In general, $Rdf = (n_1 - 1) + (n_2 - 1) + \ldots + (n_t - 1) = N - t$. For this example, if any two variety means are compared using a t-test, the degrees of freedom are

$$N - t = 20 - 4 = 16$$

To test whether varieties A and B have significantly different yields find

$$t = \frac{\text{difference in means}}{SED} = \frac{28.6 - 20.4}{2.10} = 3.90 \quad \text{on 16 df}$$

From tables $t_{(16, 2.5\%)} = 2.120$. The calculated value of 3.90 is greater than 2.120. Hence you reject the hypothesis that varieties A and B give equal yields, at the 5% level on a two-tailed test. There is evidence that variety B gives higher yields than variety A. You reach this conclusion in spite of the overlap in the confidence intervals.

Because of the equal replication, you use the same *SED* value for comparing varieties A and C. For this comparison,

$$t = \frac{22.40 - 20.40}{2.10} = 0.95 \quad \text{on 16 df}$$

As this *t*-value is less than 2.120, we have insufficient evidence to conclude that varieties A and C give different yields.

You should also check that the *t*-value for comparing varieties A and D is 2.19. What does this show?

9.6.2 One-tailed *t*-Tests

If in advance of the experiment you had evidence variety B would give a greater yield than A, a one-tailed test would be appropriate. In this case the table value for testing at the 5% level is $t_{(N-1, 5\%)} = t_{(16, 5\%)} = 1.746$. The conclusions are the same as before. Remember, if the difference is significant on a two-tailed test, it is bound to be significant on a one-tailed test. The reverse is not necessarily true. Hence you should not carry out one-tailed tests unless you have a very good reason for doing so.

9.6.3 Least Significant Difference Analysis

In both the *t*-tests just conducted, the same *SED* value is used. This is because the treatments (varieties) were equally replicated. If the experiment had contained many varieties to be compared with A, it would be tedious to carry out many repetitive *t*-tests. As the same *SED* value is used in all the tests, you can declare two means significantly different at the 5% level on a two-tailed test if

$$\frac{\text{Difference in means}}{SED} \text{ is greater than the } t\text{-table value of } t_{(N-1, 2.5\%)}$$

In other words, two means are significantly different if the difference in means is greater than $t_{(N-1, 2.5\%)} \times SED$. This is called the **least significant difference at the 5% level**:

$$LSD_{5\%} = t_{(N-1, 2.5\%)} \times SED$$

For this example, $LSD_{5\%} = 2.120 \times 2.10 = 4.45$.

THE COMPLETELY RANDOMISED DESIGN   121

A and B are significantly different as the difference in means is 8.20, which is greater than 4.45. A and C are not significantly different as the difference in means is only 2.00, which is less than 4.45. B and D are not significantly different as the difference in means is 3.60, which is less than 4.45.

Caution: The LSD should only be used for comparisons planned in your experimental objectives. No more than $t-1$ should made. Some statisticians recommend that LSDs should not be used if the F-test in the ANOVA table is not significant and some say LSDs should never be used. These points are discussed in greater detail in Section 9.6.7.

9.6.4 Confidence Intervals for the Difference between Two Treatment Means

Often it is known that treatments have different effects. In this case the SED calculated by pooling all the data can be used to find confidence intervals for the difference between two treatment means.

A 95% confidence interval is *difference in means* $\pm LSD_{5\%}$
A 95% confidence interval for the difference between means of varieties A and B is 8.20 ± 4.45

The lower limit is $8.20 - 4.45 = 3.75$ and the upper limit is $8.20 + 4.45 = 12.65$. So we are 95% confident that the mean yield of variety B exceeds that of variety A by between 3.75 and 12.65 kg/plot.

9.6.5 Confidence Intervals and Hypothesis Testing

Note that the 95% confidence limits of 3.75 and 12.65 do not include zero. This shows that the two treatment means are significantly different at the 5% level on a two-tailed test. If a 95% confidence interval for the difference between two treatment means includes zero, these two means are not significantly different at the 5% level on a two-tailed test. For example, if you find a 95% confidence interval for variety C mean minus variety A mean, the answer is -2.45 to $+6.45$. This interval includes zero so we conclude the means do not differ significantly at the 5% level.

9.6.6 Confidence Intervals for Individual Treatment Means

The pooled standard error for an individual treatment mean is $SEM = \sqrt{s_p^2/r}$, where r is the number of replications of the treatment. For this example,

$$SEM = \sqrt{\frac{s_p^2}{r}} = \sqrt{\frac{11.025}{5}} = 1.485$$

A 95% confidence interval for a particular treatment mean is *mean* $\pm t_{(N-t, 2.5\%)} \times SEM$.

A 95% confidence interval for the mean of variety B is

$$28.60 \pm (2.120 \times 1.485) = 28.60 \pm 3.15$$

The lower limit is $28.60 - 3.15 = 25.45$ and the upper limit is $28.60 + 3.15 = 31.75$. We are 95% certain that the mean yield for B lies between 25.45 and 31.75 kg/plot.

Caution: In the calculation of *SED* and *SEM* you are pooling data from all the treatments. This is only valid if the variation within each treatment is similar. If you cannot assume equal variation you may obtain valid results by transforming the data before analysis. If treatments differ greatly in their variability it is not wise to pool the data. In this case each treatment could be analysed separately. For each treatment you could find a separate standard error and a 95% CI for the mean. For variety B, the sample mean is 28.60 and the sample variance, s^2 is 16.965 (Table 9.4). Hence the standard error, *SE*, is

$$SE = \sqrt{\frac{s^2}{r}} = \sqrt{\frac{16.965}{5}} = 1.842$$

This compares with 1.485 obtained by **pooling the data** from all the treatments. Using data from B only, a 95% CI for the true mean of B is

$$\bar{x} \pm t_{(r-1, 2.5\%)} \times SE = 28.6 \pm (2.776 \times 1.842) = 28.6 \pm 5.11 = (23.49 \text{ to } 33.71)$$

Note how much wider this interval is than 25.45 to 31.75 which was calculated from the pooled data. Using the pooled method the degrees of freedom are $(N - t)$, 16 in this example. Using only the data for the treatment under consideration the degrees of freedom are $r - 1 = 4$. Hence a considerable amount of precision is lost by not pooling. However, pooling data can lead to the wrong conclusions if the variability within each (or some) treatments is markedly different.

The procedure of finding separate confidence intervals for each treatment using the above formula would still be invalid if the variability within each treatment is markedly non-normally distributed. This might be the case if the response variable is a count which might be zero, in tens or hundreds. In this case it may be necessary to transform the data before analysis. If in doubt it is always best to examine your raw data in detail and then consult a statistician.

9.6.7 Dangers in LSD Analysis

An LSD analysis is in effect the use of *t*-tests to compare all possible pairs of means. When the treatments have a structure an LSD analysis may not be appropriate. For example, if they represent increasing levels of fertilizer, you should try to fit a response curve using regression methods (Chapters 7 and 8). Only certain comparisons are appropriate if the treatments have a factorial structure (Chapter 12). You should also be aware of the relative merits of other multiple comparison tests (Chapter 13).

In many experiments there is a control treatment and the objective may be to compare each of the other treatments with the control. In this situation an LSD analysis is valid. You may even find that the *F*-test in the ANOVA table is not significant while one or two of the individual comparisons are. Do not worry because these comparisons were planned in advance. In this case the ANOVA table was not necessary. You could calculate the *SED* and hence *LSD* directly from the separate *Sxx* values.

Note that if too many comparisons are carried out, it is highly likely that false conclusions will be made. The larger the number of treatments, the greater the chance that some pairs of treatment means will be declared significantly different even when there are no real treatment differences. In this situation the *LSD* should not be used to compare all possible pairs of treatments. It is especially misleading to compare the largest and smallest treatment means in this way. Generally, it is acceptable to use the *LSD* for particular comparisons planned in advance such as each treatment with a control.

9.7 REPORTING RESULTS

When reporting the results of an analysis of variance you should always quote:

- The treatment means
- The residual degrees of freedom and
- The *P*-value (the exact value or in the form $P < 0.05$).

If the treatments are equally replicated you should report the pooled *SED*. In the case of unequal replications, the residual mean square (*RMS*) should be quoted. This is often called the pooled variance (s_p^2). Its square root, the pooled standard deviation, may be quoted instead. LSDs are often not acceptable in publications.

When there is no structure in the treatments and no natural order, an appropriate graphical representation is a bar chart. Each bar represents a treatment mean and the *SED* value can be inserted. If appropriate, the *LSD* value at the 5% level can be inserted instead. This enables the reader to judge at a glance which treatments differ from which.

In cases where the means themselves are of interest, as opposed to differences between means, it is more appropriate to insert error bars at the tops of the individual bars of the bar chart. These are usually plus and minus one standard error. The standard error for each treatment is calculated using the data for that treatment, ignoring the data from the other treatments. An example is Figure 9.1. For a given treatment the formula is $SE = s/\sqrt{n}$ where n is the number of replications of that treatment. If the standard deviations of the treatments are similar, a pooled standard error of a treatment mean (*SEM*) can be quoted.

If confidence intervals are required for individual treatment means, it is best to use the square root of the pooled residual mean square ($\sqrt{RMS} = s_p$), also called the pooled standard deviation, instead of different *s*-values for each treatment. This is satisfactory if the conditions for pooling are valid. When the treatment represents increasing levels of a factor, a line graph is often a more appropriate graphical representation.

Remember that statistical analysis is only valid if certain assumptions are made:

- You should always check the assumptions. If some assumptions are not valid when applied to the raw data, they may be valid after transforming the data. Assumptions and transformations are discussed in more detail in Chapter 14.

9.8 THE COMPLETELY RANDOMISED DESIGN — UNEQUAL REPLICATION

This can happen by choice or chance. Some treatments may be more important than others and need more replication. A control may have more replication than new, expensive treatments. In biological research some experimental units are frequently destroyed by pests and diseases, so the results table has missing values. In the case of a one-way analysis of variance this presents no problems for the analysis. However, it makes treatment comparisons more tedious as there is no single appropriate SED or LSD value for all comparisons.

Example 9.2

Table 9.7 shows Example 9.1 with four missing values (denoted by asterisks). To carry out the analysis using a computer the treatment (variety) codes should be entered in the first column and the yields in the second column of a data sheet. There should be 16 rows (do not enter rows for the missing data). If you do have 20 rows, with asterisks denoting missing values, check that your program ignores the rows with asterisks because some programs may use estimates of the missing values in subsequent calculations.

To obtain the ANOVA table by hand, the methods described in Section 9.5.6 can be used. As an exercise delete from Table 9.5 the rows corresponding to the missing data and recalculate the entries in columns three to eight. Confirm that the sums of squares of the values in columns five, six and seven are $TotalSS = 347.92$, $TreatSS = 214.92$ and $RSS = 133.00$ respectively. Now fill in the ANOVA table.

Table 9.7. Data of Example 9.1 minus 4 values

Treatment	Variety A	Variety B	Variety C	Variety D
	*	24.1	25.9	23.9
	17.3	30.3	18.4	21.7
	21.2	*	*	24.8
	*	26.4	21.9	28.2
	16.1	34.8	22.6	26.4
Total	54.6	115.6	88.8	125.0
Replication (r)	3	4	4	5
Mean	18.20	28.90	22.20	25.00

THE COMPLETELY RANDOMISED DESIGN

Table 9.8. ANOVA table for data in Table 9.7

Source	DF	SS	MS	VR
Treatments	3	214.92	71.640	6.46
Residual	12	133.00	11.083	
Total	15	347.92		

9.8.1 Summary of Calculations

- Total degrees of freedom = $N - 1 = 16 - 1 = 15$
- Treatment degrees of freedom = $t - 1 = 4 - 1 = 3$
- Residual degrees of freedom = $(N - 1) - (t - 1) = 15 - 3 = 12 = N - t$
- Residual sum of squares $(ResidSS) = TotalSS - TreatSS = 347.92 - 214.92 = 133.00$
- The Mean Squares are obtained by dividing the sums of squares by the corresponding degrees of freedom.
- The Variance Ratio (VR) is obtained by dividing the treatment mean square ($TMS = 71.640$) by the residual mean square ($RMS = 11.083$).

The Residual Sum of Squares ($ResidSS$) could be found directly by calculating Sxx separately for each treatment and adding the results (Section 9.5). The VR is 6.46 on (3, 12) df. From the 5% F-table (Appendix 4a) find $F_{(3, 12, 5\%)} = 3.49$. As the VR of 6.46 is larger, reject the null hypothesis of equal treatment effects at the 5% level and also reject at the 1% level because $F_{(3, 12, 1\%)} = 5.95$.

9.8.2 Comparison of Treatment Means — Unequal Replication

This is not so straightforward as you cannot find a common SED and LSD that apply to all comparisons. You must consider each comparison separately.

Comparison of treatments 1 and 2 from Table 9.7

Step 1. Find the SED.

$$SED = \sqrt{RMS\left(\frac{1}{n_1} + \frac{1}{n_2}\right)} = \sqrt{11.083\left(\frac{1}{3} + \frac{1}{4}\right)} = 2.543$$

This is similar to the formula for SED in Section 6.2 where only two treatments were compared. In this case RMS replaces s_p^2. RMS is a pooled estimate of σ^2 based on all four treatments.

Step 2. Find the LSD value at the 5% level of significance:

$$LSD_{(5\%)} = t_{(12, 2.5\%)} \times SED$$
$$= 2.179 \times 2.543 = 5.541$$

Note: The 12 is the residual degrees of freedom. It is the pooled degrees of freedom from all four treatments.

Treatment 1 has 3 values, so its df = 2 Treatment 2 has 4 values, so its df = 3 Treatment 3 has 4 values, so its df = 3 Treatment 4 has 5 values, so its df = 4 Hence the residual degrees of freedom = 2 + 3 + 3 + 4 = 12

Step 3. Find the difference between the means for treatments 1 and 2:

$$\bar{x}_2 - \bar{x}_1 = 28.90 - 18.20 = 10.70$$

Step 4. Compare this difference with the LSD found in step 2.

The difference between the means (10.70) is greater than the $LSD_{(5\%)}$ value of 5.541, so conclude that treatments 1 and 2 differ significantly at the 5% level.

Note: An alternative to this test is to find a 95% confidence interval for the difference between the two means. The formula is *difference in means* $\pm LSD_{(5\%)}$. For this comparison the interval is $10.70 \pm 5.54 = (5.16, 16.24)$. As this interval does not include zero conclude that these two treatment means are significantly different at the 5% level.

Comparison of treatments 2 and 4 from Table 9.7

Step 1. Find the *SED*:

$$SED = \sqrt{RMS\left(\frac{1}{n_2} + \frac{1}{n_4}\right)} = \sqrt{11.083\left(\frac{1}{4} + \frac{1}{5}\right)} = 2.233$$

Step 2. Find the *LSD* at the 5% level:

$$LSD_{(5\%)} = t_{(12, 2.5\%)} \times SED$$
$$= 2.179 \times 2.233 = 4.866$$

Step 3. Find the difference between the means for treatments 2 and 4:

$$\bar{x}_4 - \bar{x}_2 = 25.00 - 28.90 = -3.90$$

Step 4. Compare this difference (3.90 in magnitude) with the *LSD* (5%) value of 4.866. As 3.90 is less than 4.866 conclude that treatments 2 and 4 do not differ significantly at the 5% level. Alternatively confirm that the 95% confidence interval is $-3.90 \pm 4.87 = (-8.77, 0.97)$. As this interval includes zero we conclude that these two means do not differ significantly at the 5% level. In a similar way we can compare any other pair of treatments, and also make comparisons at the 1% and 0.1% levels as well as at the 5% level.

Caution: If some treatments differ very much from others with respect to the variation within them, it is not a good idea to pool the within-treatment variation for all the treatments. The ANOVA approach is not valid if you cannot assume σ^2 is the same for each treatment. A transformation may help (Chapter 14). Remember to **consider the 'background variation'**.

THE COMPLETELY RANDOMISED DESIGN   127

Output 9.2 Analysis of variance and pairwise comparisons for Example 9.2 (Minitab output)

```
Analysis of Variance for Y
Source    DF        SS         MS         F         P
Variety    3      214.9       71.6       6.46     0.007
Error     12      133.0       11.1
Total     15      347.9

                                    Individual 95% CIs For Mean
                                    Based on Pooled StDev
Level    N      Mean      StDev     -------+---------+---------+---------
A        3     18.200     2.666     (------*------)
B        4     28.900     4.693                              (------*------)
C        4     22.200     3.076             (------*------)
D        5     25.000     2.467                     (------*-----)
                                    -------+---------+---------+---------
Pooled StDev =   3.329                    18.0      24.0      30.0

Fisher's pairwise comparisons

    Family error rate = 0.184
Individual error rate = 0.0500

Critical value = 2.179

Intervals for (column level mean) − (row level mean)

               A             B            C

    B      −16.241
            −5.159

    C       −9.541        1.570
             1.541       11.830

    D      −12.098       −0.966       −7.666
            −1.502        8.766        2.066
```

9.8.3 Using a Computer to Make Pairwise Comparisons

Output 9.2 shows the output produced by Minitab to compare the means in Example 9.2. Notice that the P-value is 0.007. This shows that we can reject the null hypothesis that the yields are independent of the variety at the 1% level of significance. The pooled StDev is 3.329. This is the square root of the Residual Mean Square (11.083). It is also known as the residual standard deviation (s_p).
Hence

$$s_p = \sqrt{RMS} = \sqrt{11.083} = 3.329$$

In Output 9.2 showing Fisher's pairwise comparisons, the critical value of 2.179 is the *t*-table value on 12 df. The table shows all possible pairs of means being compared. For each comparison the two values shown are the 95% confidence limits for the difference between the two means. For example, you are 95%

confident that the B mean exceeds the A mean by between 5.159 and 16.241. This interval does not include zero so you conclude the A mean (18.2) and the B mean (28.9) are significantly different. In general, an interval that includes zero shows that the two corresponding means are not significantly different. Hence the comparisons which are not significantly different are A versus C, B versus D and C versus D.

Caution: The Fisher's pairwise comparison procedure is equivalent to an LSD analysis so the warnings given in Section 9.6.7 apply. Under the null hypothesis of no treatment differences, the chance of finding an individual comparison significant at the 5% level is 0.05. However, when making all possible pairwise comparisons, the chance of at least one significant difference is much higher. In this example it is 0.184. This is called the family error rate. This error rate can be reduced to 0.05 by using Tukey's pairwise comparisons. Other methods of making multiple comparisons are given in Chapter 13.

9.9 DETERMINATION OF NUMBER OF REPLICATES PER TREATMENT

The amount of replication required to detect a given difference between the population means depends on the within-treatment variation, σ^2 which is not known before the experiment. A pilot experiment or a previous study may provide an estimate in the form of the residual mean square (RMS). Section 6.7 gives details of how to use this estimate. If you have too few replicates you may not detect 'real' treatment differences, while too many will use more resources than necessary. If you have no prior information you should plan for at least 12 residual degrees of freedom. This implies that $N - t \geqslant 12$, so to compare five treatments $(r \times 5) - 5$ should be at least 12, which implies r at least 4. There is more discussion on this topic in Section 10.10.

9.9.1 Further Discussion on Replication

Replication allows an estimate of experimental error and randomisation reduces the bias that could occur if a particular treatment was applied to the 'best' units. However, beware of pseudoreplication. Suppose you wish to compare three treatments affecting plant growth. An unsatisfactory design would be to have three pots each with four plants, with each pot receiving a different treatment. You may think you have four replications per treatment, and that the principles of randomisation apply if the 12 plants are randomly assigned to the pots. However, the performances of the four plants within a pot are not independent as each plant will affect the growth of its neighbours. Furthermore, you cannot be sure that the environmental conditions in the pots are identical apart from the treatments; differences in the means could be due to other factors. You should have several pots per treatment and use the pot means in your analysis. A further reduction in experimental error may be achieved by randomly changing the positions of the pots at regular intervals.

Suppose in a field experiment you only have one plot per treatment and take several random samples per plot. It is not valid to use the samples as replicates. The experimental unit is the whole plot to which the treatment was applied. The sample measurements give information on within-plot variation (the sampling error) but do not provide an estimate of experimental error which can only be obtained from the variation between whole plots treated alike. As the sampling error is usually less than the experimental error, it is likely that using the within-plot samples as replicates that the conclusion will be reached that there are significant treatment differences when there are not (a Type 1 error). Also you may confound treatment effects with environmental effects because differences between whole plots may not be entirely due to treatment differences.

In field experiments with more than one plot per treatment, the researcher may be tempted to further increase replication by taking several random samples per plot. This will give information on within-plot variation, but the plot means should be used in the analysis to compare treatments. The plot is the unit of measurement used to compare treatment effects. For example, if ten plots are assigned to each treatment, and a sample of twenty plants is taken from each plot and all are measured, the number of replications per treatment is ten, not 200.

The same situation can occur in growth cabinet experiments. You may have three cabinets each maintained at a different temperature and containing 10 plants each. Although the 30 plants are randomly assigned to the cabinets and randomly placed within them, any differences in the mean performance may not be due to the temperature differences. The interpretation of the results using a one-way analysis of variance is only valid if you can assume that temperature is the only environmental difference between the cabinets. Ideally, you should have several cabinets at each temperature but usually resources do not allow this. An alternative is to repeat the experiment several times, randomising the allocation of temperatures to cabinets. The cabinets means for each temperature should then be used as replicates.

Another erroneous method used to increase replication is as follows. Consider an experiment to compare storage methods on the quality of onions. Fifty onion bulbs are randomly assigned to four storage cabinets, each having a different atmospheric concentration of carbon dioxide. Two bulbs are randomly removed from each cabinet for sucrose analysis. This gives only four residual degrees of freedom. In order to increase these the researcher chops up the two bulbs from each treatment, pools the material and makes sucrose determinations using 10 samples. These cannot be regarded as 10 replications per treatment as they include within-onion variation which is likely to be much smaller than the variation between onions treated alike. The pooling of the material has lost information on the between-onion variation. To make valid treatment comparisons the means from the pooled material should be used, but in this case there is effectively no replication. Ideally at least four bulbs should be removed from each storage cabinet for the sucrose analysis. Each bulb should be chopped but the material not pooled, and the means of the within-bulb readings used to compare methods in a one-way

130 PRACTICAL STATISTICS AND EXPERIMENTAL DESIGN FOR PLANT AND CROP SCIENCE

analysis of variance. The growth cabinets problem described in the last paragraph may also apply to this example.

Exercise 9.1
In order to test the yielding abilities of five different wheat varieties, 20 equal-sized plots were allocated to a uniform site. Each variety was planted in four plots, the distribution of the varieties being random. The yields of grain in kg/plot were measured, the results being as follows:

Variety	1	2	3	4	5
Mean yield	26.50	21.50	25.75	18.75	22.50
Corrected sum of squares	13.00	5.00	14.75	8.75	21.00

(a) Complete the analysis of variance table and test whether these data show significant differences in the mean yields of the five strains. (b) Compare the strain means using a least significant difference analysis at the 5% and 1% levels of significance. (c) Calculate the coefficient of variation.

Exercise 9.2
An experiment was carried out to investigate the effect of nitrogen on the protein percentage of spring wheat, cv Axona. The experiment was laid out as a completely randomised design with four replications per treatment. The results were as follows:

Treatment (N kg/ha)	0	40	80	120	160	200
Mean % protein	8.70	9.08	9.58	10.04	10.84	9.66

Total corrected sum of squares = 18.4868

Complete the analysis of variance table and use the results of your analysis to comment on the relationship between % protein and nitrogen level.

Answers to Exercise 9.1
(a) Add the corrected sums of squares to obtain $ResidSS = 62.5$. Find the variety totals and calculate $VarietySS = 161.5$. Then complete the ANOVA table.

Source	df	SS	MS	VR
Variety	4	161.5	40.375	9.69 on (4, 15)df
Residual	15	62.5	4.167	
Total	19	224.0		

From tables, $F_{(4, 15, 5\%)} = 3.06$, $F_{(4, 15, 1\%)} = 4.89$, $F_{(4, 15, 0.1\%)} = 8.25$. The calculated VR is greater than 8.25 so the result is significant at the 0.1% level ($P < 0.001$). This means that the observed variation in the variety means could occur with a chance of less than 0.1% if all the varieties have equal yielding ability, assuming the experiment was completely randomised.

(b) From tables, $t_{(15, 2.5\%)} = 2.131$, $t_{(15, 0.5\%)} = 2.947$. Also $SED = 1.443$ so $LSD_{5\%} = 3.075$ and $LSD_{1\%} = 4.25$. A least significant difference analysis involves comparing all possible pairs of means. If the number of treatments is large such an analysis is cumbersome, misleading and should be avoided. In this case the significant differences at the 5% level are: Variety 1 versus 2, 4 and 5; Variety 2 versus 3; Variety 3 versus 4 and 5; Variety 4 versus 5.

(c) The coefficient of variation (CV%) is independent of the units of measurement. It is found by dividing the square root of the residual mean square by the overall mean and multiplying by 100 to express the result as a percentage. For this example

$$CV\% = \frac{\sqrt{RMS}}{\text{grand mean}} \times 100 = \frac{\sqrt{4.167}}{23} \times 100 = 8.9\%$$

Answers to Exercise 9.2

Find the treatment totals and hence $TreatSS = 11.2024$. The ANOVA table is as follows:

Source	df	SS	MS	VR
Nitrogen	5	11.2024	2.2405	5.54 on (5, 18)df
Residual	18	7.2844	0.4047	
Total	23	18.4868		

From tables $F_{(5, 18, 5\%)} = 2.77$, $F_{(5, 18, 1\%)} = 4.25$, $F_{(5, 18, 0.1\%)} = 6.81$. Reject the null hypothesis that changes in nitrogen fertilizer level have no effect on protein% at the 1% level ($P < 0.01$). $SED = 0.450$. Draw a graph of mean yield versus nitrogen level and discuss the nature of the response curve. See Section 13.4 for details of how to test for a linear trend.

Chapter 10

The Randomised Block Design

10.1 INTRODUCTION

The main disadvantage of the completely randomised design (CRD) is revealed if it is used to assess treatment effects on an experimental site which is not uniform. Experimental units (plots) treated alike may have a large variation in their yields because plots having the same treatments can be widely scattered. The estimate of the error variance may be so great that large differences between treatment means may not be declared significant as it could be argued that observed differences are really due to random background variation. A particular treatment can be assigned by chance to plots in the most fertile part of the experimental site. A false conclusion may then be made about the response to this treatment.

These disadvantages can mainly be overcome by using the principle of blocking. For example, if it is suspected there is systematic variation in soil conditions across the site due to a gradual increase in pH or soil fertility, you can allow for this by dividing the experimental area into blocks at right angles to the gradient. The treatments would then be randomly assigned to plots within each block. In a greenhouse there is likely to be a light gradient so blocks could consist of rows of pots such that each row has a different light level.

In order to compare five treatments (T_1, T_2...T_5) using six replications per treatment, you could form six blocks each consisting of five equal sized plots. The blocks would be separated so that environmental conditions vary between them. The five treatments would be randomly assigned to the five plots within each block. Figure 10.1 shows a typical layout for two of the blocks. The spaces between the plots are to allow access and to reduce the possibility of neighbouring plots affecting each other. For more details of the practical problems which can occur in field experiments see Dyke (1988).

This experimental design is called the **randomised block design** (RBD) but is often called the randomised complete block design (RCBD) because each block contains a complete set of treatments. This is to distinguish it from incomplete block designs (Section 15.5) where each block contains less than the full set of treatments.

As each treatment occurs once within each block, treatments can be compared within blocks, and so block-to-block (gradient) variation does not affect the treatment comparisons. The advantage of the RBD over the CRD depends on the plot-to-plot variation within blocks being smaller than the block-to-block variation. Ideally, the experimenter should have acquired some prior knowledge of the site in order to ensure that blocks are positioned correctly. If the site is patchy with no obvious gradients the blocks should be positioned so that each is within a uniform patch.

If blocks are formed without using prior information of site variation, the variation within blocks could be greater than the variation between blocks. In this case blocking would be unsuccessful and the treatments would be compared with less precision than if a CRD had been used. If the site is uniform, a CRD should be used because the loss of the residual (error) degrees of freedom due to blocks will not be compensated for by the reduction of error variance.

10.1.1 Randomisation

Treatments must be assigned to the plots at random within each block. One way of achieving the randomisation for a given block is to generate random numbers using a calculator and then rank them. The order of treatments in the block is determined by the ranks. For example, in block one of Figure 10.1 the random numbers may have been 0.352, 0.190, 0.656, 0.205, 0.559. The ranks are 3, 1, 5, 2, 4 and these determine the assignment of treatments as shown. This procedure is repeated independently for the other blocks. Alternatively, you could use a computer package to generate a random design automatically.

10.1.2 Advantages of Blocking

- Experimental error is usually reduced by forming blocks of similar plots. Precision of treatment comparisons is increased by blocking the effects of site variability. Allowance can be made for a slope, a fertility gradient or changes in pH by arranging blocks at right angles to the gradient (Figure 10.1).
- For some field trials it is wise to drill one block at a time. When testing treatments such as row width and seed spacing, each block may take several hours to complete. It may only be possible to finish blocks 1 and 2 on the first day. A period of rainfall may then prevent progress for several days. Although a break in the drilling programme should be avoided if possible, it is important to remain patient until soil conditions are again suitable for efficient machine operations. In this situation the influence of unplanned drilling date effects can be separated from the row width and spacing effects when the block analysis is completed.
- When sampling cannot be completed in one day, it should be carried out block by block. If more than one person is employed, each should sample a different block. The time and person effects are thus part of the block effect and thus do not affect the precision of treatment comparisons.

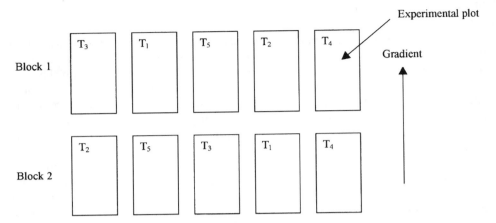

Figure 10.1. Two typical blocks showing the direction of a slope, fertility or pH gradient

- If the experimental site is too small to accommodate the complete experiment, several sites can be used with one or more blocks at each site. However, there may be some problems if each site represents a different soil type. The RBD assumes no interaction between blocks and treatments which occurs if the best treatment on one soil is not the best on another. Interaction inflates the residual variation and reduces the precision of the treatment comparisons. It can only be estimated and allowed for if there is more than one plot for each treatment in each block.
- Block designs are very popular for trials imposed onto commercial sites which have been tramlined and areas between tramlines can be used as blocks. Treatments may be applied to equal sized plots by hand or with small plot machines. Overall crop protection can be achieved with farm sprayers which use the tramline network in the field.

Note: Blocks do not have to be groups of plots in a field. They can be a row or group of pots in a greenhouse or growth cabinet.

10.1.3 Disadvantages of Blocking

- Missing values cause problems with the statistical analysis as block effects become interwoven with treatment effects. Differences between treatment means are partly due to block differences if there are missing values.
- If plots are uniform over the whole site, the RBD is less efficient than the CRD.

Table 10.1. The field layout and plot yields (t/ha)

Block 1	V_2 9.8	V_4 9.5	V_3 7.3	V_1 7.4
Block 2	V_3 6.1	V_2 6.8	V_1 6.5	V_4 8.0
Block 3	V_3 6.4	V_2 6.2	V_4 7.4	V_1 5.6

Table 10.2. Derandomised version of Table 10.1

	V_1	V_2	V_3	V_4	Total	Mean
Block 1	7.4	9.8	7.3	9.5	34.0	8.50
Block 2	6.5	6.8	6.1	8.0	27.4	6.85
Block 3	5.6	6.2	6.4	7.4	25.6	6.40
Total	19.5	22.8	19.8	24.9	87.0	
Mean	6.50	7.60	6.60	8.30		7.25

- If the number of treatments is large, it may not be possible to locate uniform blocks which are large enough to contain a complete set of treatments. One solution is to use an incomplete block design (Section 15.4).

Example 10.1

Suppose you wish to compare three new varieties of wheat (V_2, V_3 and V_4) with a standard variety (V_1). The experimental area could be divided into three blocks each containing four plots of equal size. In practice you should use more than three blocks but we use three here because we want to illustrate the principles without too much calculation. A discussion on this block replication is given in Section 10.10. Table 10.1 shows the layout and yields, and Table 10.2 the yields arranged in a Blocks × Variety table.

Yields mainly decrease from block 1 to block 3. This could have been due to a fertility gradient at right angles to the blocks. Is there sufficient evidence to argue that the new varieties are significantly different from the standard (V_1)?

For each variety there is a large variation in yields which is mainly due to the block effects. As an exercise first analyse these data ignoring blocks, and assume the design had been completely randomised.

10.2 THE ANALYSIS IGNORING BLOCKS

Using one of the methods described in Chapter 9, complete the analysis of variance to obtain Table 10.3. From F-tables (Appendix 4a) find $F_{(3,8,5\%)} = 4.066$. As the calculated $VR\ (F) = 1.45$ conclude there is very little evidence that the varieties give

Table 10.3. Analysis of variance table of the yields in Table 10.2 ignoring blocks

Source	DF	SS	MS	VR
Treatment (variety)	3	6.630	2.210	1.45
Residual (error)	8	12.180	1.5225	
Total	11	18.810		

Table 10.4. Yields adjusted for block effects

	V_1	V_2	V_3	V_4	Mean
Block 1	6.15	8.55	6.05	8.25	7.25
Block 2	6.90	7.20	6.50	8.40	7.25
Block 3	6.45	7.05	7.25	8.25	7.25
Mean	6.50	7.60	6.60	8.30	
Sxx	0.285	1.365	0.735	0.015	

different yields. In the absence of blocking there is insufficient evidence that the new varieties perform differently from the standard variety.

10.3 THE ANALYSIS INCLUDING BLOCKS

Carry out the analysis taking into account the block effects. For each yield we find the block effect which is the block mean minus the grand mean. The yield adjusted for the block effect is then the original yield minus the block effect. For example, the block 1 effect is $8.50 - 7.25 = 1.25$. Hence the adjusted yields for block 1 are 6.15, 8.55, 6.05 and 8.25. These are the original yields minus 1.25. Table 10.4 shows all the adjusted yields.

- The adjusted block means are all the same and equal to the grand mean of 7.25. This restriction reduces the degrees of freedom. Within each block there are only 3 degrees of freedom $(t-1)$ in general; within each treatment (variety) there are 2 degrees of freedom $(r-1)$ in general. These two restrictions together imply that the degrees of freedom used for the comparison of treatment means are $2 \times 3 = 6$ or $(r-1)(t-1)$ in general. This is the number of cells remaining in the Table 10.4 after crossing out the last variety and the last block.
- The adjusted variety means are the same as before but the variation within each treatment is much reduced. This reduction in variation is the result of 'removing' the block differences so that **treatments can be compared more precisely**. The variation in the treatment means is unaffected by 'removing' the block differences. This is because each treatment mean is equally affected by the block differences. The sum of the Sxx values in Table 10.4 gives the residual sum of squares of 2.40 which is much smaller than the corresponding value of 12.18 when blocking is ignored.

10.4 USING THE COMPUTER

To analyse these data by computer you should enter the variety codes in column one, the block codes in column two and the yields in column three of a spreadsheet (first

THE RANDOMISED BLOCK DESIGN

Output 10.1 Analysis of variance of wheat yield data from Example 10.1 (Minitab output)

```
Analysis of Variance for yield

Source    DF       SS        MS        F        P
Block      2    9.7800    4.8900    12.22    0.008
Variety    3    6.6300    2.2100     5.52    0.037
Error      6    2.4000    0.4000
Total     11   18.8100

Means

Variety    N      yield
1          3     6.5000
2          3     7.6000
3          3     6.6000
4          3     8.3000
```

three columns of Table 10.5). Output 10.1 shows the ANOVA table and variety means produced by Minitab.

The output shows that the total variation in yields is 18.81. This is the sum of the squares of the deviations of the 12 yields from their overall mean. Of this, 9.78 is due to block differences, 6.63 to variety differences and 2.4 is unexplained. The P-value for variety is 0.037. Because this is less than 0.05 there is evidence that the variety means are significantly different after allowing for block effects. The small P-value for blocks indicates that blocking was successful in increasing the precision of the treatment comparisons. Having found evidence that the varieties differ, you should now explore the assumptions which make the analysis valid (Section 10.6).

10.5 THE EFFECT OF BLOCKING

Table 10.3 is the analysis of variance table of Example 10.1 ignoring blocks. The treatment sum of squares was 6.63 and the residual sum of squares was 12.18. The VR of 1.45 on 3 and 8 degrees of freedom was not significant. After allowing for blocks, the residual sum of squares is reduced to 2.40 (Output 10.1). This reduction $(12.18 - 2.40 = 9.78)$ is the block sum of squares and is large enough to make the VR of 5.52 on 3 and 6 degrees of freedom significant. The VR for blocks is 12.22 on 2 and 6 degrees of freedom. This is highly significant ($P = 0.008$) and indicates large differences between blocks. This is not surprising because the experiment was designed using prior knowledge of a site gradient. A small VR for blocks may have indicated that a uniform site had been chosen, or that the blocks were not positioned correctly. This would have resulted in conditions within blocks being less uniform than conditions between them.

However, blocking may not always be efficient especially on a uniform site or when blocks are not formed at right angles to a gradient. Suppose, in this example, that $BlockSS$ was only 2.04. This would make $ResidSS = 10.14$ and $RMS =$

10.14/6 = 1.69. In this case, the reduction in *ResidSS* due to blocking would not compensate for the loss in residual degrees of freedom from 8 to 6. The value of *RMS* would actually increase from 1.5225 (Table 10.3) to 1.69 and the VR would reduce from 1.45 to 1.31 (2.21/1.69).

10.6 THE RANDOMISED BLOCKS MODEL

This section gives a demonstration by example of the theory upon which the RBD analysis is based. This approach avoids too much mathematical notation while allowing an insight into the analytical method. The procedure depends on the assumption that each plot yield is made up of an overall mean, a treatment effect, a block effect and an unexplained part, called the residual or error.

Using Example 10.1 fit

Yield = overall mean + block effect + treatment effect + residual, thus
Yield = fitted value + residual, where
Fitted value = overall mean + block effect + treatment effect, and
Residual = yield − fitted value

Table 10.5 shows how the above terms are obtained for Example 10.1, together with the sum of squares.

- The overall mean is estimated by the grand mean (GM).
- The block effect is estimated as BE = (block mean − grand mean) = (BM− GM).

Table 10.5. Table showing components of yield

Treatment (variety)	block	Yield Y	GM	BM	TM	Y−GM	TM−GM TE	BM−GM BE	R	F
1	1	7.4	7.25	8.50	6.50	0.15	−0.75	1.25	−0.35	7.75
1	2	6.5	7.25	6.85	6.50	−0.75	−0.75	−0.4	0.4	6.1
1	3	5.6	7.25	6.40	6.50	−1.65	−0.75	−0.85	−0.05	5.65
2	1	9.8	7.25	8.50	7.60	2.55	0.35	1.25	0.95	8.85
2	2	6.8	7.25	6.85	7.60	−0.45	0.35	−0.4	−0.4	7.2
2	3	6.2	7.25	6.40	7.60	−1.05	0.35	−0.85	−0.55	6.75
3	1	7.3	7.25	8.50	6.60	0.05	−0.65	1.25	−0.55	7.85
3	2	6.1	7.25	6.85	6.60	−1.15	−0.65	−0.4	−0.1	6.2
3	3	6.4	7.25	6.40	6.60	−0.85	−0.65	−0.85	0.65	5.75
4	1	9.5	7.25	8.50	8.30	2.25	1.05	1.25	−0.05	9.55
4	2	8.0	7.25	6.85	8.30	0.75	1.05	−0.4	0.1	7.9
4	3	7.4	7.25	6.40	8.30	0.15	1.05	−0.85	−0.05	7.45
Sum		87.0	87.0	87.0	87.0	0	0	0	0	87.0
Sum of squares		649.56	630.75	640.53	637.38	18.81	6.63	9.78	2.4	647.16
				CF		TotalSS	TreatSS	BlockSS	ResidSS	

- The treatment effect is estimated as TE = (treatment mean − grand mean) = (TM − GM).
- The fitted yield predicted by the model is estimated as

$$F = GM + (BM - GM) + (TM - GM) = BM + TM - GM$$

- The residual is the difference between the observed yield (Y) and the fitted value (F). It is estimated as $R = Y - F = Y - BM - TM + GM$.

As an exercise recreate this table using a computer spreadsheet package. It can be seen that for each yield, $Y = GM + BE + TE + R = F + R$. For block one, variety two $9.8 = 7.25 + 1.25 + 0.35 + 0.95$. For each yield, the residual is $R = Y - BM - TM + GM$. For block one, variety two the residual is $9.80 - 8.50 - 7.60 + 7.25 = 0.95$.

10.6.1 The Residuals

Think of the residuals as the yields after adjusting for the grand mean, the block effects and the treatment effects. Their sum is always zero (Table 10.5). Also within each block and treatment they sum to zero (Table 10.6). Their sum of squares is the **Residual Sum of Squares (ResidSS)**.

These residuals are not independent because they are constrained to sum to zero within each row and each column. You could cross out the last row and the last column (B_3 and V_4) of Table 10.6 and still find the eliminated residuals by using these constraints. The number of independent residuals is the number of residuals left after crossing out the last row and last column; in this case $(3 - 1) \times (4 - 1) = 2 \times 3 = 6$. In general it is $(r - 1) \times (t - 1)$ where $r =$ number of blocks and $t =$ number of treatments.

The number of independent residuals is called the **Residual Degrees of Freedom (Rdf)**. For Example 10.1 the sum of the squares of the residuals is $ResidSS = 2.400$ and they have 6 degrees of freedom. i.e. $Rdf = 6$. This is given as the Error SS in Output 10.1.

The residuals represent the variation between the yields after removing the Grand Mean, the Block effects and the Treatment effects. If you divide *ResidSS* by its degrees of freedom you obtain the **Residual (or Error) Mean Square (RMS)** which is an estimate of the variance σ^2 of the population from which the residuals are assumed to belong. For the statistical tests associated with the RBD to be valid, it is assumed that the residuals are independent of each other and come from a Normal

Table 10.6. Residuals for Example 10.1

	V_1	V_2	V_3	V_4	Mean
Block 1	−0.35	0.95	−0.55	−0.05	0
Block 2	0.40	−0.40	−0.10	0.10	0
Block 3	−0.05	−0.55	0.65	−0.05	0
Mean	0	0	0	0	

distribution with variance σ^2. You can obtain some idea of whether these assumptions are true by studying the residuals. If there is a pattern, for example if some treatments or blocks have much larger (in magnitude) residuals than others or if there is a sequence of positive residuals followed by a sequence of negative residuals, there may be systematic effects other than those due to treatments and blocks. Some very large residuals could be the result of not aligning blocks at right angles to an environmental gradient. Perhaps the conditions within each block were not uniform or they were uniform in some blocks and not others. If you use a computer for your analysis, it is always a good idea to ask for a printout of the residuals so that the basic assumptions can be checked. The residuals have a mean of zero and should be normally distributed and have no relationship with the estimated or **fitted y-values** if the significance tests in the analysis of variance are to be valid. You should check these assumptions by looking at a histogram and a normal probability plot of the residuals, and a plot of the residuals versus the fitted values. Most statistical software packages allow you to do this easily. Figure 10.2 shows that there is no relationship between the residuals and fitted values. The histogram is symmetrical but the residuals do not appear normally distributed. The normal plot is not straight which also indicates a deviation from normality. These results are inconclusive as they are based on a very small set of data.

10.6.2 Partitioning the Variance

According to the RBD model each yield is made up of an overall effect, a variety effect, a block effect, plus a random error. For example, the yield of 6.5 in variety 2 block 1 can be partitioned as

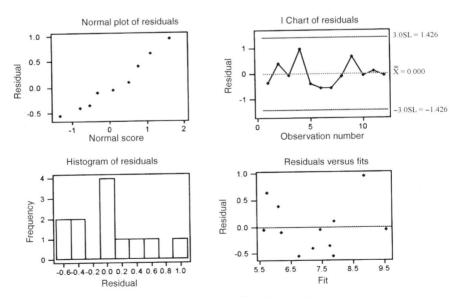

Figure 10.2. Residual diagnostics for Example 10.1 (Minitab)

THE RANDOMISED BLOCK DESIGN

$$Y = GM + BE + TE + R \quad 9.80 = 7.25 + 1.25 + 0.35 + 0.95$$

The deviation of this yield from the grand mean is $Y - GM = 9.80 - 7.25 = 2.55$. This deviation is made up of the variety 2 effect (0.35), the block 1 effect (1.25), and the residual (0.95).

In general, $(Y - GM) = BE + TE + R = (BM - GM) + (TM - GM) + R$. It can also be shown (proof not given) that

$$\Sigma(Y - GM)^2 = \Sigma(BM - GM)^2 + \Sigma(TM - GM)^2 + \Sigma R^2$$

This equation can be summarised as **Total Sum of Squares = Block Sum of Squares + Treatment Sum of Squares + Residual Sum of Squares** or in shorthand notation as

$$TotalSS = BlockSS + TreatSS + ResidSS$$

In Table 10.5, *TotalSS* (18.81) represents the total variation in the yields. The sum of squares for block, treatment and residual add up to the total sum of squares $(9.78 + 6.63 + 2.40 = 18.81)$. Thus the total variation in these data can be attributed to these three sources. More than half of the variation is due to block differences (9.78) and very little (2.40) is due to residual. The residual sum of squares (*ResidSS*) represents the average variation between the plot yields after allowing for the effects of treatment and block. It is often called the **error sum of squares**.

10.7 USING A HAND CALCULATOR TO FIND THE SUMS OF SQUARES

Use your calculator in SD or STATS mode. Make sure you know how to obtain the standard deviation. The relevant key is either marked σ_{n-1} or s. In what follows SD means sample standard deviation.

Step 1. Enter all N numbers. Press the n key to make sure you have entered all the numbers. Obtain the SD, square the result and multiply by $N - 1$ to obtain *TotalSS*. This is the corrected sum of squares of all N values. (*TotalSS* = Sxx for all N values.)

Step 2. To find *BlockSS* clear the results of previous calculations. Enter the r block totals and obtain their SD, square the result, and multiply by $r - 1$ (the block degrees of freedom). This gives the corrected sum of squares of the block totals. Divide this result by t (the number of treatments) to get *BlockSS*.

Step 3. To find *TreatSS* clear the results of previous calculations. Enter the t treatment totals and obtain their SD, square the result, and multiply by $t - 1$ (the treatment degrees of freedom). This gives the corrected sum of squares of the treatment totals. Divide this result by r (the number of blocks) to get *TreatSS*.

Step 4. Find the residual sum of squares by subtraction:

$$ResidSS = TotalSS - BlockSS - TreatSS$$

Once the SS column has been completed, the rest of the ANOVA table (Table 10.7) can be filled in easily.

Table 10.7. ANOVA table for a randomised blocks design

Source	DF	SS	MS	VR
Blocks	$r-1$	$BlockSS$	$BMS = BlockSS/(r-1)$	BMS/RMS
Treatments	$t-1$	$TreatSS$	$TMS = TreatSS/(t-1)$	TMS/RMS
Residual	$(r-1) \times (t-1)$	$ResidSS$	$RMS = ResidSS/Rdf$	
Total	$N-1$	$TotalSS$		

DF = Degrees of Freedom   SS = Sums of Squares r = number of blocks
MS = Mean Square   VR = Variance Ratio t = number of treatments

10.8 COMPARISON OF TREATMENT MEANS

Some authors do not recommend carrying out this step if the VR is not significant. However, if the main aim of the experiment is to compare a particular treatment, say a control treatment, with each of the others it is still valid to carry out a Least Significant Difference analysis even if the VR is not significant.

10.8.1 LSD Analysis (using Results of Example 10.1)

Step 1. Calculate the standard error of the difference between two treatment means (SED). This can be obtained using the RMS value from the analysis of variance table:

$$SED = \sqrt{\frac{2 \times RMS}{r}} = \sqrt{\frac{2 \times 0.400}{3}} = 0.516$$

Step 2. Find the Least Significant Difference (LSD). Start with the 5% level:

$$LSD_{5\%} = t_{(Rdf, 2.5\%)} \times SED$$

where Rdf is the Residual degrees of freedom $= (r-1) \times (t-1) = 6$

$$= t_{(6, 2.5\%)} \times SED = 2.447 \times 0.516 = 1.263 \text{ t/ha}$$

Step 3. Compare differences between the means with the LSD value. In this example variety 1 is the control treatment so we may wish to compare varieties two, three and four with variety one. The following table shows the relevant differences.

$V_2 - V_1$ $7.60 - 6.50 = 1.10$
$V_3 - V_1$ $6.60 - 6.50 = 0.10$
$V_4 - V_1$ $8.30 - 6.50 = 1.80$

Of these differences only 1.80 is greater than the $LSD_{5\%}$ value of 1.26 so we conclude that only variety four gives a yield significantly different from the control (V_1).

We could go on and find $LSD_{1\%}$. Check that it comes to 1.91. At this level none of the other varieties are significantly different from V_1.

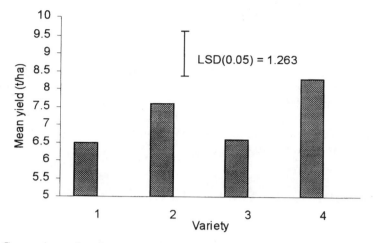

Figure 10.3. Comparison of variety means for Example 10.1

A visual presentation of this analysis is shown in Figure 10.3 which shows the variety means together with the 5% LSD value.

If your main interest is in comparing certain treatments, for instance the control versus each of the others, there is no need to complete an ANOVA table. Nevertheless this table is almost always output by a computer package so you should know what all the entries mean. Furthermore, it provides an easy way to find the *SED* using the *RMS* value in the formula in step 1 above.

10.8.2 Estimation

Some researchers are interested not in comparing treatments but in estimating the population means for each treatment. To do this find the pooled standard error of a treatment mean:

$$SEM = \sqrt{\frac{RMS}{r}}$$

This could be used to find a 95% confidence interval for the true mean of a particular treatment provided the conditions for pooling are valid.

10.8.3 Standard Errors and Confidence Intervals

A 95% CI for μ is $\bar{x} \pm t_{(Rdf, 2.5\%)} \times SEM$. In this example, $SEM = \sqrt{0.400/3} = 0.365$ is found by pooling the data from all the treatments. The mean for treatment 2 (V_2) is 7.60 so a 95% CI for the true mean yield of V_2 is

$$7.60 \pm (2.447 \times 0.365) = 7.60 \pm 0.89 = (6.71,\ 8.49)$$

We are thus 95% certain that the mean yield for V_2 lies between 6.71 and 8.49 t/ha.

10.9 REPORTING THE RESULTS

The advice given concerning reporting the results of a one-way analysis of variance apply here. See Section 9.7 for details.

Some computer packages report individual standard errors or standard deviations for each treatment ignoring blocks. While these are valid to report in a CRD, they are misleading in a RBD because they incorporate block differences. The individual standard errors calculated from Table 10.2 are 0.520, 1.111, 0.361, 0.624. The standard errors after allowing for block effects are 0.218, 0.477, 0.350, 0.050 and can be calculated from Table 10.4.

10.10 DECIDING HOW MANY BLOCKS TO USE

At least two blocks are required. With only one block there is no replication and no estimate of residual (error) variation can be obtained. Thus, there is no way of telling whether observed differences in yield are due to treatment or other effects. However, even if several blocks are used the analysis of variance may show that the treatment means are not significantly different ($P > 0.05$). This is because (a) there are no treatment differences or (b) the treatments which do have different effects on yield are declared non-significant due to large background variation. Although this variation can be partly taken into account by blocking some will always remain. There will always be variation between yields of plots treated alike after allowing for block effects. This is summarised by the residual mean square (RMS) in the ANOVA table. Its square root is the standard deviation of the residuals. A large value for RMS requires large observed differences in treatment means for a significant VR to be obtained. The VR reflects the ratio of the variation in treatment means to the background variation. Whether a particular value for this ratio is significant or not (for a given number of treatments) depends on the number of blocks used. For example, suppose you have four treatments and the VR is 4.2. Table 10.8 shows two skeleton ANOVA tables, (a) assuming three blocks and (b) assuming five blocks. The residual degrees of freedom are 6 and 12 respectively. For (a) the 5% F-table value is $F_{(3, 6, 5\%)} = 4.76$ and for (b) it is $F_{(3, 12, 5\%)} = 3.49$. Hence the calculated VR of 4.2 is not significant with three blocks but it is with five.

As a rule of thumb it is recommended that in field trials the number of blocks should be sufficient to give at least 12 residual degrees of freedom. In general $Rdf = (t - 1) \times (r - 1)$ where $t = $ number of treatments and $r = $ number of blocks. In more controlled environments, such as greenhouses where the background variation is likely to be less, Rdf can be smaller. With too small a number of blocks, the results may be inconclusive while too large a number implies that more resources than necessary may have been used.

10.10.1 Detection of a Specified Difference

Application of a new treatment will only be recommended if the benefits outweigh the extra cost of using it compared to the standard. It is proposed to carry out a

Table 10.8. Skeleton ANOVA tables for a RBD with four treatmemts and (a) three blocks, (b) five blocks

(a)

Source	DF
Blocks	2
Treatments	3
Residual	6
Total	11

(b)

Source	DF
Blocks	4
Treatments	3
Residual	12
Total	19

RBD experiment to compare several promising new treatments with a control. Suppose a difference of δ or more is necessary before a new treatment can be recommended. How many replications (blocks) are required such that a difference in two means of δ or more will be declared significantly different at the 5% level? Recall that the formula for the LSD is $LSD_{5\%} = t_{(Rdf, 2.5\%)} \times SED$. You require this to be less than δ. The difficulty is that in advance of the experiment you cannot calculate LSD as the t-value depends on the number of blocks and the SED depends on the number of blocks and the background variation. The background variation is denoted by σ^2 (Section 10.6.1) and is estimated by RMS. The true value of SED is given by

$$SED = \sqrt{\frac{2 \times \sigma^2}{r}}$$

You require r such that $t \times SED$ is less than δ. This implies that r must be greater than $(2 \times t^2 \times \sigma^2)/\delta^2$.

Example 10.2

In Example 10.1 the mean yield for variety two exceeded that for variety one by 1.10 t/ha. This difference was found not to be significant. Suppose a difference as great as 0.8 t/ha is of economic significance. How many blocks would be needed in a repeat experiment to have a good chance of declaring such a difference statistically different? As a first approximation take t to be 2.0. An estimate of σ^2 from the previous experiment is $RMS = 0.40$ (Output 10.1). Hence we require r to be at least $(2 \times 4 \times 0.40)/0.8^2 = 5$.

This is only an approximation as with five blocks and four treatments the residual degrees of freedom is $4 \times 3 = 12$ and the t-value is 2.179. Confirm that with this value of t the number of blocks required would be at least 5.94. As r must be an integer, six blocks are needed. The residual degrees of freedom increase to 15 and the t-value decreases to 2.131. Repeating the calculation still shows that six blocks are required so that is the number to use.

In some cases the number of replications required for the stated degree of precision would make the experiment too expensive to carry out. Yet an experiment with too few replications would be a waste of time and resources.

10.10.2 Using the Coefficient of Variation

Suppose the criterion is to recommend a new variety if it outperforms the average of those in the trial by at least by $q\%$. In this case δ would be $(q/100 \times \mu)$ where μ is the population average yield of all the varieties. The coefficient of variation (CV) expressed as a percentage should be $(\sigma/\mu \times 100)$. It can be estimated from previous similar experiment by dividing the square root of the residual mean square by the grand mean and expressing the result as a percentage. For Example 10.1 its value is

$$CV = \frac{\sqrt{RMS}}{GM} \times 100 = \frac{\sqrt{0.400}}{7.25} \times 100 = 8.72\%$$

Using the CV the formula for estimating the minimum value of r is $(2 \times t^2 \times (CV)^2)/q^2$.

For example, if you require a good chance of detecting a variety whose yield differs by 10% from the average and estimate the CV as 8.72% you will need r to be at least $(2 \times 4 \times 8.72^2)/10^2 = 6.08$ as a first approximation. Using $r = 7$ the residual degrees of freedom are 18 assuming four varieties in the experiment. The corresponding t-value would be 2.101. On repeating the calculation with this value the minimum r is still seven so this is the number of blocks required.

10.11 PLOT SAMPLING

Frequently, in a field experiment it is not possible to measure the yield of the whole plot and so samples must be taken. In particular, soil measurements have to be estimated for the whole plot by taking representative soil samples. Each plot should be sampled the same number of times. The individual measurements can be used to assess the within-plot variation but the plot means should be used to compare the treatments in the analysis of variance. This is because the plots are the units to which the treatments have been applied and the within-plot variation gives no information on the variation between plots receiving the same treatment. The same situation can occur in greenhouse or laboratory pot experiments when several plants are grown in each pot. Suppose three treatments are being compared and five pots are used for each treatment and each pot contains four plants. This can be a CRD if the environmental conditions are uniform. However, if there is a light or temperature gradient a RBD should be used with five rows of pots arranged at right angles to the gradient. Each of the three pots in a row should receive a different treatment at random. Measurements are made on each plant several days after the application of the treatments. For the purpose of comparing treatments there are only five replications per treatment and the pot means should be used. The Rdf for comparing treatment means are 8 $(14 - 4 - 2)$ and not 53 $(59 - 4 - 2)$.

Exercise 10.1

An experiment was carried out to compare the effects of three cultivation systems (A, B and C) with traditional seedbed preparation (D) on the total dry matter yield of kale. The trial was laid out in five randomised blocks. Plot yields in t/ha and the experimental layout are given below.

Block 1	C 5.5	A 5.4	B 6.9	D 7.2
Block 2	D 5.8	B 5.3	C 5.6	A 4.1
Block 3	D 6.9	A 5.3	B 6.6	C 4.5
Block 4	A 5.0	B 7.2	C 6.1	D 7.0
Block 5	B 6.2	C 5.7	A 4.8	D 5.8

(a) Test the null hypothesis that the cultivation systems all have the same effect on yield.
(b) Compare systems A, B and C with D.

Answers to Exercise 10.1

The first step is to form a blocks times treatments table of yields including the block totals, treatment totals and treatment means as follows:

| Block | Treatment | | | | Total |
	A	B	C	D	
1	5.4	6.9	5.5	7.2	25
2	4.1	5.3	5.6	5.8	20.8
3	5.3	6.6	4.5	6.9	23.3
4	5	7.2	6.1	7	25.3
5	4.8	6.2	5.7	5.8	22.5
Total	24.6	32.2	27.4	32.7	116.9
Mean	4.92	6.44	5.48	6.54	

The next step is to complete the analysis of variance table. You should confirm that the ANOVA table is as follows:

Source	DF	SS	MS	VR
Blocks	4	3.4370	0.8593	3.344
Treats	3	9.1295	3.0432	11.845
Residual	12	3.0830	0.2569	
Total	19	15.6495		

From tables, $F_{(3, 12, 0.1\%)} = 10.80$.
- As VR for treatments is greater than this, we reject the null hypothesis that the four cultivation systems all have the same effect on yield at the 0.1% level ($P < 0.001$).

To compare A, B and C with D, we find the SED and multiply it by the 2.5% t-table value on the residual degrees of freedom to obtain the 5% LSD value:

$$SED = \sqrt{\frac{2 \times RMS}{r}} = \sqrt{\frac{2 \times 0.2569}{5}} = 0.321$$

$$LSD_{5\%} = t_{(12, 2.5\%)} \times SED = 2.179 \times 0.321 = 0.698$$

The mean differences between A and D, B and D, C and D are compared to the LSD value. If any difference is greater than the LSD, we conclude this cultivation system gives yields which are significantly different from D.

The means for A, B, C and D are 4.92, 6.44, 5.48 and 6.54 respectively. The mean differences from D are as follows:

A − D	B − D	C − D
−1.62	−0.1	−1.06

Both the A and C means are significantly different from the D mean but the B mean is not. A bar chart of the means with a vertical line to represent the LSD will illustrate this conclusion.

Chapter 11

The Latin Square Design

11.1 INTRODUCTION

Where there is soil heterogeneity, possibly due to a gradient (Figure 10.1), the randomised block design is an improvement over the completely randomised design. This is because each block is be made up of similar plots and blocks are arranged at right angles to the gradient. If there are gradients in two directions, their effects can be allowed for by using a Latin square design. This design uses double blocking and is a row and column design. Each treatment occurs once in each row and once in each column. For five treatments (A, B, C, D and E), Figure 11.1 shows a possible field layout, and Table 11.1 the skeleton ANOVA indicating how the degrees of freedom are partitioned.

In this design the number of rows = number of columns = number of treatments = t. Thus the number of replicates per treatment is t, and the total number of plots is $N = t^2$. These restrictions make the single Latin square a design of limited value. If the number of treatments is four or less, the residual degrees of freedom are too small to allow sufficient precision of treatment comparisons (Table 11.2). However, if you have many treatments, the Latin square design may be difficult to manage. For example, an experiment with ten treatments would require

A	B	C	D	E
B	C	D	E	A
C	D	E	A	B
D	E	A	B	C
E	A	B	C	D

Figure 11.1. A possible layout for a Latin square design

Table 11.1. Skeleton ANOVA table for a 5×5 Latin square design

Source	DF
Rows	$r - 1 = 4$
Columns	$c - 1 = 4$
Treatments	$t - 1 = 4$
Residual	12 (by subtraction)
Total	$N - 1 = 24$

Table 11.2. Skeleton analysis of variance tables for 3×3 and 4×4 Latin Square Designs

A 3×3 square

Source	DF
Rows	2
Columns	2
Treatments	2
Residual	2
Total	8

A 4×4 square

Source	DF
Rows	3
Columns	3
Treatments	3
Residual	6
Total	15

100 plots. Each treatment would have to be replicated ten times, which may be impossible with the resources available. Furthermore, the precision of treatment comparisons may be far greater than is required. As a working rule, you should design your experiments to give you at least 12 degrees of freedom for residual. A 5×5 Latin square gives 12 residual degrees of freedom.

It is possible to use 3×3 or 4×4 squares if the experiment includes more than one of them. The results from the several squares are then combined into a single analysis.

The Latin square analysis is not valid if there is interaction between the blocking factors and the treatments. No interaction implies that the relationship between the treatment yields is the same for each row and for each column.

11.1.1 Uses of the Latin Square

- This design is useful for controlling two sources of variation.
- It can be used along a fertility gradient. The following diagram shows a possible field plan with four treatments.

Block 1: | A | B | C | D |

Block 2: | B | C | D | A |

Block 3: | C | D | A | B |

Block 4: | D | A | B | C |

Assuming the gradient goes from left to right, each treatment occurs at each level of gradient within a block. The analysis is carried out assuming the blocks are the 'rows' and the positions within the blocks are the 'columns'.
- The Latin square design is very popular for animal experiments. For nutrition trials on dairy cattle, only a few cows may be available for financial reasons. If five diets are being compared, each can be fed to five cows in a different order, say three weeks on each diet. Each 'column' could represent a different cow and each 'row' a different period of lactation. This is more economical than using a CRD with, say, three cows on each diet. In this case, 15 cows would be required to give 10 degrees of freedom for residual. However, one disadvantage of giving several diets to the same cow is that there may be carryover effects. The effects of one diet may remain when the next diet is given.

11.2 RANDOMISATION

Figure 11.2 shows a 4 × 4 Latin square where the treatment order within each row follows an alphabetical pattern. Before carrying out an experiment, the design should be randomised with the restriction that each treatment occurs once within each row and once within each column. The randomisation should be carried out as follows (Figures 11.3 and 11.4):

(1) Randomise the order of the rows.
(2) Randomise the order of the columns.
(3) Randomly assign the treatment to the letters.

- Draw a square with the letters in alphabetical order (Figure 11.2).
- Randomise the order of the rows. Suppose the first four random numbers generated by your calculator are 0.950, 0.854, 0.319, 0.697. Their ranks are 4, 3, 1, 2. Thus the fourth row is put first, the third row is put second, the first row is put third, and the second row is put fourth (Figure 11.3).
- Randomise the columns. Suppose the next four random numbers generated by your calculator are 0.638, 0.861, 0.147, 0.936. Their ranks are 2, 3, 1, 4. Thus column 2 is put first, column 3 is put second, column 1 is put third, column 4 is

A	B	C	D
B	C	D	A
C	D	A	B
D	A	B	C

Figure 11.2. Systematic arrangement for a 4 × 4 Latin square

D	A	B	C
C	D	A	B
A	B	C	D
B	C	D	A

Figure 11.3. The 4 × 4 Latin square of Figure 11.2 after randomising the rows

A	B	D	C
D	A	C	B
B	C	A	D
C	D	B	A

Figure 11.4. The 4 × 4 Latin square of Figure 11.2 after randomising the rows and columns

kept in fourth place (Figure 11.4). Each letter occurs once in each row and once in each column.
- The treatments are randomly assigned to the letters. Suppose the next four random numbers are 0.032, 0.724, 0.556, 0.176. Their ranks are 1, 4, 3, 2. The assignment is

Letter	A	B	C	D
Treatment	T_1	T_4	T_3	T_2

Note: Some statistical packages, including Genstat, enable you to obtain a computer-generated design.

Example 11.1
Table 11.3 shows the field layout and yields (kg/plot) of four varieties of wheat arranged in a 4 × 4 Latin square. Table 11.4 gives the variety totals and means.
 Data were arranged as shown in the first four columns of Table 11.5 and Output 11.1 shows an analysis using Genstat.

Table 11.3. Layout and yields (kg/plot) for Example 11.1

Row	Column				Total
	1	2	3	4	
1	16.6 C	13.9 D	18.0 B	21.9 A	70.4
2	12.6 B	17.4 A	17.9 C	16.5 D	64.4
3	9.3 D	16.2 C	18.4 A	15.5 B	59.4
4	15.7 A	11.3 B	10.5 D	13.1 C	50.6
Total	54.2	58.8	64.8	67.0	244.8

Table 11.4. Variety totals and means for Example 11.1

	A	B	C	D
Totals	73.4	57.4	63.8	50.2
Means	18.35	14.35	15.95	12.55

11.3 INTERPRETATION OF COMPUTER OUTPUT

- After allowing for variation due to rows and columns the *P*-value for variety (F pr) is 0.009. This is the probability of obtaining a VR(v.r.) at least as great as 10.26 if the varieties give the same yield. You can think of the *P*-value as the probability that the observed differences in the variety means are due to chance and would therefore conclude that these yield differences are almost certainly due to other factors.
- As the VRs of 7.42 and 3.57 are much greater than 1.0 they indicate that the Latin square design was effective in increasing the precision of treatment comparisons in this experiment.
- The 5% LSD value of 2.66 indicates that varieties A and C each give yields which are significantly from that of D. Note that the LSD should only be used for comparisons planned in advance. If variety D is a standard and A, B and C are promising new varieties, the above comparisons would have been the main experimental objective. The LSD value also shows that the B and C means are not significantly different from each other. It would be safer to just quote the SED (1.087), the *P*-value (0.009) and the R*df* (6) when presenting the means in a scientific paper so the reader can carry out further tests. A bar chart showing the variety means together with a vertical line showing the SED or LSD value would be a suitable way of presenting the results graphically.
- Genstat emphasises the nature of the row and column effects by its use of strata. The design has a block structure and a treatment structure. Before the treatments are applied, the plots are arranged in rows and columns. These are called strata and have **random effects**. The treatments are applied to the plots according to the

Output 11.1 Genstat output for the Latin square — data from Example 11.1

```
***** Analysis of variance *****
```

Variate: Yield

Source of variation	d.f.	s.s.	m.s.	v.r.	F pr.
Row stratum	3	52.620	17.540	7.42	
Column stratum	3	25.340	8.447	3.57	
Row.Column stratum					
Variety	3	72.760	24.253	10.26	0.009
Residual	6	14.180	2.363		
Total	15	164.900			

```
***** Tables of means *****
```

Variate: Yield

Grand mean 15.30

Variety	A	B	C	D
	18.35	14.35	15.95	12.55

*** Standard errors of differences of means ***

Table	Variety
rep.	4
d.f.	6
s.e.d.	1.087

*** Least significant differences of means (5% level) ***

Table	Variety
rep.	4
d.f.	6
l.s.d.	2.660

***** Stratum standard errors and coefficients of variation *****

Variate: Yield

Stratum	d.f.	s.e.	cv%
Row	3	2.094	13.7
Column	3	1.453	9.5
Row.Column	6	1.537	10.0

design restrictions and have **fixed effects**. To understand this, imagine the experiment is repeated. It would have the same treatments (fixed) but different rows and columns (random). For this reason the row and column means are normally of little interest.

- The row standard error (s.e. = 2.094) is found by dividing the row mean square (17.54) by the number of rows (4) and taking the square root. The column standard error (1.453) is found similarly. The row.column standard error (1.537) is the square root of the residual mean square. The corresponding coefficients of variation are obtained by dividing the s.e. by the grand mean (15.30) and multiplying by 100. The cv% of 10.0 is the one normally quoted in publications as it represents the background variation after allowing for the effects of row, column and treatment.

Note: If you use Minitab or SAS you should use the general linear model procedure. This is because the design is not balanced in the sense that only particular row, column, treatment combinations are present.

11.4 THE LATIN SQUARE MODEL

The procedure depends on the assumption that each plot yield is made up of an overall mean, a row effect, a column effect, a treatment effect and an unexplained part, called the residual or error. As in the CRD and RBD, it is assumed that the residuals come from a normal distribution with a mean of zero and an unknown variance σ^2 which is estimated by the residual mean square in the ANOVA table. They are also assumed to be independent of row, column and treatment effects.

Table 11.5 shows, for Example 11.1, how the model terms are made up for all the 16 yields. It also shows the sum of squares in the last row.

Yield (Y) = overall mean + row effect + column effect + treatment effect + residual

Table 11.5. Table showing the model components of yield

Row	Col.	Treat.	Y	GM	RM	CM	TM	Y−GM	RE	CE	TE	R	F
1	1	3	16.6	15.3	17.6	13.55	15.95	1.3	2.3	−1.75	0.65	0.1	16.5
2	1	2	12.6	15.3	16.1	13.55	14.35	−2.7	0.8	−1.75	−0.95	−0.8	13.4
3	1	4	9.3	15.3	14.85	13.55	12.55	−6	−0.45	−1.75	−2.75	−1.05	10.35
4	1	1	15.7	15.3	12.65	13.55	18.35	0.4	−2.65	−1.75	3.05	1.75	13.95
1	2	4	13.9	15.3	17.6	14.7	12.55	−1.4	2.3	−0.6	−2.75	−0.35	14.25
2	2	1	17.4	15.3	16.1	14.7	18.35	2.1	0.8	−0.6	3.05	−1.15	18.55
3	2	3	16.2	15.3	14.85	14.7	15.95	0.9	−0.45	−0.6	0.65	1.3	14.9
4	2	2	11.3	15.3	12.65	14.7	14.35	−4	−2.65	−0.6	−0.95	0.2	11.1
1	3	2	18	15.3	17.6	16.2	14.35	2.7	2.3	0.9	−0.95	0.45	17.55
2	3	3	17.9	15.3	16.1	16.2	15.95	2.6	0.8	0.9	0.65	0.25	17.65
3	3	1	18.4	15.3	14.85	16.2	18.35	3.1	−0.45	0.9	3.05	−0.4	18.8
4	3	4	10.5	15.3	12.65	16.2	12.55	−4.8	−2.65	0.9	−2.75	−0.3	10.8
1	4	1	21.9	15.3	17.6	16.75	18.35	6.6	2.3	1.45	3.05	−0.2	22.1
2	4	4	16.5	15.3	16.1	16.75	12.55	1.2	0.8	1.45	−2.75	1.7	14.8
3	4	2	15.5	15.3	14.85	16.75	14.35	0.2	−0.45	1.45	−0.95	0.15	15.35
4	4	3	13.1	15.3	12.65	16.75	15.95	−2.2	−2.65	1.45	0.65	−1.65	14.75
Total			244.8	244.8	244.8	244.8	244.8						244.8
Sum of Squares								164.9	52.62	25.34	72.76	14.18	

Write this as   yield = fitted value + residual, hence
Fitted value (F) = overall mean + row effect + column effect + treatment effect, and
Residual (R) = yield − fitted value
The overall mean is estimated by the grand mean (GM)
The row effect is estimated as RE = (row mean − grand mean) = (RM − GM)
The column effect is estimated as CE = (column mean − grand mean) = (CM − GM)
The treatment effect is estimated as TE = (treatment mean − grand mean) = (TM − GM)

The fitted value (F) is estimated as

$$F = GM + RE + CE + TE = GM + (RM - GM) + (CM - GM) + (TM - GM)$$
$$= RM + CM + TM - (2 \times GM)$$

This is the yield predicted by the model. The residual is the difference between the observed yield (Y) and the fitted value (F). It is estimated as

$$R = Y - F = Y - RM - CM - TM + (2 \times GM)$$

Check that for each row in Table 11.5, F = GM + RE + CE + TE and R = Y − F. You should also check that Y = F + R = GM + RE + CE + TE + R and hence that

$$(Y - GM) = RE + CE + TE + R = (RM - GM) + (CM - GM) + (TM - GM) + R$$

It can also be shown that

$$\Sigma(Y - GM)^2 = \Sigma(RM - GM)^2 + \Sigma(CM - GM)^2 + \Sigma(TM - GM)^2 + R^2$$

This means that the total variation in the yields is made up of the variation between the rows, plus the variation between the columns, plus the variation between the treatments, and the residual variation. Write this as *TotalSS = RowSS + ColSS + TreatSS + ResidSS*.
The last row of Table 11.5 verifies this relationship for Example 11.1 as

$$164.90 = 52.62 + 25.34 + 72.76 + 14.18$$

11.5 USING YOUR CALCULATOR

You have already seen how the sums of squares are calculated using a computer spreadsheet (Table 11.5). As an alternative we present a method using your calculator in SD mode.
  To find *TotalSS*:

- Enter all $N = 16$ yields into your calculator in SD mode.
- Press the *n* key to make sure you have entered 16 numbers.
- Press the SD key (labelled σ_{n-1} or *s*), square the result and multiply by $N - 1 = 15$.
- You should obtain 164.90.

THE LATIN SQUARE DESIGN

To find the other sums of squares:

- Enter the relevant totals.
- Press the SD (σ_{n-1} or s) key and square the result.
- Multiply by (the number of totals entered -1) and
- Divide by the number of yields making up each total.

For example, to find *RowSS*:

- Enter the four row totals.
- Press the SD key and square the result.
- Multiply by 3 and divide by 4.
- You should obtain 52.62.

When you have calculated these sums of squares, find *ResidSS* using the formula

$$ResidSS = TotalSS - RowSS - ColumnSS - TreatSS$$
$$ResidSS = 164.90 - 52.62 - 25.34 - 72.76 = 14.18$$

You can now complete the ANOVA table and confirm the results of Output 11.1. If you do not have access to a computer you will not be able to calculate the *P*-values and will need to use *F*-tables. From tables,

$$F_{(3,6,5\%)} = 4.76, \ F_{(3,6,1\%)} = 9.78 \ \text{and} \ F_{(3,6,0.1\%)} = 23.71$$

As the *VR* of 10.26 is greater than 9.78 and not as large as 23.71, the null hypothesis that all four varieties give equal yields is rejected at the 1% level but not at the 0.1% level. Hence the *F*-tables show that the *P*-value for treatments is between 0.01 and 0.001. Output 11.1 gives the exact value of 0.009. Hence there is very strong evidence that the varieties give different yields. The SED and LSD values given in the output are calculated from the ANOVA table as follows:

$$SED = \sqrt{\frac{2 \times RMS}{r}} = \sqrt{\frac{2 \times 2.363}{4}} = 1.087$$

$$LSD_{5\%} = t_{(6, 2.5\%)} \times SED = 2.447 \times 1.087 = 2.66$$

A 95% confidence interval for the difference between the A and D population means is

$$(18.35 - 12.55) \pm LSD_{5\%} = 5.80 \pm 2.66 = (3.14, \ 8.46)$$

We are thus 95% confident that the mean yield for variety A exceeds that for variety D by between 3.14 and 8.46 kg/plot.

Exercise 11.1
An experiment was designed to test the effects of five levels of phosphate fertiliser (P_1, P_2, P_3, P_4 and P_5) on the yield of potato tuber dry matter. The experiment was laid out in the field as a 5×5 Latin square, and the following yields (kg DM/plot) were obtained:

P_2 62.3	P_3 61.3	P_5 62.5	P_1 63.8	P_4 75.0
P_4 64.1	P_5 68.4	P_1 62.9	P_2 66.2	P_3 77.4
P_3 69.2	P_1 55.8	P_2 67.8	P_4 71.3	P_5 74.8
P_5 65.0	P_2 68.7	P_4 69.8	P_3 76.0	P_1 70.9
P_1 63.3	P_4 75.0	P_3 69.3	P_5 78.0	P_2 75.4

Analyse these data and report your conclusions.

Answers to Exercise 11.1
The analysis of variance table is

Source	DF	SS	MS	VR	F prob
Rows	4	147.81	36.95	3.56	
Columns	4	350.23	87.56	8.44	
Treatments	4	196.74	49.18	4.74	0.006
Residual	12	124.54	10.38		
Total	24	819.31			

The treatment means are

Treatment	P_1	P_2	P_3	P_4	P_5
Mean	63.34	68.08	70.64	71.04	69.74

The SED value is 2.037 and the 5% LSD value is 4.439. The coefficient of variation is 4.7%.

The treatment means are significantly different (VR = 4.74, $P = 0.006$). The LSD shows that P_1 gives significantly different yields from the other levels of P, and these are not significantly different from each other. The high VR values for rows and columns suggest that the Latin square design was effective in reducing experimental error.

Chapter 12

Factorial Experiments

12.1 INTRODUCTION

Chapters 9 and 10 considered only single-factor experiments. Example 10.1 compared the yields of several varieties of wheat. All other factors were held constant; there was one planting date, seed rate, level and type of fertiliser, etc. As a result the highest yielding variety could only be recommended in conditions similar to those used in this experiment. If a different rate of fertiliser had been used, another variety may have been the highest yielding.

For most research programmes in applied biology there are several environmental and treatment factors affecting the results. For example, if an agronomist wished to assess the economic potential of a 'new' crop plant, he or she would be required to produce data on different varieties, seed rates, and agrochemical requirements. One approach would be to test individual crop husbandry inputs—one trial to assess optimum seeding density, a second response to nitrogen fertiliser, a third the value of plant growth regulators . . . and so on. The list is long. This is clearly time consuming and expensive, and gives no measure of possible interactions between different factors. In statistical terms it may also give a poor estimate of standard error.

In a factorial experiment the treatments are combinations of two or more levels of two or more factors. A factor is a classification or categorical variable which can take one or more values called levels.

Examples of factors are:

Variety	The levels are the particular varieties of a species used in the experiment.
Fertiliser	The levels are the different kinds of fertiliser used.
Concentration	The levels are the different values of concentration of, for example, a fertiliser or another chemical.
Spacing	The levels are the different row spacings, or seed spacings in a row.

When designing a factorial trial it is worth remembering that ambitious plans are often impossible to manage in a scientific way under field, glasshouse and laboratory conditions. Although it is tempting to include numerous levels of each factor, in practice it is often acceptable to have wide differences between levels of many agrochemical inputs. For example, it would be unwise to test the nitrogen fertiliser response of several varieties of a new crop within the range zero up to 200 kg N/ha using increasing increments of 10 kg N/ha. In an **initial** experiment a control and levels of 50, 100, 150 and 200 kg N/ha would probably be adequate.

It is also debatable whether different sowing dates can be regarded as factor levels. As change of sowing date results in a range of environmental differences which affect establishment and harvest date, it may be wiser to avoid a factorial design when assessing this important factor, and regard each sowing date as a separate trial.

It is also important to check that the correct statistical procedures are employed. Varying concentrations of ammonium nitrate (fertiliser), Fusilade (herbicide), Baytan (fungicide), etc. can be used in a field or glasshouse experiment. However, it may be unwise to use a pooled residual mean square for estimating SED in order to compare the above agrochemicals which belong to distinct chemical groupings.

12.2 ADVANTAGES OF FACTORIAL EXPERIMENTS

Suppose you wish to compare the response of three varieties to five rates of nitrogen fertiliser using either a factorial or separate experiments for nitrogen and variety. For the nitrogen experiment, you could use a randomised block design with four blocks and one variety. For the variety experiment, you could use a randomised block design with seven blocks each containing the same nitrogen level. The skeleton ANOVA tables are shown in Table 12.1.

For these two experiments, the number of blocks would be chosen to ensure at least 12 residual degrees of freedom for estimating the error variance. The nitrogen experiment would require 20 plots and the variety experiment 21 plots (41 plots altogether). The main disadvantage of using separate experiments is that you cannot test for a possible interaction between the factors nitrogen and variety. There would be an interaction if the response to nitrogen is not the same for the three varieties (Figures 12.1 and 12.2).

In Figure 12.1, the change in yield as the nitrogen level is increased is the same for each variety. The response curves are parallel. Also, the differences in yields for the three varieties are independent of the level of nitrogen fertiliser. When there is no interaction, you can summarise in your results the main effects of nitrogen and

Table 12.1. Skeleton ANOVA tables for separate Nitrogen and Variety experiments

Source	DF	Source	DF
Blocks	3	Blocks	6
Nitrogen	4	Variety	2
Residual	12	Residual	12
Total	19	Total	20

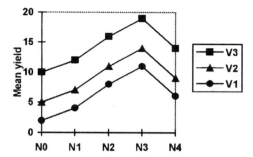

Figure 12.1. Example of no interaction

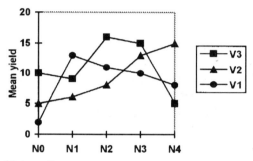

Figure 12.2. Example of interaction

variety. The differences in the overall nitrogen means are the same as the differences in the nitrogen means for any particular variety. In this case a rate of nitrogen fertiliser can be recommended which will be suitable for all the varieties used in the trial. Similarly, the differences in the overall variety means are the same as the differences in the variety means for each level of nitrogen fertiliser if there is no interaction. In this case, the highest yielding variety at a particular nitrogen level will be the highest-yielding variety at all the other nitrogen levels.

In Figure 12.2, the differences between the variety yields depend on the level of nitrogen fertiliser. Also, the change in yield as the nitrogen level is increased depends

Block 1

N_3V_1	N_3V_3	N_5V_2
N_1V_3	N_1V_2	N_5V_3
N_4V_2	N_4V_1	N_2V_1
N_4V_3	N_1V_1	N_2V_3
N_3V_2	N_5V_1	N_2V_2

Block 2

N_3V_2	N_4V_2	N_2V_2
N_4V_1	N_1V_2	N_1V_1
N_3V_1	N_4V_3	N_1V_3
N_2V_1	N_5V_1	N_2V_3
N_5V_2	N_3V_3	N_5V_3

Figure 12.3. Possible layout for a randomised blocks factorial experiment

Table 12.2. Skeleton ANOVA table for design of Example 12.1

Source	DF
Blocks	$r - 1 = 1$
Nitrogen	$n - 1 = 4$
Variety	$v - 1 = 2$
$N \times V$	$4 \times 2 = 8$
Residual	14
Total	$N - 1 = 29$

on the variety. The optimal nitrogen level for maximum yield depends on the variety used. When there is interaction, you cannot discuss main effects. You should investigate the nitrogen mean yields for each variety or compare the variety mean yields at each level of nitrogen.

Example 12.1 A factorial experiment to compare five nitrogen levels (N_1, N_2, ... N_5) for three varieties (V_1, V_2, and V_3)

This is an example of a 5×3 factorial experiment where Factor A is Nitrogen at 5 levels and Factor B is Variety at 3 levels. There are 15 treatments, giving all possible combinations of the five levels of nitrogen and the three levels of variety. They are:

N_1V_1 N_1V_2 N_1V_3 N_2V_1 N_2V_2 N_2V_3 N_3V_1 N_3V_2 N_3V_3 N_4V_1 N_4V_2 N_4V_3
N_5V_1 N_5V_2 N_5V_3

The design could be a completely randomised design, a randomised block design, a split-plot design (Chapter 16) or some other, but it is unlikely that a Latin square (Chapter 11) would be used.

Suppose the design is a randomised block design with only two blocks. Figure 12.3 shows a possible layout and Table 12.2 the corresponding skeleton ANOVA table.

Partitioning of treatment degrees of freedom

In this factorial experiment, there are 15 treatments and so the treatment degrees of freedom are $t - 1 = 14$. They are split into 4 for nitrogen, 2 for variety and $4 \times 2 = 8$ for the nitrogen × variety interaction. The factorial is advantageous because in the two separate experiments a total of 41 plots are required and the residual degrees of freedom are only 12. There is no test for interaction. The factorial experiment with two blocks requires a total of 30 plots to provide 14 residual degrees of freedom. With fewer plots you achieve greater residual degrees of freedom, and can also test for interaction.

If there is no interaction there is hidden replication for testing the main effects

In this experiment, although there are only two replications per treatment, each nitrogen level is replicated six times. You can see from the plan that six of the plots contain N_1, six contain N_2, etc. In the nitrogen experiment with 12 residual

FACTORIAL EXPERIMENTS   163

degrees of freedom, each nitrogen level was replicated only three times. Hence, if there is no interaction, the effective replication for comparing nitrogen levels is six.

If there is no interaction, the effective replication for comparing varieties in the factorial experiment is 10. This is because there are 30 plots altogether and each of the three varieties occurs in 10 plots (Figure 12.3). In the variety experiment, each variety occurs in only seven plots.

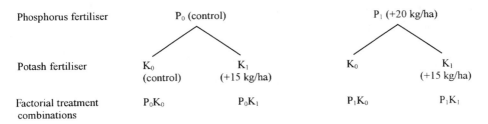

Figure 12.4. Treatment combinations for Example 12.2

12.3 MAIN EFFECTS AND INTERACTIONS

The concepts of main effect and interaction are easier to understand when there are only two factors each at two levels. Such experiments are called 2 × 2 factorial experiments.

Example 12.2
Consider an experiment to investigate the effect of phosphorus and potassium fertilisers on the yield of a crop. The two factors are called P and K. The two levels of P could be zero and 20 kg/ha. The two levels of K could be zero and 15 kg/ha. The four treatment combinations are (Figure 12.4):

Treatment 1 P_0, K_0 No fertiliser (control)
Treatment 2 P_0, K_1 15 kg/ha of potassium
Treatment 3 P_1, K_0 20 kg/ha of phosphorus
Treatment 4 P_1, K_1 20 kg/ha of phosphorus and 15 kg/ha of potassium.

Three possible results of this experiment are summarised in Figures 12.5–12.7 which show positive, negative and no interaction respectively. The mean yields are in arbitrary units. The design may be a CRD, a RBD or some other. It is assumed that each treatment is replicated r times so that the points on the graphs represent the means of r yields.

Positive interaction (Figure 12.5)

The effect of P in the absence of K $= (18 - 10) = 8$
The effect of P in the presence of K $= (26 - 12) = 14$
The effect of K in the absence of P $= (12 - 10) = 2$
The effect of K in the presence of P $= (26 - 18) = 8$

Thus the response to P depends on whether K is present. Similarly, the response to K depends on whether P is present. This interaction is shown to be positive as follows.

The increase in yield due to adding P alone is $(18 - 10) = 8$. The increase in yield due to adding K alone is $(12 - 10) = 2$. If the two fertilisers were acting independently, you would expect the yield to increase by $(8 + 2) = 10$ units when both are applied. If fact, the increase is $(26 - 10) = 16$, which is 6 more than expected. Thus, the interaction is positive. The interaction effect is $+6$. This can be also be calculated in two ways:

Interaction effect = (effect of P in the presence of K) − (effect of P in the absence of K) = $14 - 8 = 6$
Interaction effect = (effect of K in the presence of P) − (effect of K in the absence of P) = $8 - 2 = 6$

The main effect of K is $(19 - 14) = 5$. The main effect of P is $(22 - 11) = 11$. However, because of the interaction, it would be misleading to only quote these main effects in a report. The effect of P depends on whether K is present, and the effect of K depends on whether P is present.

Negative interaction (Figure 12.6)

You can see that P increases yield by 8 units if K is absent, but decreases yield by 2 units when K is present. Similarly, K increases yield by 6 units when P is absent, but decreases yield by 4 units when P is present. It would be very misleading to say that the main effect of K is $(15 - 14) = 1$ unit.

Interaction effect = (effect of P in the presence of K) − (effect of P in absence of K) = $(14 - 16) - (18 - 10) = -2 - 8 = -10$

No interaction (Figure 12.7)

The effect of P in the absence of K   = $(18 - 10) = 8$
The effect of P in the presence of K   = $(22 - 14) = 8$
The effect of K in the absence of P   = $(14 - 10) = 4$
The effect of K in the presence of P   = $(22 - 18) = 4$

Thus the response to P is independent of whether K is present. Similarly, the response to K is independent of whether P is present. This is an example of no interaction.

The increase in yield due to adding P alone is $(18 - 10) = 8$. The increase in yield due to adding K alone is $(14 - 10) = 4$. If the two fertilisers were acting independently, you would expect the yield to increase by $(8 + 4) = 12$ units when both are applied. If fact, the increase is $(22 - 10) = 12$, which is as expected. This shows that the two fertilisers are acting independently. There is no interaction.

Interaction effect = (effect of P in the presence of K) − (effect of P in the absence of K) = $8 - 8 = 0$

Interaction effect = (effect of K in the presence of P) − (effect of K in the absence of P) = 4 − 4 = 0

The main effect of K is $(18 − 14) = 4$. The main effect of P is $(20 − 12) = 8$. Because the interaction is zero, it is correct to quote these main effects in a report. The effect of P does not depend on whether K is present or absent, and the effect of K does not depend on whether P is present or absent. When there is no interaction, the lines on the graph of mean yield are parallel.

	P_0	P_1	Mean
K_1	12	26	19
K_0	10	18	14
Mean	11	22	

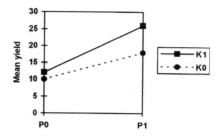

Figure 12.5. Results showing positive interaction

	P_0	P_1	Mean
K_1	16	14	15
K_0	10	18	14
Mean	13	16	

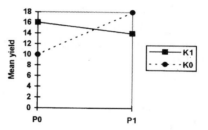

Figure 12.6. Results showing negative interaction

	P_0	P_1	Mean
K_1	14	22	18
K_0	10	18	14
Mean	12	20	

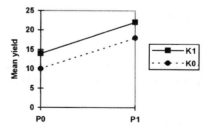

Figure 12.7. Results showing no interaction

12.4 VARIETIES AS FACTORS

Consider a 2 × 2 factorial experiment in which the effect of nitrogen fertiliser is to be determined for two varieties. The two factors are

Nitrogen at 2 levels, N_0 and N_1
Variety at two levels, V_1 and V_2

166   PRACTICAL STATISTICS AND EXPERIMENTAL DESIGN FOR PLANT AND CROP SCIENCE

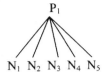
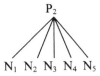
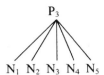

The treatment combinations are:

P_1N_1	P_2N_1	P_3N_1
P_1N_2	P_2N_2	P_3N_2
P_1N_3	P_2N_3	P_3N_3
P_1N_4	P_2N_4	P_3N_4
P_1N_5	P_2N_5	P_3N_5

Figure 12.8. Treatment combinations for the experiment of Example 12.3

The treatment combinations are

$T_1 = N_0, V_1$ $T_2 = N_0, V_2$ $T_3 = N_1, V_1$ $T_4 = N_1, V_2$

The results can be displayed in tables and graphs similar to those discussed for the PK experiment (Example 12.2). However, there is a slight difference in the meaning of the term interaction. In the PK experiment, you can talk about P and K both being present and both being absent, so the interaction is a measure of the increase in yield when both are present over and above the sum of the yield increases when each is applied alone. In the variety experiment you cannot speak of the factor variety being present or absent. In this case, there is interaction if the change in yield using nitrogen depends on variety. The interaction can be illustrated using graphs similar to those used above, except that the horizontal axis should show N_0 and N_1, one of the lines should represent V_1 and the other should represent V_2.

12.5 ANALYSIS OF A RANDOMISED BLOCKS FACTORIAL EXPERIMENT WITH TWO FACTORS

Example 12.3
A factorial trial was carried out in order to determine the response of **one variety** to three levels of a new plant growth regulator (PGR), denoted by P_1, P_2, P_3 and five levels of nitrogen fertiliser (N_1, N_2, N_3, N_4, N_5). This is a 3×5 factorial experiment. The 15 treatment combinations are shown in Figure 12.8.

A randomised block design with three replications was adopted and the 15 treatments were randomly assigned to the plots within each block. This experiment can be used to explore the following objectives:
(1) The main effect of nitrogen fertiliser on the yield of oilseed rape when averaged over levels of the plant growth regulator (P)

(2) The main effect of P on the yield of oilseed rape when averaged over levels of nitrogen (N).
(3) Interaction — if the response curves to nitrogen fertiliser are not the same for the three levels of P. This is equivalent to saying that the differences between the PGR yields are not the same at each nitrogen level. If the increase in yield in going from one nitrogen level to the next is not the same for all PGRs this indicates interaction.

If there is a significant interaction it is misleading to compare the main effects of N or P. You should compare the effects of P at each level of N or the effects of N at each level of P. Plot yields are given in Table 12.3.

To facilitate the calculations of sums of squares by hand the plot yields are arranged in Table 12.3. Each column represents a different treatment combination. Table 12.4 gives the 15 treatment totals arranged by growth regulator and nitrogen. Each value in the body of Table 12.4 is the **sum** of three yields, one from each block. Level 1 of the PGR (P_1) gave the highest yield at all nitrogen levels except for N_5, the yields having increased with nitrogen level up to N_4. A similar pattern of response was recorded for P_3. However, for the second level of PGR an increased yield was also achieved using the highest level of fertiliser. This suggests an interaction between P and N which can be tested using the analysis of variance table (Output 12.1).

12.5.1 Using the Computer

You are advised to enter the data in 45 rows and 4 columns. If you enter the yields according to the order of columns in Table 12.3, the first column will be 15 ones, followed by 15 twos and 15 threes. This column contains the codes for PGR (P). The second column will be the sequence 3 ones, 3 twos, 3 threes, 3 fours and 3 fives repeated 3 times. This column contains the codes for nitrogen (N). The third column will be the sequence 1, 2, 3 repeated 15 times. This column contains the codes for blocks (B). The fourth column will contain the yields (Y). They must be entered according to the codes in the first three columns. Table 12.5 gives the first six and the last six rows as they should be entered in the spreadsheet. As an exercise you should enter the remaining 33 rows into this table. An advantage of entering data in standard order with factor codes is that this format can be used by alternative statistical packages.

12.5.2 Interpretation of Computer Output (Output 12.1)

The F pr. column shows that the interaction is significant at the 0.1% level ($P < 0.001$). The interaction SED is 0.114 and should be used with the two-way table of means. It can be used to compare two PGR means at the same level of nitrogen, and two nitrogen means at the same level of PGR. The corresponding 5% LSD value is 0.234. It suggests that for the first nitrogen level there are no significant differences between the levels of PGR.

Table 12.3. Plot yields (kg) for Example 12.3

Block	P$_1$					P$_2$					P$_3$					Total
	N$_1$	N$_2$	N$_3$	N$_4$	N$_5$	N$_1$	N$_2$	N$_3$	N$_4$	N$_5$	N$_1$	N$_2$	N$_3$	N$_4$	N$_5$	
B$_1$	0.9	1.2	1.3	1.8	1.1	0.9	1.1	1.3	1.6	1.9	0.9	1.4	1.3	1.4	1.2	19.3
B$_2$	0.9	1.3	1.5	1.9	1.4	0.8	0.9	1.5	1.3	1.6	1.0	1.2	1.4	1.5	1.1	19.3
B$_3$	1.0	1.2	1.4	2.1	1.2	0.8	0.9	1.1	1.1	1.5	0.7	1.0	1.4	1.4	1.3	18.1
Total	2.8	3.7	4.2	5.8	3.7	2.5	2.9	3.9	4.0	5.0	2.6	3.6	4.1	4.3	3.6	56.7

Table 12.4. P × N table of totals for Example 12.3

	N$_1$	N$_2$	N$_3$	N$_4$	N$_5$	Total
P$_1$	2.8	3.7	4.2	5.8	3.7	20.2
P$_2$	2.5	2.9	3.9	4.0	5.0	18.3
P$_3$	2.6	3.6	4.1	4.3	3.6	18.2
Total	7.9	10.2	12.2	14.1	12.3	56.7

For the main effect of PGR, v.r. = 4.32 and F pr = 0.023. This *P*-value of 0.023 is less than 0.05 and suggests significant differences between the overall PGR means (1.347, 1.220 and 1.213). The 5% LSD value of 0.1048 is less than the difference between 1.347 and 1.220 (0.127) and suggests that P$_1$ gives significantly higher yields than P$_2$. However, at the fifth level of nitrogen the PGR means are 1.233, 1.667, 1.200 showing that P$_1$ gives lower yields than P$_2$. This illustrates a danger of interpreting main effects in the presence of interaction. The main effect SED values of 0.0511 and 0.0660 should not be used to compare the main effect means of PGR and nitrogen respectively, because of the significant interaction.

The residual coefficient of variation is 11.12%. This is acceptable for many field experiments and indicates little background variation in plot yields when compared with the overall mean. The coefficient of variation for blocks is low (3.7%). This, together with the low v.r. of 1.63 suggests that blocking was not very successful in this experiment.

12.5.3 Notes on the Calculations

- **Degrees of freedom**

df for Blocks (B)	= number of blocks − 1	= 3 − 1 = 2
df for PGR (P)	= number of levels of PGR − 1	= 3 − 1 = 2
df for Nitrogen (N)	= number of nitrogen levels − 1	= 5 − 1 = 4
df for P × N	= (df for nitrogen) × (df for PGR)	= 4 × 2 = 8
df for Total	= Total number of yields − 1	= 45 − 1 = 44

  df for Residual are found by subtraction (44 − 2 − 2 − 4 − 8 = 28)

Output 12.1 Genstat analysis of Example 12.3

```
***** Analysis of variance *****

Variate: Yield

Source of variation    d.f.      s.s.      m.s.      v.r.    F pr.

Block stratum             2    0.06400   0.03200     1.63

Block.*Units* stratum
P                         2    0.16933   0.08467     4.32    0.023
N                         4    2.49022   0.62256    31.73   <0.001
P.N                       8    1.01511   0.12689     6.47   <0.001
Residual                 28    0.54933   0.01962

Total                    44    4.28800


***** Tables of means *****

Variate: Yield

Grand mean 1.260


     P              1        2        3
                 1.347    1.220    1.213

     N              1        2        3        4        5
                 0.878    1.133    1.356    1.567    1.367

     P     N        1        2        3        4        5
     1           0.933    1.233    1.400    1.933    1.233
     2           0.833    0.967    1.300    1.333    1.667
     3           0.867    1.200    1.367    1.433    1.200


*** Standard errors of differences of means ***

Table                     P        N        P
                                            N
rep.                     15        9        3
d.f.                     28       28       28
s.e.d.               0.0511   0.0660   0.1144


*** Least significant differences of means (5% level) ***

Table                     P        N        P
                                            N
rep.                     15        9        3
d.f.                     28       28       28
l.s.d.               0.1048   0.1353   0.2343


***** Stratum standard errors and coefficients of variation *****

Variate: Yield

Stratum                d.f.     s.e.     cv%

Block                     2    0.0462    3.7
Block.*Units*            28    0.1401   11.1
```

Table 12.5. Spreadsheet entry for data of Example 12.3

P	N	B	Y
1	1	1	0.9
1	1	2	0.9
1	1	3	1.0
1	2	1	1.2
1	2	2	1.3
1	2	3	1.2
3	4	1	1.4
3	4	2	1.5
3	4	3	1.4
3	5	1	1.2
3	5	2	1.1
3	5	3	1.3

- **Sums of squares**
  An explanation is given in Section 12.5.6 and their method of calculation by a hand calculator in Section 12.5.7.
- **Mean square**
  The entries in the mean square column (m.s.) are obtained by dividing the entries in the sum of squares column (s.s.) by the corresponding degrees of freedom (d.f.).
- **Variance ratio**
  The four variance ratio values (v.r.) are obtained by dividing the corresponding mean square by the residual mean square ($RMS = 0.01962$). The greater the variance ratio the stronger the evidence of an effect.
- **P-values**
  The F. pr column shows the P-values. This is the probability of obtaining a v.r. greater than the calculated value assuming no effect. The greater the v.r., the smaller the P-value. A P-value smaller than 0.05 is generally regarded as evidence of an effect.
- **Standard errors of difference (SED)**
  The standard error of the difference between two overall P means is

$$SED_P = \sqrt{\frac{2 \times RMS}{r \times n}} = \sqrt{\frac{2 \times 0.0196}{15}} = 0.0511$$

where r = no. of reps (blocks), $r \times n$ = no. of yields for each level of PGR, and n = no. of levels of nitrogen.
The standard error of the difference between two overall N means is

$$SED_N = \sqrt{\frac{2 \times RMS}{r \times p}} = \sqrt{\frac{2 \times 0.0196}{9}} = 0.0660$$

where $r \times p$ = no. of yields at each level of nitrogen, and p = no. of levels of PGR. The standard error of the difference between two overall P × N combination means is

$$SED_{P \times N} = \sqrt{\frac{2 \times RMS}{r}} = \sqrt{\frac{2 \times 0.0196}{3}} = 0.1143$$

where r is the number of yields for each combination mean.
- **Least significant differences**
Corresponding to each of the three SED values there are three LSD values. To calculate the 5% LSD values given in Output 12.1, multiply the corresponding SED values by $t_{(28, 2.5\%)} = 2.048$ where 28 is the residual degrees of freedom.
- **Residual Coefficient of Variation**

$$CV = \frac{\sqrt{RMS}}{Grand\ mean} \times 100 = \frac{0.1401}{1.260} \times 100 = 11.12\%$$

The square root of the residual mean square (0.1401) is often called the residual standard error and is shown in the last line of the computer output.
- **Block Coefficient of Variation**

$$CV = \frac{\sqrt{Block\,MS/(n \times p)}}{Grand\ mean} \times 100 = \frac{\sqrt{0.032/15}}{1.260} \times 100 = \frac{0.0462}{1.260} \times 100 = 3.7\%$$

The square root of the block mean square divided by the number of units per block (0.0462) is called the block standard error and is shown in the penultimate line of the output.

12.5.4 Use of Statistical Tables

If you have calculated the ANOVA table by hand (Section 12.5.7) you will have to compare the variance ratio (VR) values with values found in the F-tables (Appendix 4).

- From the F-tables, $F_{(8, 28, 5\%)} = 2.29$, $F_{(8, 28, 1\%)} = 3.23$ and $F_{(8, 28, 0.1\%)} = 4.70$ so the interaction VR of 6.47 is very highly significant ($P < 0.001$).
- $F_{(2, 28, 5\%)} = 3.34$, $F_{(2, 28, 1\%)} = 5.45$ so the VR for P of 4.32 is significant at the 5% but not at the 1% level ($P < 0.05$).
- The VR for blocks of 1.63 is not significant as it is less than $F_{(2, 28, 5\%)} = 3.34$.
- $F_{(4, 28, 5\%)} = 2.71$, $F_{(4, 28, 1\%)} = 4.07$ and $F_{(4, 28, 0.1\%)} = 6.25$ so the VR for N of 31.73 is very highly significant ($P < 0.001$).

12.5.5 Presentation of Results

You can summarise the results by presenting a two-way table of means (Table 12.6), and drawing graphs of the mean yields versus nitrogen level for each level of PGR (Figure 12.9). For publication in scientific journals the main effect and interaction standard errors of differences are also presented. Normally, the number of decimal places given for these standard errors is one greater than the number given for each mean. The main effect SED values should not be used to compare the main effect means in the presence of significant interaction.

Suppose you wish to compare the levels of PGR at a particular nitrogen level. The appropriate SED is $SED_{P \times N} = 0.114$ and the LSD at the 5% level is 0.234. The SED value is presented in Figure 12.9 as a vertical line so visual comparisons can be made between the treatment means. You could present the LSD (0.05) value of 0.234 but this is usually not recommended by referees of scientific papers.

From the graph and table of means you see that there are no significant differences between the mean yields at levels N_1 or N_3 but P_1 is significantly different from both P_2 and P_3 at nitrogen level N_4. You can use the same SED to compare the nitrogen yields within each level of PGR.

Because the levels of the nitrogen factor represent increasing values of a quantitative variable, an attempt should be made to fit straight lines or curves separately for each level of PGR (Section 13.4.4).

12.5.6 The Model

Example 12.3 uses a RBD where the treatments are the 15 combinations of P and N. Hence we can present the model as in Section 10.6 using Tables 12.7 and 12.8.

Yield = overall mean + block effect + treatment effect + residual
Yield = fitted value + residual, hence
Fitted value = overall mean + block effect + treatment effect, and
Residual = yield − fitted value

- The overall mean is estimated by the grand mean (GM)
- The block effect is estimated as BE = (block mean − grand mean) = (BM − GM)

Table 12.6 P × N Table of means for Example 12.3

	N_1	N_2	N_3	N_4	N_5	Means
P_1	0.93	1.23	1.40	1.93	1.23	1.35
P_2	0.83	0.97	1.30	1.33	1.67	1.22
P_3	0.87	1.20	1.37	1.43	1.20	1.21
Means	0.88	1.13	1.36	1.57	1.37	

$SED_P = 0.051$ $SED_N = 0.066$ $SED_{P \times N} = 0.114$

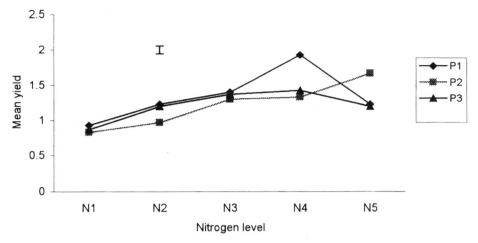

Figure 12.9. Mean yield versus nitrogen level for each level of PGR

- The treatment effect is estimated as TE = (treatment mean − grand mean) = (TM − GM)
- The fitted yield predicted by the model is estimated as

$$F = GM + (BM - GM) + (TM - GM) = BM + TM - GM$$

- The residual is the difference between the observed yield (Y) and the fitted value (F). It is estimated as R = Y − F = Y − BM − TM + GM

You should now be able to construct a spreadsheet similar to Table 10.5 and confirm the values for TotalSS, BlockSS, TreatSS and ResidSS as 4.288, 0.064, 3.675 and 0.549 respectively. Table 12.7 shows the first and last six rows for this example.

- The P effect is estimated as PE = (P mean − grand mean) = (PM − GM)
- The N effect is estimated as NE = (N mean − grand mean) = (NM − GM)
- The interaction effect is estimated as IE = (treatment effect − P effect − N effect)

$$= (TM - GM) - (PM - GM) - (NM - GM)$$
$$= (TM - PM - NM + GM)$$

A table can now be constructed with columns for TE, PE, NE and IE. Table 12.8 shows the first and last six rows of such a table. As an exercise you should complete the table and confirm that the sums of squares of the entries for PE, NE and IE are 0.1693, 2.4902 and 1.0151 respectively. These are the sums of squares for P, N and the interaction respectively. Note that they add up to TreatSS = 3.675.

It can be shown (proof not given) that

$$\Sigma(TM - GM)^2 = \Sigma(PM - GM)^2 + \Sigma(NM - GM)^2 + \Sigma(TM - PM - NM + GM)^2$$

Table 12.7. Fitting the model Yield = overall mean + block effect + treatment effect + residual for Example 12.3

Treat	Block	Yield	GM	BM	TM	Y−GM	TM−GM	BM−GM	R	F
		Y					TE	BE		
1	1	0.9	1.26	1.2867	0.9333	−0.36	−0.3267	−1.62	−0.06	0.96
1	2	0.9	1.26	1.2867	0.9333	−0.36	−0.3267	−1.62	−0.06	0.96
1	3	1	1.26	1.2067	0.9333	−0.26	−0.3267	−1.52	0.12	0.88
2	1	1.2	1.26	1.2867	1.2333	−0.06	−0.0267	−1.32	−0.06	1.26
2	2	1.3	1.26	1.2867	1.2333	0.04	−0.0267	−1.22	0.04	1.26
2	3	1.2	1.26	1.2067	1.2333	−0.06	−0.0267	−1.32	0.02	1.18
14	1	1.4	1.26	1.2867	1.4333	0.14	0.1733	−1.12	−0.06	1.46
14	2	1.5	1.26	1.2867	1.4333	0.24	0.1733	−1.02	0.04	1.46
14	3	1.4	1.26	1.2067	1.4333	0.14	0.1733	−1.12	0.02	1.38
15	1	1.2	1.26	1.2867	1.2	−0.06	−0.06	−1.32	−0.0267	1.2267
15	2	1.1	1.26	1.2867	1.2	−0.16	−0.06	−1.42	−0.1267	1.2267
15	3	1.3	1.26	1.2067	1.2	0.04	−0.06	−1.22	0.1533	1.1467
Sum of Squares						4.288	3.6747	0.064	0.549	
						TotalSS	TreatSS	BlockSS	ResidSS	

This equation can be summarised as **Treatment Sum of Squares = Pgr Sum of Squares + Nitrogen Sum of Squares + Interaction Sum of Squares** or in shorthand notation as

$$TreatSS = PSS + NSS + (P \times N)SS$$

Hence the interaction sum of squares can be found by subtraction:

$$InteractionSS = TreatSS - PSS - NSS$$

12.5.7 Using your Calculator

Look at Output 12.1. We now show how to calculate the sums of squares column using your calculator in SD mode. In what follows the σ_{n-1} key may be marked s on your calculator.

TotalSS = 4.288
This is found as follows. Enter all the 45 yields shown in Table 12.3. Press the n key to check you have entered 45 numbers. Press the σ_{n-1} key, square the result and multiply by $N - 1 = 44$ to obtain 4.288.

BlockSS = 0.064
Enter the three block totals. Press the n key to ensure you have entered three numbers. Press the σ_{n-1} key, square the result, multiply by (number of blocks − 1) = 2, and divide by (the number of yields making up each block total) = 15 to get 0.064.

Table 12.8. Partitioning of the treatment effect and sum of squares for Example 12.3

Treat	P	N	Yield Y	TM	PM	NM	TM−GM TE	PM−GM PE	NM−GM NE	TM−PM−NM+GM IE
1	1	1	0.9	0.9333	1.3467	0.8778	−0.3267	0.0867	−0.3822	−0.0311
1	1	1	0.9	0.9333	1.3467	0.8778	−0.3267	0.0867	−0.3822	−0.0311
1	1	1	1	0.9333	1.3467	0.8778	−0.3267	0.0867	−0.3822	−0.0311
2	1	2	1.2	1.2333	1.3467	1.1333	−0.0267	0.0867	−0.1267	0.0133
2	1	2	1.3	1.2333	1.3467	1.1333	−0.0267	0.0867	−0.1267	0.0133
2	1	2	1.2	1.2333	1.3467	1.1333	−0.0267	0.0867	−0.1267	0.0133
⋮										
14	3	4	1.4	1.4333	1.2133	1.5667	0.1733	−0.0467	0.3067	−0.0867
14	3	4	1.5	1.4333	1.2133	1.5667	0.1733	−0.0467	0.3067	−0.0867
14	3	4	1.4	1.4333	1.2133	1.5667	0.1733	−0.0467	0.3067	−0.0867
15	3	5	1.2	1.2	1.2133	1.3667	−0.06	−0.0467	0.1067	−0.12
15	3	5	1.1	1.2	1.2133	1.3667	−0.06	−0.0467	0.1067	−0.12
15	3	5	1.3	1.2	1.2133	1.3667	−0.06	−0.0467	0.1067	−0.12
Sums of Squares							3.6747 TreatSS	0.1693 PgrSS	2.4902 NitrogenSS	1.0151 InteractionSS

PSS = 0.169
Enter the 3 pgr totals found in Table 12.4. Check you have entered three numbers. Press the σ_{n-1} key, square the result, multiply by (the number of pgr levels − 1) = 2, and divide by (the number of yields in each pgr level) = 15 to get 0.169. Note that each pgr × nitrogen total in Table 12.4 is made up of three yields.

NSS = 2.490
Enter the five nitrogen totals found in Table 12.4. Check you have entered five numbers. Press the σ_{n-1} key, square the result, multiply by (the number of nitrogen levels − 1) = 4, and divide by (the number of yields in each nitrogen level) = 9 to get 2.490. Note that each pgr nitrogen total in Table 12.4 is made up of three yields.

TreatSS = 3.675
Enter the 15 treatment totals found in the last row of Table 12.3. Check you have entered 15 numbers. Press the σ_{n-1} key, square the result, multiply by (the number of treatments − 1) = 14, and divide by (the number of yields in each treatment) = 3 to get 3.675.

(P × N)SS = 1.016
For a factorial experiment with two factors, A and B, the interaction sum of squares is the treatments sum of squares − the factor A sum of squares − the factor B sum of squares. Hence

$$(P \times N)SS = TreatSS - PSS - NSS = 3.675 - 0.169 - 2.490 = 1.016$$

ResidualSS = 0.549
$ResidSS = TotalSS - BlocksSS - TreatSS = 4.288 - 0.064 - 3.675 = 0.549$

12.6 GENERAL ADVICE ON PRESENTATION

When presenting results graphically, a line graph similar to Figure 12.9 is appropriate when the levels of one of the factors is quantitative. These levels are shown on the horizontal axis. Separate lines are drawn for each level of the other factor. If the levels of both factors represent categories, a multiple bar chart is suitable for displaying the combination means. The following example illustrates this.

Example 12.4
Factor A is variety with levels V_1, V_2 and V_3, and factor B is method of tillage with levels M_1, and M_2. Table 12.9 shows a two-way table of mean yields.

Figures 12.10 and 12.11 are two ways of illustrating the information in Table 12.9. Figure 12.10 shows that Method 1 is best for Variety 1, but Method 2 is best for Variety 3. A vertical line showing the interaction SED could be included. Figure 12.11 emphasises differences between the varieties. Variety 1 performs best when Method 1 is used, while Variety 3 performs best when Method 2 is used. If the interaction is not significant bar charts showing the overall variety means and overall method means are sufficient. However, graphs similar to the above are always useful as they are usually easier to interpret than tables of means.

Table 12.9. Two-way table of means for Example 12.4

	V_1	V_2	V_3
M_1	7.8	6.2	5.1
M_2	4.8	5.6	7.0

12.7 EXPERIMENTS WITH MORE THAN TWO FACTORS

When there are more than two factors the analysis becomes increasingly complex.

12.7.1 A Three-factor Experiment

Factor A has a levels. Suppose it is *Variety* and there are four varieties in our experiment, so $a = 4$. Factor B has b levels. Suppose it is *Spacing* and there are five spacing levels in our experiment, so $b = 5$. Factor C has c levels. Suppose it is *Sowing Date* and there are three sowing dates in our experiment, so $c = 3$. The number of treatment combinations is $abc = 4 \times 5 \times 3 = 60$.

We need at least two replications per treatment. Hence, at least 120 plots are required. You can now appreciate the problems associated with factorial experiments when there are more than two factors.

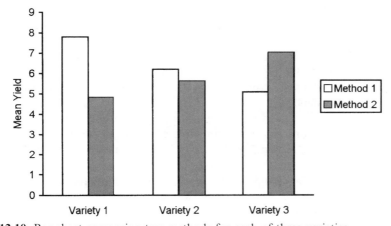

Figure 12.10. Bar chart comapring two methods for each of three varieties

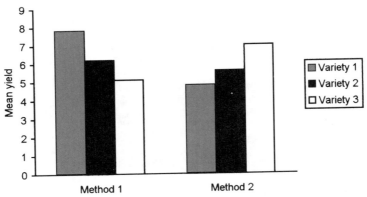

Figure 12.11. Bar chart comparing three varieties for each of two methods

Table 12.10. Skeleton ANOVA table for a typical randomised block factorial experiment with three factors

Source	DF
Blocks	$r - 1 = 1$
Variety	$a - 1 = 3$
Spacing	$b - 1 = 4$
Date	$c - 1 = 2$
Variety × Spacing	$(a - 1)(b - 1) = 12$
Variety × Date	$(a - 1)(c - 1) = 6$
Spacing × Date	$(b - 1)(c - 1) = 8$
Variety × Spacing × Date	$(a - 1)(b - 1)(c - 1) = 24$
Residual	By subtraction $= 59$
Total	$N - 1 = 119$

Table 12.10 shows the skeleton ANOVA table assuming the experiment is laid out in two randomised blocks ($r = 2$).

When you look at computer output for this type of experiment, you should first check whether the three-factor interaction is significant. If it is, you need to present the three-way table of means and the A × B × C *SED* value in your report. This table is also useful when there is no interaction.

If the A × B × C interaction is not significant, you should look to see if any of the two-factor interactions are significant. For example, if the A × B interaction is significant, you need to present the two-way table of means and the A × B *SED* value.

If, in addition, the A × C and B × C interactions are not significant, you should look to see if the main effect of C is significant. If it is, you should present the one-way table of the factor C means, together with the factor C *SED* value. Overall means can be misleading in the presence of interactions.

12.7.2 More than Three Factors

Factorial experiments with more than three factors have the following disadvantages:

- Three-factor and higher-order interactions are very difficult to interpret.
- The number of plots needed for a complete replicate may be prohibitively large.
- If a RBD design is proposed, it may not be possible to find a block large enough to accommodate all the plots required.
- The plots in a large block are unlikely to be uniform, and this will increase the experimental error variance.

12.7.3 2^n Experiments

A 2^n experiment is a factorial experiment in which each factor has two levels. For example, if you wish to study the effects of several fertiliser additives on the yield of a crop, you could carry out an experiment in which the treatments are all possible combinations of the presence and absence of each additive. If there are four additives, A, B, C and D, the number of treatments is $2^4 = 16$. If a RBD is used, each block would consist of 16 plots. A 2^5 experiment in randomised blocks requires a block size of 32 plots. Even if a block this large can be found, it is unlikely that it will be uniform. The problem is worse for a 2^6 experiment, which requires 64 plots per block. A partial solution to this problem is to use the principle of confounding.

12.8 CONFOUNDING

Within one replicate of a 2^6 experiment, there are 64 plots, one for each of the 64 treatment combinations. The block size is reduced if half of the treatments are put into one block, and the other half into another block. A pair of blocks is needed for a complete replicate. If each treatment is replicated three times, three pairs of blocks are required. The block size is reduced to 16 if a complete replicate of the 64 treatments is assigned to 4 blocks of 16 plots each. The advantage of confounding is that block size is only a fraction of the number of treatments. The disadvantage is that some effects are confounded with blocks.

The allocation of treatments to blocks should be such that all main effects can be estimated, and only higher-order interactions, or those that are thought to be negligible, are confounded with blocks. If an effect is confounded with blocks, this means that it is not possible to tell whether the difference in means is due to blocks or to the effect in question. For example, if a complete replicate is in two blocks, and those treatments with A at level 1 are in block 1, and those treatments with A at level 2 are in block 2, the block effect is the A effect. We say that factor A is confounded with blocks.

We will not give details here of how to decide which treatments to put in which blocks. For a full account see Cochran and Cox (1957).

12.9 FRACTIONAL REPLICATION

In a 2^6 experiment there are 32 plots with the first level of factor A and 32 plots with the second level of A. The statement is true of the other five factors. If the experiment contains only one replicate, each main effect mean is estimated with an effective replication of 32. This seems adequate replication. However, there is only one replicate of each of the 64 treatments, so there is no estimate of the residual variation. Two replicates would require 128 plots.

What is often done in this situation is to have only one replicate and assume that some of the highest-order interactions are negligible and use them as experimental error. The number of plots can be further reduced by having only a fraction of the treatment combinations in the experiment. This is fractional replication.

A half replicate of a 2^6 experiment requires only 32 plots instead of 64. Care must be taken on the choice of treatment combinations to include. With the proper choice, each main effect is **aliased** with a high-order interaction. If the high-order interaction is assumed negligible, the observed effect can be assumed to be the main effect. Several high-order interactions are combined and used as the error variance. A quarter replicate of a 2^7 experiment requires only 32 plots instead of the 128 that would be required for a full replicate.

The advantage of fractional replication is that a large number of factors can be investigated in the same experiment with a relatively small number of plots. The disadvantage is the results are subject to misinterpretation. An observed difference in means may be due to a main effect or a high-order interaction.

Experiments with large numbers of factors each at two levels should be regarded as exploratory. They should be carried out to discover which factors may be important. Having isolated a few important factors, two- or three-factor experiments can be carried out with each factor at more than two levels. In this way, the important factors can be studied in greater depth. For further details on 2^n experiments, confounding and fractional replication see Cochran and Cox (1957).

Exercise 12.1

An experiment was carried out to investigate the effect of row spacing on the yield of three varieties (V_1, V_2, V_3) of dwarf French beans. The spacings were: $S_1 = 20$ cm. $S_2 = 40$ cm. $S_3 = 60$ cm. $S_4 = 80$ cm.

This was a factorial experiment in which factor A was Variety at three levels, and factor B was spacing at four levels. These 12 treatment combinations were arranged in a randomised complete block design with five blocks, so that there were 60 plots altogether. Within each block, the treatment combinations were randomly assigned to the plots. The yields in kg per plot are summarised in the following table:

	V_1				V_2				V_3				Block Totals
	S_1	S_2	S_3	S_4	S_1	S_2	S_3	S_4	S_1	S_2	S_3	S_4	
Block 1	25	27	30	31	30	27	24	20	27	28	33	35	337
Block 2	20	23	26	28	28	26	24	21	28	31	35	36	326
Block 3	19	20	23	23	27	25	22	20	23	30	35	34	301
Block 4	21	22	23	25	29	27	25	24	24	27	29	30	306
Block 5	22	24	21	23	30	28	26	23	25	28	32	34	316
Total	107	116	123	130	144	133	121	108	127	144	164	169	1586

Analyse these data and report your conclusions.

Answers to Exercise 12.1

An analysis of variance shows a significant interaction ($VR = 15.62$ on 6 and 44 df, $P < 0.001$). The two-way table of means is as follows:

	S_1	S_2	S_3	S_4
V_1	21.40	23.20	24.60	26.00
V_2	28.80	26.60	24.20	21.60
V_3	25.40	28.80	32.80	33.80

Due to the significant interaction the marginal means are not presented. $SED_{V \times S} = 1.284$ can be used to compare the variety means at each level of spacing. These results can be illustrated by drawing a graph showing the mean yields against the spacing levels for each variety separately. The main conclusion is that the mean yields of V_1 and V_3 increase with spacing while that of V_1 decreases with spacing. With the knowledge that the spacing levels are 20, 40, 60 and 80 cm you can obtain separate regression lines. The slopes for V_1, V_2 and V_3 are 0.076, -0.120, 0.146 respectively. To find out how to test whether these slopes are significantly different see Section 13.4.4.

Chapter 13

Comparison of Treatment Means

13.1 INTRODUCTION

Data from many experiments in applied biology are analysed using analysis of variance. If the results indicate treatment differences, the next step is to discover where these differences lie. There are many methods used to compare treatment means, but unfortunately most of them are subject to misinterpretation. The ideal solution is to decide which comparisons are of interest at the design stage of the experiment. These should be independent, no more in number than the treatment degrees of freedom, and have practical meaning in the context of the treatment structure. If other comparisons are found to be significant, they should ideally be investigated in a further experiment.

13.2 TREATMENTS WITH NO STRUCTURE

In some experiments, such as variety trials, there is no structure; a trial may include many new varieties. If there is no prior information on relative yield performance, comparative tests may be required of all possible pairs of means. If there are 20 new varieties there are $(20 \times 19)/2 = 190$ comparisons to be made. It is tempting to summarise the results by making multiple pairwise comparisons. This practice may be misleading and some statisticians would advise you to avoid it altogether for the following reasons.

If there are t varieties in a trial, there are $t(t-1)/2$ possible pairwise comparisons. If two means are randomly chosen, and the difference is greater than the 5% LSD value, you would reject the null hypothesis of equal variety effects at the 5% level. If the corresponding two population variety means are really the same, there is a 5% chance of declaring them different using this procedure; the Type I error is 5% (Section 5.4).

If the difference between the two means is less than the 5% LSD value when in fact there is a real difference between them, a Type II error is committed and important new varieties could be overlooked. Most plant breeders try to reduce the Type I error because they do not want to recommend new varieties that are not really as good as they seem. However, if the Type I error is reduced to 1% by using the 1% LSD value, the Type II error is automatically increased with the consequence that important new varieties may be missed. The probability of the Type II error is denoted by β. The power of the test is $1 - \beta$ and is the chance of declaring two means significantly different when the true means are really different. The power of the test can be increased without increasing the Type I error if the replication is increased.

It is quite acceptable to use the LSD test for comparisons planned in advance. Do not use it for comparisons suggested by the data. For example, if it is used to compare the largest mean versus the smallest mean the chance is much larger than 5% that this particular difference will be found significant if the null hypothesis is true. Other tests have been devised which control the Type I error for comparisons suggested by the differences in means.

Note: If the 5% LSD procedure is used to compare all possible pairs of means from the same experiment, the probability of committing at least one Type I error is much greater than 5%. For one comparison, there is a 5% chance of declaring two randomly selected means significantly different. This is the **comparisonwise Type I error rate**. You would expect 5% of your comparisons to be significant if there were no real variety differences. In general, this rate gives the proportion of all differences in means that are expected to be found significant when there are no real differences.

Suppose there are six treatments in your experiment. If all $(6 \times 5)/2 = 15$ possible pairs of means are compared using the 5% LSD procedure, you would expect to get $15 \times 0.05 = 0.75$ (1 to the nearest whole number) significant differences if there were no real differences. The probability of obtaining at least one significant difference is $1 - (1 - 0.05)^{15} = 0.54$. This is (approximately) the **experimentwise Type I error rate**. It gets larger when there are more treatments in the experiment. If there are only two, the comparisonwise and experimentwise error rates are identical. The case against using the LSD test for comparing all possible pairs of means is the high experimentwise Type I error rate. Other tests have been devised which control this. They are not so powerful as the LSD test; they fail to detect as many differences that are real. Before describing some of these we will illustrate the LSD procedure with an example, and then use the same example to illustrate some alternative tests. The method of presentation is similar for the results of all the tests.

Example 13.1
Six varieties are replicated four times using a randomised block design. The variety means are:

Variety	V_1	V_2	V_3	V_4	V_5	V_6
Mean	50.8	69.2	25.0	89.0	76.2	74.0

and the residual mean square (RMS) in the ANOVA table is 144.54 on 15 degrees of freedom. The standard error of the difference between two means is

$$SED = \sqrt{\frac{2 \times RMS}{r}} = \sqrt{\frac{2 \times 144.54}{4}} = 8.501$$

13.2.1 Least Significant Difference

The least significant difference at the 5% level is

$$LSD\ (5\%) = t_{(15, 2.5\%)} \times SED = 2.131 \times 8.501 = 18.12$$

Similarly $LSD\ (1\%) = 2.947 \times 8.501 = 25.05$, and $LSD\ (0.1\%) = 4.073 \times 8.501 = 34.63$. We now describe three methods of presenting the results.

Method 1

Table 13.1 shows all the mean differences. A single asterisk (*) indicates the difference is greater in magnitude than the 5% LSD value and is therefore significant at the 5% level. A double asterisk (**) indicates significance at the 1% level, and a treble asterisk (***) denotes significance at the 0.1% level. Any differences less than the 5% LSD value are not significant (NS).

There are $(6 \times 5)/2 = 15$ possible comparisons. This method of presentation would become very cumbersome if the number of means was large.

Method 2

A space-saving presentation uses the underscore method. First, the means are ranked from smallest to largest, as in Table 13.2. A separate diagram is needed for each level of significance. We show only the one for the 5% level in Figure 13.1.

Table 13.1. Mean differences and their significance

	V_2	V_3	V_4	V_5	V_6
V_1	18.4*	−25.8**	38.2***	25.4**	23.2*
V_2		−44.2***	19.8*	7.0 NS	4.8 NS
V_3			64.0***	51.2***	49***
V_4				−12.8 NS	−15.0 NS
V_5					−2.2 NS

Table 13.2. Varieties ranked according to their means

Rank	1	2	3	4	5	6
Variety	V_3	V_1	V_2	V_6	V_5	V_4
Mean	25.0	50.8	69.2	74.0	76.2	89.0

25.0 (V$_3$)   50.8 (V$_1$)   69.2 (V$_2$)   74.0 (V$_6$)   76.2 (V$_5$)   89.0 (V$_4$)
 <u>$$</u> <u>$$</u>

Figure 13.1. Significant differences using the LSD (5%) method

Step 1. Represent these means on a horizontal axis not necessarily drawn to scale.

Step 2. Add the LSD of 18.12 to the lowest mean of 25.0 to obtain 43.12. This is less than the next mean (50.8), indicating that all the other varieties are significantly different from V$_3$ at the 5% level. Do not underscore.

Step 3. Now start at 50.8 and add the LSD of 18.12. This gives 68.92. This is less than 69.2 so V$_1$ is significantly different from all the other varieties. Do not underscore.

Step 4. Next, start at 69.2 and add 18.12 to obtain to 87.32. This is greater than 74.0 and 76.2. Draw a horizontal line underscoring V$_2$, V$_6$ and V$_5$. This indicates that these varieties do not differ significantly.

Step 5. Now add 18.12 starting at 74.0. This is greater than 76.2 and 89.0. Draw a line underscoring V$_6$, V$_5$ and V$_4$. This shows that these means are not significantly different.

As V$_6$ and V$_4$ are not significantly different, there is no need to add the LSD to the V$_5$ mean. The difference between the V$_5$ and V$_6$ means is less than the difference between the V$_6$ and V$_4$ means, so V$_5$ and V$_6$ do not need a separate underscore to indicate that they are not significantly different. Any two means underscored by the same line are not significantly different (Figure 13.1).

Method 3

This method uses underscoring to assign letters. The treatment (variety) means are arranged in order of magnitude and letters are added such that any two means having the same letter in common are not significantly different. The SAS presentation for this example is shown in Output 13.1. In scientific papers lower-case letters are usually employed.

Note that SAS has ranked the means from largest to smallest. The mean square error $(MSE) = 144.54$ is the residual (error) mean square (RMS). T = 2.13 is $t_{(15, 2.5\%)}$, the t-value on the residual degrees of freedom (df = 15) and is used to calculate the LSD of 18.12 at the 5% level. Using Output 13.1 you can see many differences are significant using the LSD procedure. Some protection is afforded by using the **Fisher's protected LSD** (FPLSD) procedure. This is the same as the ordinary LSD procedure, but no comparisons are made if the F-test used to test the null hypothesis of equal treatment effects shows a non-significant result. There are cases where the overall F-test is not significant but there is at least one comparison shown to be significant using the LSD.

13.2.2 Tukey's Honestly Significant Difference test (HSD) (Also Called Tukey's Studentised Range Test)

A criticism of the LSD is that if the comparisonwise error rate is fixed at, say 5%, the experimentwise error rate increases as the number of treatments increases. The HSD

Output 13.1 SAS output for LSD comparisons

```
                        The SAS System
                  Analysis of Variance Procedure

                    T tests (LSD) for variable: YIELD

NOTE:  This test controls the type I comparisonwise error rate not the
       experimentwise error rate.

                    Alpha = 0.05   df = 15   MSE = 144.5433
                          Critical Value of T = 2.13
                       Least Significant Difference = 18.12

           Means with the same letter are not significantly different.

              T     Grouping       Mean      N     VARIETY

                        A         89.000     4     V4
                        A
              B         A         76.200     4     V5
              B
              B         A         74.000     4     V6
              B
              B                   69.200     4     V2

                        C         50.800     4     V1

                        D         25.000     4     V3
```

test controls the experimentwise error rate. In repeated experiments, it controls the proportion of experiments in which at least one difference in means is found to be significant when there are no real differences. For any given level of significance a minimum significant difference (MSD) is calculated. It is greater than the corresponding LSD. Whereas the LSD does not depend on the number of treatments, the MSD increases with the number of treatments. We illustrate the procedure with Example 13.1, and as before use a 5% significance level.

Step 1. Consult the 5% tables of the Studentised range (Appendix 6). Find Q. This is the value in the row corresponding to the residual degrees of freedom, and the column corresponding to the number of means. For this example, consult row 15 and column 6 to obtain $Q = 4.595$.

Step 2. Calculate

$$MSD = Q \times \sqrt{\frac{RMS}{r}} = 4.595 \times \sqrt{\frac{144.54}{4}} = 4.595 \times 6.011 = 27.62$$

Step 3. Any differences greater than 27.62 are significant. The results are presented in Figure 13.2 and Output 13.2.

Note that the MSD value (27.62) is much greater than the LSD value (18.12). Hence fewer differences are significant using this test. V_2 is not significantly different from V_4 or V_1, whereas these differences were significant using the LSD.

25.0 (V₃)   50.8 (V₁)   69.2 (V₂)   74.0 (V₆)   76.2 (V₅)   89.0 (V₄)

Figure 13.2. Significant differences using Tukey's HSD test at the 5% level

Output 13.2 SAS output for Tukey's test

```
                          The SAS System
                     Analysis of Variance Procedure

         Tukey's Studentized Range (HSD) Test for variable: YIELD

NOTE:  This test controls the type I experimentwise error rate, but
       generally has a higher type II error rate than REGWQ.

              Alpha = 0.05   df = 15   MSE = 144.5433
              Critical Value of Studentized Range = 4.595
                 Minimum Significant Difference = 27.62

        Means with the same letter are not significantly different.

              Tukey Grouping       Mean     N    VARIETY

                           A      89.000    4    V4
                           A
                   B       A      76.200    4    V5
                   B       A
                   B       A      74.000    4    V6
                   B       A
                   B       A      69.200    4    V2
                   B
                   B              C  50.800    4    V1
                                  C
                                  C  25.000    4    V3
```

The disadvantage of Tukey's test is that it fixes the experimentwise error rate to a low value. As the number of treatments increases, MSD increases because Q increases. Hence, the power of the test to detect real differences decreases and too few differences are declared significant. If used in variety trials important new varieties may be missed.

13.2.3 Student–Newman–Keuls Test (SNK)

This is a multiple-range test and is intermediate in its power between the LSD and the HSD tests. For comparing the largest versus the smallest mean it is equivalent to Tukey's test. For comparing pairs of means adjacent in magnitude, it is equivalent to the LSD test.

Step 1. Arrange the t means in order of magnitude as before (Table 13.2).
Step 2. Compute the MSD $t - 1$ times; corresponding to $t, t - 1, \ldots 2$ means. In this example, there are six means, so find the MSD for 6, 5, 4, 3, and 2 means.

Use columns 6 to 2 inclusive in the 5% tables of the Studentised range (Appendix 6). Use row 15 as there are 15 residual degrees of freedom. Write down the six values of Q and hence calculate the six values of MSD using the formula in the last section. For six means the HSD is 27.62 as found for Tukey's test. Call this SNK(6). For five means the HSD = $4.367 \times 6.011 = 26.25$. Call this SNK(5). You should calculate the other SNK values (critical ranges) and confirm the entries in Table 13.3. Note that for two means, SNK(2) = LSD(5%) = 18.12.

Step 3. Compare the highest mean with the lowest using SNK(t). If the range is less than SNK(t), none of the differences are significant and no more comparisons are made. One line would underscore all the means. In this example, the range is $89.0 - 25.0 = 74$. This is not less than 27.62 so these two means are significantly different and more testing is required.

Step 4. Compare the two ranges of $t - 1$ adjacent ranked means with SNK($t - 1$). In this example, these ranges are $(76.2 - 25.0) = 51.2$ and $(50.8 - 89.0) = 38.2$. Both these ranges are greater than the SNK(5) value of 26.25. Hence V_3 is significantly different from V_5 and V_1 is significantly different from V_4.

Step 5. Continue in this way until all significant pairwise differences have been found. If at any stage a certain range is not significant, there is no need to compare means within that range. For this example, the ranges of four adjacent means are $74.0 - 25.0 = 49$, $76.2 - 50.8 = 25.4$, and $89.0 - 69.2 = 19.8$. As 19.8 is less than $24.50 = $ SNK(4), the four means, V_2, V_6, V_5 and V_4 are not significantly different and can be underscored.

Output 13.3 shows the SNK values which it calls critical ranges. More declarations of significance are made with this test than using Tukey's test, but fewer than with the LSD test.

13.2.4 Duncan's Multiple Range Test (DMRT)

This test has been very popular in the past for multiple comparisons but it has now fallen out of favour and most statisticians warn against its use. It is very similar to the SNK test, except that the critical ranges are smaller. These ranges are calculated by the same formula as for the SNK test but a special table is used instead of the Studentised range table. It is a more powerful test than SNK (it has a larger probability of detecting differences). Conversely, it has a large probability of one or more incorrect rejections. It is the same as the LSD test for comparing two adjacent means after ranking them. Output 13.4 shows how much smaller the critical ranges are than for the SNK test. Note also that the results are the same as for the LSD test

Table 13.3. Critical ranges for the SNK test

No. of means	2	3	4	5	6
Q from tables	3.014	3.674	4.076	4.367	4.595
SNK	18.12	22.08	24.50	26.25	27.62

Output 13.3 SAS output for the SNK test

```
                        The SAS System
                  Analysis of Variance Procedure

            Student-Newman-Keuls test for variable: YIELD

NOTE:  This test controls the type I experimentwise error rate under
       the complete null hypothesis but not under partial null
       hypotheses.

                   Alpha = 0.05   df = 15   MSE = 144.5433

Number of Means         2          3          4          5          6
Critical Range   18.120038  22.081809  24.501939  26.251296  27.620372

         Means with the same letter are not significantly different.

                SNK Grouping        Mean     N     VARIETY

                      A            89.000    4     V4
                      A
                      A            76.200    4     V5
                      A
                      A            74.000    4     V6
                      A
                      A            69.200    4     V2

                      B            50.800    4     V1

                      C            25.000    4     V3
```

in this example (Output 13.1). This is often true and is one of the reasons for its lack of popularity. It does not control the Type I experimentwise error rate sufficiently.

13.2.5 Waller–Duncan's Bayes MSD Test

In this test, all possible pairs of means can be compared by calculating a minimum significant difference (MSD) whose value depends on the result of the F-test in the analysis of variance. The higher the variance ratio (VR or F-ratio), the smaller the critical MSD, thereby increasing the power of the test. If the F-ratio is small, indicating little overall treatment differences, the MSD is increased, making it less likely that two means will be declared significantly different. The advantage of this test is that it is almost as powerful as the ordinary LSD, but does not suffer from having such a high probability of an experimentwise Type I error. For this example the F-ratio is 14.37 and the LSD is 18.12 (Output 13.1). On carrying out the Waller test using SAS (output not shown) the MSD value is 17.25. This is smaller than the LSD, reflecting the large F-ratio. However, the difference is not great enough to alter the conclusions arrived at using the LSD test.

After deliberately altering some of the yields in Example 13.1 and rerunning the SAS program (data and output not shown) we obtained an F-ratio of 1.86 with a corresponding P-value of 0.162. You would therefore conclude that there are no significant variety differences with respect to yield. The revised variety means were:

Output 13.4 SAS output for Duncan's Multiple Range test

```
                          The SAS System
                    Analysis of Variance Procedure
              Duncan's Multiple Range Test for variable: YIELD

   NOTE:  This test controls the type I comparisonwise error rate, not
          the experimentwise error rate

                    Alpha = 0.05   df = 15   MSE = 144.5433

  Number of Means         2        3        4        5        6
  Critical Range       18.12    18.99    19.54    19.91    20.17

        Means with the same letter are not significantly different.

              Duncan Grouping        Mean      N     VARIETY

                           A       89.000      4      V4
                           A
                     B     A       76.200      4      V5
                     B     A
                     B     A       74.000      4      V6
                     B
                     B             69.200      4      V2

                           C       50.800      4      V1

                           D       25.000      4      V3
```

Variety	V_3	V_1	V_2	V_4	V_6	V_5
Mean	50.2	57.6	58.3	60.4	66.5	68.7

However, an LSD value of 14.74 was obtained which would lead you to conclude that V_3 is significantly different from V_5 and V_6. The Waller output gave a MSD value of 18.52 and this is larger than the LSD to reflect the small F-ratio. Using this test, none of the mean differences are declared significant, a conclusion consistent with the F-test.

13.2.6 Summary of Multiple-comparison Tests

The number of significant differences between two means for Example 13.1 were as follows: LSD (10), Tukey (5), SNK (9), DMR (10), WDB (10).

You are probably wondering which of the above tests you should use. You are in good company as statisticians cannot agree among themselves. For simplicity, Fisher's protected LSD has many proponents, but the Waller–Duncan procedure is gaining favour. For more discussion on this controversial topic you should read Carmer and Walker (1982) who discuss the relative merits of some of the tests used for multiple comparisons. Some computer packages give you the option of carrying out the tests we have mentioned as well as others. You should avoid trying all the options and using only the one that gives you the result you are seeking. Ideally you should plan the comparisons in advance. They should be independent of each other and number no more than the number of treatments minus one.

13.2.7 Least Significant Increase (LSI)

In many plant breeding trials new varieties are compared with standard varieties. The purpose is to find new acquisitions that give higher yields or quality than the standards. You lose interest if a new variety gives a significantly lower result. A criterion that can be used to test whether each new variety is significantly higher yielding than a standard is the LSI procedure which is the one-sided version of the LSD procedure. It consists of a series of one-tailed t-tests. If each variety is equally replicated, a single LSI value can be used. For the tests at the 5% level, the formula is:

$$LSI = t_{(Rdf, 5\%)} \times \sqrt{\frac{2 \times RMS}{r}} = t_{(Rdf, 5\%)} \times SED$$

where Rdf is the residual degrees of freedom and RMS is the residual mean square from the ANOVA table.

In Example 13.1, six varieties are replicated four times in a randomised block design. Suppose you wish to test at the 5% level whether any of the varieties outyield the standard (V_1). The standard error of the difference between two means (SED) = 8.501, thus the least significant increase at the 5% level is:

$$LSI\,(5\%) = t_{(15, 5\%)} \times SED = 1.753 \times 8.501 = 14.90$$

Any variety with a mean yield greater than $\bar{x}_s + LSI$ (where $\bar{x}_s$ is the mean of the standard) significantly outyields the standard. In this example, $\bar{x}_s + LSI = 50.8 + 14.9 = 65.7$. Hence varieties V_2, V_4, V_5 and V_6 significantly outyield the standard.

13.2.8 One-sided Confidence Intervals

Confidence intervals give more information than significance tests. For Example 13.1, 95% upper confidence intervals for the increases in mean yields over that of the standard can be given. Find the mean increase and subtract the LSI (5%) value. For V_2, the mean increase is $(69.2 - 50.8) = 18.4$. Subtract 14.9 to obtain 3.5. Hence you are 95% confident that V_2 outyields V_1 by at least 3.5 units.

13.3 TREATMENTS WITH STRUCTURE (FACTORIAL STRUCTURE)

Example 13.2
Consider the following set of treatments:

T_1: No fertiliser applied to variety A
T_3: No fertiliser applied to variety B
T_5: Nitrogen applied to variety A

T_7: Nitrogen applied to variety B

T_2: Potassium applied to variety A
T_4: Potassium applied to variety B
T_6: Nitrogen and potassium applied to variety A
T_8: Nitrogen and potassium applied to variety B

These eight treatments possess a 2^3 factorial structure (Chapter 12). However, this is often ignored by researchers when analysing data and presenting results. A presentation after applying a LSD analysis to the means is given in Table 13.4. This does not take into account the factorial structure.

Suppose a RBD with four blocks had been chosen and that the ANOVA table (Table 13.5) was obtained. This is only a preliminary analysis because the factorial treatment structure is ignored.

From tables, $F_{(7, 21, 0.1\%)} = 5.56$. As the calculated VR is 16.3 you conclude there are very highly significant differences between the treatments. After obtaining the SED it is possible to compare all 28 possible pairs of treatments. However, this is not appropriate for this example.

Because Example 13.2 was designed to find the effect of nitrogen and potassium on the yields of the two varieties, the factorial structure should determine the method of analysis and presentation of results. Table 13.6 shows an appropriate presentation of the treatment means.

The SED value of 0.42 can be used to compare any two means in Table 13.6. However, due to the factorial structure other comparisons using other SED values may be more appropriate depending on which, if any, interactions are significant. The interactions, and if appropriate, the main effects should be tested. This is carried out by partitioning the treatment sum of squares and degrees of freedom as in Table 13.7.

Table 13.4. An inappropriate presentation for the results of Example 13.2

Treatment	Mean yield	
T_1	4.3 a	
T_2	4.7 a b	
T_3	4.8 a b	
T_4	7.4	e
T_5	5.2 b c	
T_6	5.8 c d	
T_7	6.4 d	
T_8	7.4	e

Treatments with the same letter in common are not significantly different (P = 0.05).

Table 13.5. Preliminary ANOVA table

Source	DF	SS	MS	VR
Blocks	3	27.55	9.18	
Treatments	7	41.12	5.87	16.3
Residual	21	7.59	0.36	
Total	31	76.26		

Raw data not given.

Table 13.6. Presentation of treatment means from a three-factor experiment (Example 13.2)

	Mean yield (t/ha)			
	Variety A		Variety B	
	N_0	N_1	N_0	N_1
K_0	4.3	5.2	4.8	6.4
K_1	4.7	5.8	7.4	7.4

SED = 0.42

Table 13.7. ANOVA table showing partitioning of the treatment sum of squares and degrees of freedom for Example 13.2

Source	DF	SS	MS	VR
Variety (V)	1	18.00	18.00	50.0***
Nitrogen (N)	1	6.48	6.48	18.0***
V × N	1	0.08	0.08	0.22
Potassium (K)	1	10.58	10.58	29.4***
K × V	1	3.38	3.38	9.4***
K × N	1	0.98	0.98	2.7
K × V × N	1	1.62	1.62	4.5*
Treatment	7	41.12	5.87	16.3

The values of VR for the main effects and interactions are calculated by dividing the corresponding MS values by the RMS (0.36) from Table 13.5 and comparing with F-tables on 1 and 21 degrees of freedom.

Note: The three-factor interaction is significant and this can be interpreted several ways. You can conclude that K and V interact differently at each level of N, or that N and V interact differently at each level of K, or that the interaction between K and N depends on the variety. You can see from Table 13.6 that the effect of N is similar at both levels of K for variety A. For variety B, the effect of N is fairly large in the absence of K, but has no effect in the presence of K. The SED for testing these effects is the SED for comparing any two means in Table 13.6. It is calculated as

$$SED_{V \times K \times N} = \sqrt{\frac{2 \times RMS}{r}} = \sqrt{\frac{2 \times 0.36}{4}} = 0.42$$

- The ANOVA table suggests that there are very large main effects, but it can be misleading to quote the overall main effect means when there are significant interactions.
- There are various ways the results can be presented graphically. Figure 13.3 shows two possible bar charts. The first emphasises the effects of K for the variety × nitrogen combinations while the second emphasises the effects of N for the variety × potassium combinations.

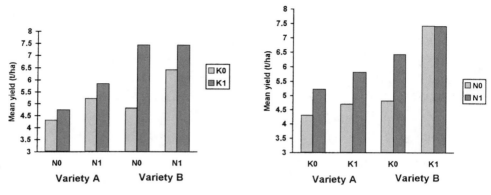

Figure 13.3. Two ways of presenting Table 13.6 graphically

- If the three-factor interaction had not been significant, you would present the K × V table of means (Table 13.8) as this interaction was highly significant. The SED for comparing two of these means is

$$SED_{K \times V} = \sqrt{\frac{2 \times RMS}{8}} = 0.30$$

(Each K × V mean is the average of 8 original yields.)
- For variety A, the potassium effect is small (0.5), whereas for variety B, it is much larger (1.8).
- There was no two-factor interaction with N, so if the three-factor interaction had not been significant, the overall nitrogen means (Table 13.9) would also have been presented. The SED for comparing these two means is

$$SED_N = \sqrt{\frac{2 \times RMS}{16}} = 0.21$$

(Each N mean is the average of 16 original yields.)

In a three-factor experiment with factors A, B and C there are three one-way tables of means, three two-way tables and one three-way table. Each of these seven tables has its own SED value. All these tables with their SED values and LSD values (if appropriate) can be obtained with Genstat. As you will appreciate, there are many permutations of the ways of presenting the results in tables and graphs. The tables

Table 13.8. Potassium × Variety table of means for Example 13.2

	K_0	K_1
Variety A	4.75	5.25
Variety B	5.6	7.4

SED = 0.30.

Table 13.9. Overall nitrogen means for Example 13.2

Without nitrogen	With nitrogen
5.3	6.2

SED = 0.21.

you present will depend on which main effects and interactions are significant. However, beware of using main effect SEDs to compare main effect means in the presence of significant interaction. For further discussion see Baker (1980) and Morse and Thompson (1981).

13.4 TREATMENTS WITH STRUCTURE (LEVELS OF A QUANTITATIVE FACTOR)

The treatments in an experiment could be different levels of a fertiliser such as nitrogen or potassium, planting depths, or row spacings, etc. Thus, a factorial experiment may contain a mixture of quantitative and qualitative factors. For example, in a 2 × 5 factorial experiment, the two levels of factor A may be two different varieties, and the five levels of factor B may be five different concentrations of nitrogen fertiliser. In this case, the main interest would be in discovering if the response to nitrogen varied with variety. A graph showing the two response curves would be appropriate. You may wish to fit straight lines or curves describing the responses. We begin by studying a single-factor experiment.

13.4.1 Testing for a Linear Trend in Treatment Means

As there are several Y-values for each X, an estimate of the residual variance (σ^2) can be obtained from the preliminary analysis of variance table. This table is modified by partitioning the treatment sum of squares into a part due to linear regression and a part due to deviation from regression. The modified table is then used to test for linear trend.

Example 13.3
An experiment testing the effects of six levels of nitrogen fertiliser between 0 and 125 kg/ha was carried out on winter wheat. A randomised blocks design with four replications was used. The grain yields (t/ha) are shown in Table 13.10.

In Chapter 7 a straight line was fitted using these treatments means as the Y-values and the nitrogen levels as the X-values. In Chapter 8 a quadratic curve was fitted. However, we did not take into account the variation within the treatment means. To allow for this some researchers would carry out a regression analysis on all 24 yields as the Y-values with the X-values being the nitrogen levels each repeated four times (four Y-values for each X-value). When each treatment is equally replicated, as in this example, both of these approaches give the correct fitted equation ($\hat{Y} = 4.0714 + 0.010057X$). However, they give the wrong *t*-values for testing whether the slope is significantly different from zero. When using only the means the *t*-value is 3.06 on 4 df and the corresponding *P*-value is 0.038. The R-square value is 70% (Output 7.2). As an exercise, you should confirm that when using all 24 yields in the regression analysis the *t*-value is 4.30 on 22 df, and the *P*-value is < 0.001. The regression sum of squares is 4.4251 and the R-square value is 45.7%.

The correct approach is to use the variation in the *Y*-values for each *X* (after allowing for block effects) to estimate the residual variance (σ^2). This is the residual mean square (0.078) in the **preliminary** analysis of variance table (Table 13.11) and is used to test for a linear trend in the **modified** table shown in Output 13.5.

The treatment sum of squares ($TreatSS = 6.3200$) represents the variation in the treatment means after allowing for blocks. **Part of this variation is due to regression** ($TreatSS = RegSS + DevSS$).

The regression sum of squares ($RegSS = 4.4251$) is obtained from the regression analysis on all the 24 yields given in Table 13.10. The sum of squares due to deviation from regression ($DevSS$) represents that part of $TreatSS$ not accounted for by linear regression. It is obtained by subtraction ($TreatSS - RegSS = 1.8949$). The analysis of variance table is then modified to show the partitioning of the treatment sum of squares and various *F*-tests carried out (Output 13.5).

Interpretation

The nitrogen means are significantly different ($VR = 16.21$, $P < 0.001$). You can also confirm this by using the *F*-tables, as the calculated VR is greater than the *F*-table value of 7.567 on 5 and 15 degrees of freedom at the 0.1% level. A significant amount of the variation in the means is due to a linear trend ($VR = 56.73$, $P < 0.001$). The corresponding *t*-value is 7.53 (the square root of 56.73) on 15 df and contrasts with that obtained using only the means (3.06 on 4 df) or the 24 yields (4.30 on 22 df) in the regression analysis.

Under nitrogen contrasts in Output 13.5, the slope is shown as Lin = 0.0101. To obtain the standard error of the slope (0.00134) divide the slope (0.0101) by the *t*-value of 7.53. The ss.div value of 43750 is the corrected sum of squares of all 24 *X*-values ($Sxx = 43750$). It is also four times the Sxx value of the six separate nitrogen levels because each *X*-value is repeated four times. The standard error of the slope can also be calculated as the square root of the residual mean square divided by ss.div:

$$SE = \sqrt{\frac{0.0780}{43750}} = 0.00134$$

The R-square value is $RegSS/TreatSS = 4.42514/6.3200 = 0.700$ (70%). This is the same as that obtained in Output 7.2 when only the six means were used in the analysis. Note also that because each mean is calculated from four yields, the regression sum of squares (4.42514) is four times that shown in Output 7.2 (4×1.1063).

You can test for lack of fit to a straight line by finding $VR = DevMS/RMS$. A significant result indicates that a significant amount of the variation between the

COMPARISON OF TREATMENT MEANS

treatment means is not explained by linear regression and you should look for a different relationship. Output 13.5 shows a significant lack of fit to a straight line ($VR = 6.07$, $P = 0.004$). Figure 7.4 also suggests this.

It can be shown that $RegSS$ cannot be larger than the treatment sum of squares ($TreatSS$). That is, $RegSS \leqslant TreatSS$. If $RegSS = TreatSS$ this implies that the treatment means lie exactly on a straight line. However, even if they do, this does not necessarily imply significant regression as the treatment means (after adjusting for blocks) may be based on widely scattered values as shown by the size of the residual sum of squares. In this case the R-square calculated using $RegSS/TreatSS$ would be 100% whereas that based on all the yields ($RegSS/TotalSS$) would be quite low. For this example the two R-square values are 70% and 45.7% respectively.

Note: This method still applies to a completely randomised design. However, in the case of unequal replications the equation fitted to all the points will not be the same as that fitted to the means.

Table 13.10. Yields of winter wheat (Example 13.3)

Block	\multicolumn{6}{c}{Nitrogen level (kg/ha)}					
	0	25	50	75	100	125
1	3.6	4.8	4.4	5.3	4.8	5.0
2	4.1	5.1	5.2	5.9	5.5	5.4
3	3.2	4.0	4.6	4.6	5.2	4.8
4	3.9	3.9	4.8	5.0	5.1	4.6
Mean	3.70	4.45	4.75	5.20	5.15	4.95

Table 13.11. Preliminary ANOVA table for data given in Table 13.10

Source	DF	SS	MS	VR
Blocks	3	2.190	0.730	
Treats	5	6.320	1.264	16.21
Residual	15	1.170	0.078	
Total	23	9.680		

13.4.2 Testing for a Quadratic Trend in Treatment Means

A quadratic curve has already been fitted to the means of Example 13.3 (Section 8.3). The fitted equation was

$$\hat{Y} = 3.7054 + 0.03202X - 0.0001757X^2$$

and the regression sum of squares $= 1.55655$. As each mean is based on four values this figure is multiplied by 4 to give the quadratic model regression sum of squares

Output 13.5 Analysis of variance for testing a linear trend (Genstat output for Example 13.3)

```
***** Analysis of variance *****

Variate: yield

Source of variation    d.f.    s.s.      m.s.      v.r.      F pr.

block stratum           3     2.19000   0.73000    9.36

block. *Units* stratum
nitrogen                5     6.32000   1.26400   16.21    <0.001
   Lin                  1     4.42514   4.42514   56.73    <0.001
   Deviations           4     1.89486   0.47371    6.07     0.004
Residual               15     1.17000   0.07800

Total                  23     9.68000


*** nitrogen contrasts ***

Lin    0.0101    s.e. 0.00134    ss.div. 43750.


*** Standard errors of differences of means ***

Table           nitrogen
rep.               4
d.f.              15
s.e.d.          0.1975
```

for all 24 points as $RegSS(Quad) = 6.2262$ on 2 df. This value can also be obtained by using the computer to fit the quadratic model to the 24 points using the yields as the Y-values and the nitrogen levels as the X-values.

The deviation sum of squares from the quadratic model is

$$DevSS(Quad) = TreatSS - RegSS(Quad) = 6.3200 - 6.2262 = 0.0938$$
$$\text{on } (5 - 2) = 3 \text{ df}$$

From Output 13.5 it is seen that the linear regression model sum of squares for the 24 points is $RegSS(Lin) = 4.4251$ on 1df. Hence the extra sum of squares due to quadratic over and above linear regression is $6.2262 - 4.4251 = 1.8011$ on 1 degree of freedom.

To test if a quadratic curve is significantly better than a straight line, find

$$VR = \frac{RegSS(Quad) - RegSS(Lin)}{RMS} = \frac{1.8011}{0.0780} = 23.09 \text{ on } (1, 15) \text{ df}$$

From F-tables, $F_{(1, 15, 5\%)} = 4.543$. As 23.09 is much larger, you conclude that a quadratic curve is a better fit than a straight line. These calculations are included in Output 13.6 which shows the ANOVA table of Output 13.5 modified to include the test of the quadratic model.

Interpretation

Note that an extra row, Quad, has been added. This shows that when a quadratic model is fitted the regression sum of squares is 1.8011 greater than when a linear

Output 13.6 Analysis of variance for testing linear and quadratic trends (Genstat output for Example 13.3)

```
***** Analysis of variance *****
Variate: yield

Source of variation   d.f.      s.s.       m.s.      v.r.    F pr.
block stratum          3      2.19000    0.73000    9.36

block.*Units* stratum
nitrogen               5      6.32000    1.26400   16.21    <0.001
  Lin                  1      4.42514    4.42514   56.73    <0.001
  Quad                 1      1.80107    1.80107   23.09    <0.001
  Deviations           3      0.09379    0.03126    0.40     0.754
Residual              15      1.17000    0.07800

Total                 23      9.68000

*** nitrogen contrasts ***
Lin    0.0101      s.e. 0.00134      ss.div. 43750.
Quad  -0.000176    s.e. 0.0000366    ss.div. 5.83E+07
```

model is fitted. This increase is significant ($VR = 23.09$, $P < 0.001$). You therefore conclude that a quadratic curve is a significantly better fit than a straight line. The regression sum of squares due to the quadratic model $RegSS(Quad)$ is found by adding the s.s. values in the Lin and Quad rows ($4.4251 + 1.8011 = 6.2262$).

Note also that the entries in the Deviations row have been changed. This shows that the sum of squares due to lack of fit from the quadratic model is only 0.0938 and not significant ($VR = 0.40$, $P = 0.754$). Hence the quadratic curve adequately explains the variation in mean yields (Figure 13.4).

The treatment sums of squares (6.3200) has been partitioned into that due to the

- Linear model (4.4252)
- Extra sum of squares accounted for by the quadratic model (1.8010)
- Deviation from the quadratic model (0.0938).

The degrees of freedom are similarly partitioned ($5 = 1 + 1 + 3$).

Under nitrogen contrasts in Output 13.6, the quadratic coefficient is shown as $Quad = -0.000176$ and its standard error is 0.0000366. On dividing the former value by the latter you obtain the t-value of -4.81 on 15 df which can also be used to test whether the quadratic coefficient is significantly different from zero. This is equivalent to the test in the ANOVA table as the VR (23.09) is the square of the t-value. The corresponding t-value (when only the means are fitted) is -7.59 on 3 df (Output 8.1). You should also confirm, by fitting a quadratic model to all 24 points ignoring the treatment and block structure, that the t-value is -3.31 on 21 df.

Note: This method makes the homogeneity of variance assumption, that at each nitrogen level the variance of the residuals is constant. Figure 13.4 shows this is a valid assumption.

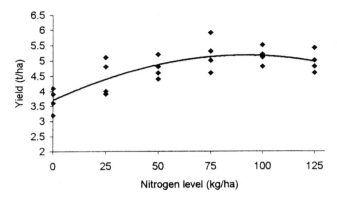

Figure 13.4. Graph of yield versus nitrogen level with fitted quadratic curve

R-square

The R-square value is obtained by dividing $RegSS(Quad) = 6.2262$ by $TreatSS = 6.3200$ to obtain 0.985 (98.5%). This compares with 70% using the linear model. This value of 98.5% reflects the fit of the treatment means to a quadratic curve. However, while this fit may be almost perfect, there may be considerable variation within treatments. To obtain the R-square value appropriate to fitting the 24 points ignoring the treatment and block structure divide $RegSS(Quad) = 6.2262$ by $TotalSS = 9.68$ to obtain 0.643 (64.3%).

13.4.3 Polynomial Fitting

In the previous two sections we showed how to fit a linear and a quadratic model (polynomials of order one and two respectively). As the treatment degrees of freedom are five, polynomials up to order four could be fitted. A fifth-order polynomial would produce a perfect fit to the treatment means as there would be no degrees of freedom left for deviation. In principle you could proceed to fit a cubic and a quartic to the data of Example 13.3. The analysis of variance table would be further modified to show how the treatment sum of squares is partitioned into sequential sums of squares. The first entry would be the regression sum of squares due to fitting the linear model. The next would be the regression sum of squares after fitting the quadratic model minus the regression sum of squares due to fitting the linear model. The third entry would be the regression sum of squares after fitting the cubic model minus the regression sum of squares due to fitting the quadratic model. The fourth entry would be the regression sum of squares after fitting the quartic model minus the regression sum of squares due to fitting the cubic model. The final entry would be the sum of squares due to deviation from the quartic model.

Note: If you have t treatments you cannot fit a polynomial of order more than $t - 1$. In particular you should not attempt to fit a quadratic curve if you only have three treatments as you would obtain a perfect fit. It is wise to have several more treatments than the order of polynomial you are considering fitting. Also be aware that the R-square value is bound to increase as higher-order polynomials are fitted. Problems of interpretation usually arise when high-order polynomials are fitted. Would you be able to justify fitting a quartic equation? It may be a good fit just because you have only six treatments.

Presentation

When response curve fitting is appropriate, comparing treatment means using LSDs is inappropriate. It may be better to include the pooled standard error of a treatment mean (Section 10.8.2), to indicate the precision of the means. A good policy is to present a graph of the data showing the individual points with the curve fitted to the treatment means (Figure 13.4). The reader can then see the variation.

13.4.4 Regression in a Two-factor Experiment

In Section 12.5 we discussed a factorial experiment with five levels of nitrogen fertiliser (N) and three levels of plant growth regulator (PGR or P). Figure 12.9 suggests that it may also be useful to test for overall linear and quadratic trends in the nitrogen means, and for the same trends at each level of P. Output 13.7 (which is a modification of Output 12.1) shows the results after testing for linear trends.

Interpretation

The extra output shows there is a significant overall linear trend in the nitrogen means ($VR = 91.35$, $P < 0.001$). Assuming the nitrogen levels are equally spaced, an increase from one level to the next is expected to increase the yield by 0.141 kg/plot when averaged over the three levels of P. However, the VR for Deviations is 11.86, indicating a significant lack of fit from a straight line. As a result you should test for a quadratic or other trend.

Due to the significant interaction between P and N ($VR = 6.47$, $P < 0.001$), you should also test for a linear trend for each level of P separately. There is significant P.Lin interaction ($VR = 5.05$, $P = 0.013$). This indicates that when straight lines are fitted to the nitrogen means for each level of P, the slopes of these lines are significantly different (Figure 12.9 also suggests this):

- The slope for P_1 is $0.141 - 0.011 = 0.130$
- The slope for P_2 is $0.141 + 0.062 = 0.203$
- The slope for P_3 is $0.141 - 0.051 = 0.090$

However, the above interpretation is misleading because the deviations of the nitrogen means about these separate lines show significant lack of fit ($VR = 6.94$, $P < 0.001$). You should try fitting separate quadratic curves.

When both factors in a factorial experiment are quantitative, simultaneous testing for linear, quadratic and possibly higher-order polynomial trends may be carried out. However, the output is complicated and interpretation is very difficult.

Output 13.7 Testing for linear trends in a factorial experiment (data from Example 12.3)

```
***** Analysis of variance *****
Variate: yield
Source of variation   d.f.      s.s.       m.s.      v.r.     F pr.
block stratum           2     0.06400    0.03200    1.63

block.*Units* stratum
P                       2     0.16933    0.08467    4.32     0.023
N                       4     2.49022    0.62256   31.73    <0.001
  Lin                   1     1.79211    1.79211   91.35    <0.001
  Deviations            3     0.69811    0.23270   11.86    <0.001
P.N                     8     1.01511    0.12689    6.47    <0.001
  P.Lin                 2     0.19822    0.09911    5.05     0.013
  Deviations            6     0.81689    0.13615    6.94    <0.001
Residual               28     0.54933    0.01962
Total                  44     4.28800

*** N contrasts ***
Lin    0.141    s.e. 0.0148    ss.div. 90.0

*** P.N contrasts ***
       P1         P2         P3
    -0.011      0.062     -0.051
```

Note: It is common for an analysis to be carried out assuming levels are equally spaced when they are not. It is also common for a quadratic trend to be tested for when there only three levels of a factor. In this case a quadratic curve is a perfect fit to the means. However, there may be significant overall lack of fit due to large variations in the values which make up the means.

13.5 TREATMENTS WITH STRUCTURE (CONTRASTS)

In an analysis of variance used to compare treatment means, a significant variance ratio (VR) indicates the presence of treatment differences. Pairs of means can be compared using an LSD analysis but this may not be appropriate (Section 13.2).

The method of contrasts can be used to compare one group of treatments with another. For example, with five treatments you may wish to compare treatment A with the average of B, C, D and E or the average of A and B with the average of C, D and E. For each contrast of interest, a sum of squares can be calculated with 1 degree of freedom and compared with the residual mean square (RMS) in the ANOVA table to determine if that contrast is significant.

This method is a special case of linear regression where the Y-values are the treatment means. The X-values are the contrast coefficients denoted by c_1, c_2, etc. They are defined to add to zero, i.e. $\Sigma c = 0$. If there are five treatments, A, B, C, D and E, each with r replications the contrast coefficients are c_1, c_2, c_3, c_4 and c_5. In the

COMPARISON OF TREATMENT MEANS

case of comparing the average of A and B with the average of C, D and E the c-values could be 3, 3, −2, −2, −2. Note that they add to zero.

In general, for five treatments each replicated r times, the Y-values are the treatment means given by

$$\frac{T_1}{r} \quad \frac{T_2}{r} \quad \frac{T_3}{r} \quad \frac{T_4}{r} \quad \frac{T_5}{r} \quad \text{where the } T\text{s are the treatment totals.}$$

The X-values are c_1, c_2, c_3, c_4 and c_5.

Using the notation of Sections 7.4 and 7.6

$$Sxx = \Sigma X^2 - \frac{(\Sigma X)^2}{5}$$

In this case $\Sigma X = \Sigma c = 0$, so $Sxx = \Sigma X^2 = \Sigma c^2$:

$$Sxy = \Sigma XY - \frac{(\Sigma X)(\Sigma Y)}{5} \quad \text{but } \Sigma X = \Sigma c = 0$$

Hence

$$Sxy = \Sigma XY = \Sigma \left(c \times \frac{T}{r} \right) = \frac{(\Sigma cT)}{r}$$

and so the slope or contrast coefficient is given by

$$\hat{\beta} = \frac{Sxy}{Sxx} = \frac{\Sigma cT}{r(\Sigma c^2)}$$

The regression sum of squares for the treatment means ($RegSS$) is

$$\hat{\beta} \times Sxy = \frac{\Sigma cT}{r(\Sigma c^2)} \times \frac{\Sigma cT}{r}$$

To find the corresponding sum of squares for the contrast based on all $5 \times r$ observations the last expression is multiplied by r to give

$$ContrastSS = \frac{(\Sigma cT)^2}{r(\Sigma c^2)}$$

Example 13.4

Suppose the following treatment totals are each based on $r = 4$ replicates:

T_1	T_2	T_3	T_4	T_5
25.1	27.3	29.2	31.3	30.8

To compare the average of treatments 1 and 2 with the average of treatments 3, 4 and 5 the contrast coefficients could be (−3, −3, +2, +2, +2) or (+3, +3, −2, −2, −2) or (+1.5, +1.5, −1, −1, −1) or any other grouping where the first two coefficients are 3/2 times the other three and the sum of the coefficients is zero. Using the second grouping,

$\Sigma c^2 = 9 + 9 + 4 + 4 + 4 = 30$ and

$\Sigma cT = 3 \times 25.1 + 3 \times 27.3 - 2 \times 29.2 - 2 \times 31.3 - 2 \times 30.8 = 25.4$

Therefore

$$Contrast SS = \frac{(\Sigma cT)^2}{r(\Sigma c^2)} = \frac{(25.4)^2}{4 \times 30} = 5.3763$$

Note that if one of the other sets of coefficients is used, for example 1.5, 1.5, -1, -1, -1, the answer is still 5.3763. To test if this sum of squares is significant, divide it by the residual mean square from the ANOVA table (not given). The resulting VR on 1 and the residual degrees of freedom is compared with F-tables in the usual way. The sum of squares for any particular contrast will be less than the treatment sum of squares. Ideally if you are interested in more than one contrast you should ensure they are orthogonal, planned in advance and are no more than the number of treatments minus one.

13.5.1 Orthogonal Contrasts

Two contrasts are orthogonal if the product of their coefficients adds to zero. If there are t treatments, there are $(t-1)$ orthogonal contrasts, one for each of the $(t-1)$ degrees of freedom for treatments. These contrasts are independent and their sums of squares add to the treatments sum of squares.

Example 13.5

The effect of five different broad-leafed weed herbicides (A to E) on the yield of barley (kg/plot) was studied in a field experiment. A randomised complete block design with four replications was used. The yields are summarised in Table 13.12, and Table 13.13 shows the preliminary analysis of variance table.

From F-tables $F_{(4, 12, 5\%)} = 3.259$. The calculated value of 10.91 is much greater so conclude that there are real treatment differences. The LSD analysis could be carried out to discover which pairs of treatments differ, but in this case a different analysis is more appropriate because the treatments were:
- A is a standard broad-leafed weed control herbicide.
- B is a newly approved chemical.
- C is the same chemical as B, with an additive for improved weed leaf penetration.
- D is the same chemical as B, with the additive and third chemical at a high concentration.
- E is the same chemical as B, with the additive and the third chemical at a low concentration.

Table 13.14 shows the contrasts of interest together with their contrast coefficients. Confirm that for each contrast the sum of the coefficients is zero and that for any pair of contrasts the sum of the products of their coefficients is also zero. Hence these form a set of **orthogonal contrasts**.

For each contrast a sum of squares can be found using the formula

$$ContrastSS = \frac{(\Sigma cT)^2}{r(\Sigma c^2)}$$

For the first contrast

$$\Sigma cT = 4 \times 257 + (-1) \times 253 + (-1) \times 283 + (-1) \times 299 + (-1) \times 265$$
$$= 1028 - 253 - 283 - 299 - 265 = -72$$

and

$$\Sigma c^2 = 16 + 1 + 1 + 1 + 1 = 20$$

Therefore

$$ContrastSS = \frac{(-72)^2}{4 \times 20} = \frac{5184}{80} = 64.8$$

You should verify that the sums of squares for the other three contrasts are 161.333, 0.1667 and 144.5. These are shown in Output 13.8 which gives the ANOVA table expanded for testing the significance of the contrasts.

Interpretation

Contrasts 2 and 4 are very highly significant (high VRs and P-values less than 0.001). Contrast 1 is significant at the 5% level ($VR = 7.62$, $P = 0.017$). Contrast 3 is not significant (small VR, P greater than 0.05). The non-significance of contrast 3 indicates that the mean of herbicide C (70.75) is not significantly different from the average of the means of herbicides D and E (70.5).

The coefficient of -0.90 for the first contrast is found as follows. This contrast compares the mean of A with the average of the other four means. The herbicide A mean is 64.25 and the mean of the other four means is 68.75. A straight line is fitted through these two means as Y-values. The two corresponding X-values are $+4$ and -1 respectively. Hence, the slope of this line is $(64.25 - 68.75)/5 = -0.9$.

For $X = +4$ there is only one Y-value (64.25). For $X = -1$, there are four Y-values (63.25, 70.75, 74,75 and 66.25). This is summarised as follows:

X	-1	-1	-1	-1	$+4$
Y	63.25	70.75	74.75	66.25	64.25

If you carry out a simple linear regression you will find the regression coefficient (slope) is -0.9 and the regression sum of squares is 16.2. As each Y-value is the mean of four numbers the contrast sum of squares (64.8) is obtained by multiplying 16.2 by 4.

The slope can also be obtained by the formula given at the beginning of Section 13.5:

$$Slope = \frac{\Sigma cT}{r(\Sigma c^2)} = \frac{-72}{4 \times 20} = -0.90$$

The standard error of the slope of the first contrast is 0.326. This is

$$\sqrt{\frac{RMS}{SSdiv1}} = \sqrt{\frac{8.50}{80.0}} = 0.326$$

where $SSdiv1 = r(\Sigma c^2) = 4 \times 20 = 80$

If you divide the slope by its standard error you get a t-value of -2.761. On squaring this you get the VR of 7.62 shown in the ANOVA table. The slopes for the other three contrasts are -1.83, 0.08 and 4.2. Similar calculations can be carried out to confirm their $SSdiv$ values and standard errors.

Note: If there are t treatments it is possible to find a set of $(t-1)$ orthogonal contrasts whose sums of squares add to the treatment sum of squares. In practice, some of these contrasts may have no meaning in the context of the experiment. Furthermore, the contrasts you are interested in may not be orthogonal. In this case their sums of squares may add to more than the treatment sum of squares. It is still possible to test for their significance individually but you must be aware that these tests will not be independent of each other. When planning an experiment you should try to ensure that the contrasts of interest are orthogonal. They do not have to be a complete set. For instance, in an experiment with six treatments, only three out of the five orthogonal contrasts may have a biological interpretation.

Output 13.8 ANOVA table showing tests of orthogonal contrasts for Example 13.5 (Genstat)

```
***** Analysis of variance *****
Variate: yield
Source of variation   d.f.        s.s.        m.s.      v.r.    F pr.
block stratum          3         99.750      33.250     3.91
block.*Units* stratum
herbicide              4        370.800      92.700    10.91   <0.001
  A vs B, C, D, E      1         64.800      64.800     7.62    0.017
  B vs C, D, E         1        161.333     161.333    18.98   <0.001
  C vs D, E            1          0.167       0.167     0.02    0.891
  D vs E               1        144.500     144.500    17.00    0.001
Residual              12        102.000       8.500
Total                 19        572.550

*** herbicide contrasts ***
A vs B, C, D, E    -0.90    s.e. 0.326    ss.div. 80.0
B vs C, D, E       -1.83    s.e. 0.421    ss.div. 48.0
C vs D, E           0.08    s.e. 0.595    ss.div. 24.0
D vs E              4.2     s.e. 1.03     ss.div. 8.00

*** Standard errors of differences of means ***
Table            herbicide
rep.                 4
s.e.d.             2.062
```

COMPARISON OF TREATMENT MEANS

Table 13.12. Yields of barley after the application of five different herbicides

	A	B	C	D	E
Block 1	64	68	71	75	69
Block 2	66	64	69	74	71
Block 3	68	60	75	78	65
Block 4	59	61	68	72	60
Total	257	253	283	299	265
Mean	64.25	63.25	70.75	74.75	66.25

Table 13.13. Preliminary ANOVA table for Example 13.5

Source	DF	SS	MS	VR
Blocks	3	99.75	33.25	3.91
Treatments	4	370.8	92.7	10.91
Residual	12	102.0	8.5	
Total	19	572.55		

Table 13.14. Table of orthogonal contrast coefficients for Example 13.5

A versus The rest	4	−1	−1	−1	−1
B versus C, D, E	0	3	−1	−1	−1
C versus D, E	0	0	2	−1	−1
D versus E	0	0	0	1	−1

A diagram is helpful for finding a complete set of orthogonal contrasts. For example, the four contrasts of Example 13.5 are shown in Figure 13.5.

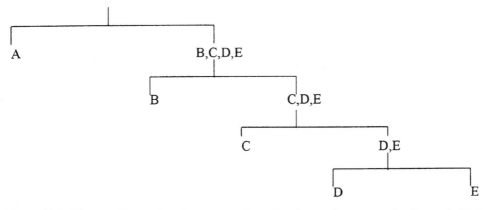

Figure 13.5. Diagram illustrating the construction of orthogonal contrasts for Example 13.5

13.5.2 Contrasts not Orthogonal

Output 13.8 was produced by Genstat with REG (herbicide; 4; cmat) as the treatment structure, where cmat is the matrix of 4 rows of coefficients given in Table 13.14. The use of REG assumes the contrasts are orthogonal. For contrasts that are not necessarily orthogonal and number less than or equal to the treatment degrees of freedom, Genstat can be used, replacing REG with COMP.

In Table 13.14, if the B versus C, D, E contrast is replaced by B, C versus D, E the coefficients become 0, 1, 1, −1, −1. This makes the second and third contrasts not orthogonal. Verify that for B, C versus D, E the contrast sum of squares is 49.000 and the variance ratio is 5.76. The sums of squares for the four contrasts now add to 258.467, which is less than the treatment sum of squares of 370.800.

The COMP method uses the same formula as the REG method to calculate the sums of squares, but uses different formulae to calculate the contrasts, their standard errors and ss.divs. They are

$$\text{Contrast} = (\Sigma cT)/r \qquad s.e. = \sqrt{\Sigma c^2 \times RMS/r} \qquad ss.div. = r/\Sigma c^2$$

The contrast part of the Genstat output using the COMP method is

```
*** herbicide contrasts ***

A versus B, C, D, E    -18.0    s.e. 6.52    ss.div. 0.200
B, C versus D, E        -7.0    s.e. 2.92    ss.div. 1.00
C versus D, E            0.5    s.e. 3.57    ss.div. 0.667
D versus E               8.5    s.e. 2.06    ss.div. 2.00
```

For example, the first contrast is $-72/4 = -18.0$, with a standard error of $\sqrt{20 \times 8.50/4} = 6.52$.

This contrast divided by its standard error is $(-18/6.52 = -2.761)$. The same answer $(-0.90/0.326 = -2.761)$ was obtained from Output 13.8 using the regression method. On squaring -2.761 we obtain 7.62, the variance ratio for testing the significance of this contrast. This is true in general because

$$\frac{\text{Contrast}}{s.e.} = \frac{(\Sigma cT)/r}{\sqrt{\Sigma c^2 \times RMS/r}} = \frac{\Sigma cT}{r\Sigma c^2} \times \frac{\sqrt{\Sigma c^2}}{\sqrt{RMS/r}} = \frac{\Sigma cT}{r\Sigma c^2} \div \sqrt{\frac{RMS}{r\Sigma c^2}}$$

which is the slope divided by its standard error using the regression method of the previous section.

On squaring this result we obtain

$$\frac{(\Sigma cT)^2}{r\Sigma c^2} \div RMS$$

This is the contrast sum of squares divided by the residual mean square and is equal to the variance ratio used to test the significance of the contrast.

Note: If the contrast coefficients for A versus the rest has been 1, −0.25, −0.25, −0.25, −0.25 instead of 4, −1, −1, −1, −1, the contrast using the COMP method would have been −4.5, which is the mean for A (64.25) minus the average of the other four means (68.75), and the standard error would have been 1.63. The ratio of the contrast to the standard error would still have been 2.761.

For further discussion of the concepts described in this chapter see Petersen (1977) and Little (1981).

Exercise 13.1

The yields (t/ha) from a randomised block experiment to compare different levels of nitrogen fertiliser on winter rapeseed are shown below. The treatments were levels of nitrogen and the levels used were 1, 2, 2.5, 3, and 3.5 × 100 kg/ha of nitrogen.

Block	X	T_1 1	T_2 2	T_3 2.5	T_4 3	T_5 3.5
1		4.28	4.68	4.78	4.89	4.69
2		4.31	4.73	4.82	4.93	4.70
3		4.43	4.75	4.89	4.91	4.80
Means	Y	4.34	4.72	4.83	4.91	4.73

(a) Carry out an analysis of variance and test whether nitrogen level affects yield.
(b) Carry out a linear regression analysis of yield on nitrogen level and use the results to partition the treatment sum of squares found in (a) into regression and deviation components. Hence test for a linear trend in the treatment means and test for a lack of fit to a linear model. Finally test whether a quadratic model fits significantly better than the linear model.

Exercise 13.2

A randomised blocks experiment with five replicates was carried out to compare the performance of four varieties of barley. This analysis of variance table is as follows:

Source	DF	SS	MS	VR
Blocks	4	25.648	6.412	
Variety	3	7.292	2.431	28.94
Residual	12	1.008	0.084	
Total	19	33.948		

There are four varieties, one old (A) and three new (B, C, D), total yields from five plots are as follows:

Variety	Total yield
A	21.7
B: New variety—cheap seed	24.5
C: A second new variety—cheap seed	27.2
D: A third new variety—expensive seed	29.8

(a) Decide upon three orthogonal contrasts, on the basis of the variety information provided.
(b) Write down a suitable matrix of contrasts coefficients and use it to calculate the sum of squares for each contrast. Verify that these sums of squares add to the variety sum of squares.
(c) Complete the modified analysis of variance table, test the significance of each contrast and report your conclusions.

Answers to Exercise 13.1

(a) The VR for the treatments is 144.5 on 4 and 8 degrees of freedom. It is very much higher then $F_{(4,8,0.1\%)} = 14.39$, so we conclude that nitrogen fertiliser level affects yield. A graph of mean yield versus nitrogen level should be drawn. This shows that a quadratic curve may fit better than a straight line. The SED for comparing two nitrogen level means is 0.0258 and the $LSD_{5\%}$ is 0.0594. The treatment sum of squares is 0.5752.

(b) The fitted line is $Y = 4.273 + 0.181X$. This indicates that for every 100 kg/ha increase in nitrogen fertiliser, the mean yield is expected to increase by 0.181 t/ha. The regression sum of squares calculated using the means is 0.1206. This is multiplied by 3 (as there are three Y-values for each X-value) to give that part of the treatment sum of squares due to regression. The complete ANOVA table is as follows:

Source	DF	SOS	MS	VR
Blocks	2	0.0216	0.0108	
Treatments	4	0.5752	0.1438	144
Reg(linear)	1	0.3618	0.3618	362
Deviation	3	0.2134	0.0711	72
Residual	8	0.0080	0.0010	
Total	14	0.6048		

From F-tables $F_{(1,8,0.1\%)} = 25.41$ hence there is a very highly significant linear trend after allowing for block effects. Also from F-tables $F_{(3,8,0.1\%)} = 15.83$ so there is also a very highly significant lack of fit indicating that not all the variation between treatments is due to linear regression.

A quadratic trend should be tested. A computer output shows that when a quadratic curve is fitted to the treatment means the fitted equation is $Y = 3.569 + 0.927X - 0.167X^2$ and the regression sum of squares is 0.18486. This is multiplied by 3 to give 0.5546, the correct quadratic regression sum of squares for all 15 points. Of this 0.3618 is due to linear regression — see the above table. Hence $0.5546 - 0.3618 = 0.1928$ is due to the increase in the quadratic fit over and above the linear fit. The corresponding VR is $0.1928/0.0010 = 193$ on (1, 8) df. Hence the quadratic fit is significantly better than the linear fit. The fitted treatments means (obtained from the quadratic equation) are 4.329, 4.754, 4.841, 4.844 and 4.763 respectively. The deviation sum of squares from the quadratic model is $0.5752 - 0.5546 = 0.0206$ on $4 - 2 = 2$ df. The corresponding mean square is $0.0206/2 = 0.0103$ and the VR is 10.3 on (2, 8) df which is significant at the 1% level. Thus, there is evidence of a lack of fit from the quadratic model.

Answers to Exercise 13.2
(a) The three orthogonal contrasts of interest are (i) Old versus New, (ii) Cheap versus Expensive, (iii) Cheap B versus Cheap C.
(b) The matrix of contrast coefficients (the c values) is

A	B	C	D	Contrast	Σc^2
3	−1	−1	−1	A versus the rest	12
0	1	1	−2	B, C versus D	6
0	1	−1	0	B versus C	2

The formula for a contrast sum of squares is

$$\text{Contrast SS} = \frac{(\Sigma cT)^2}{r(\Sigma c^2)}$$

where r is the number of replications per treatment (five in this case) and T represents the treatment totals.
For the first contrast,

$$\text{Contrast SS (1)} = \frac{(3 \times 21.7 - 24.5 - 27.2 - 29.8)^2}{5 \times 12} = 4.483$$

For the second contrast,

$$\text{Contrast SS (2)} = \frac{(24.5 + 27.2 - 2 \times 29.8)^2}{5 \times 6} = 2.080$$

For the third contrast,

$$\text{Contrast SS (3)} = \frac{(24.5 - 27.2)^2}{5 \times 2} = 0.729$$

These add to the variety sum of squares $(4.483 + 2.080 + 0.729) = 7.292$.
(c) The significance of each contrast (on 1 df) can be found by dividing by the residual mean square to obtain a variance ratio on (1, 12) df.

The following analysis of variance table shows how the variety sum of squares (on 3 df) is partitioned into the three orthogonal contrast sums of squares (each with 1 df):

Source	DF	SS	MS	VR
Blocks	4	25.648	6.412	
Variety	3	7.292	2.431	28.9
A versus the rest	1	4.483	4.483	53.4
B, C versus D	1	2.080	2.080	24.8
B versus C	1	0.729	0.729	8.68
Residual	12	1.008	0.084	
Total	19	33.948		

From F-tables:

$$F_{(1,12,5\%)} = 4.747 \qquad F_{(1,12,1\%)} = 9.330 \qquad F_{(1,12,0.1\%)} = 18.64$$

The first contrast is significant at the 0.1% level. This indicates that the average yield per plot from the new varieties, $(24.5 + 27.2 + 29.8)/(3 \times 5) = 5.43$, is significantly better than the average yield per plot from the old variety, $21.7/5 = 4.34$. The second contrast is also significant at the 0.1% level, indicating that expensive new seed gives significantly higher average yields per plot (5.96) than cheap new seed (5.17). The third contrast is significant at the 5% level. This indicates that varieties B and C give significantly different mean yields per plot (4.90 versus 5.44).

Chapter 14

Checking the Assumptions and Transformation of Data

14.1 THE ASSUMPTIONS

The analysis of variance and regression methods are based on the assumptions of independence of errors, normality, homogeneity of variance and additivity. Violation of these assumptions affects the sensitivity of the F-tests in the analysis of variance. Significance tests of mean comparisons are also affected and the level of significance may be much different from what is assumed. This can result in invalid conclusions.

Many researchers carry out analysis of variance and regression on their data without checking whether these assumptions are valid. In the past, this checking was very tedious, but now the computer can do it very straightforwardly. You should get into the habit of checking assumptions and taking the necessary action if they are seriously violated.

For each of the designs discussed so far we have assumed an underlying model in which each observed yield is made up of various effects and a random error. These errors are assumed to be independent, normally distributed with a mean of zero, and not related to any particular treatment or block. In regression models the magnitude of the errors is not supposed to depend on the X-value.

The yield predicted by the model is called the fitted value and the yield minus the fitted value is called the residual. The assumptions can be checked by analysing the residuals and their relationship to the fitted values. Details are given in the following sections. If the assumptions are seriously violated, one solution may be to transform the data and then repeat the analysis (Section 14.2). Another solution may be to carry out a non-parametric test (Chapter 19). An alternative to making a transformation when the errors do not follow a normal distribution is to use a **generalised linear model**. See McConway, Jones and Taylor (1999) for more details.

14.1.1 Independence of Errors

This is usually achieved by adequate randomisation and it is assumed that the error (residual) of an observation is not related to that of another. The yields from adjacent plots receiving the same treatment are likely to be more similar than the yields from plots far apart receiving the same treatment. You should check the experimental layout to ensure that a particular pair of treatments does not always occur in adjacent plots. If necessary you should carry out a fresh randomisation, provided the treatments have not been applied. You should not have all the plots with a given treatment next to each other. A graph of the residuals in plot order may indicate a lack of independence if there is a run of positive residuals followed by a run of negative ones.

14.1.2 Normality

A histogram of the residuals should always be obtained in order to check whether the assumption of normality is seriously violated. The computer can also be asked to produce a normal plot of the residuals. This should be a straight line if the residuals are normally distributed. Often it is not possible to tell if residuals come from a normal distribution when the amount of data is small. Fortunately, moderate departures from normality are not serious. The F-tests and t-tests usually performed with analysis of variance are **robust** tests. They are still approximately valid even when the data are not normal.

To avoid serious departures from normality when sampling from plots you should use sample means based on several within-plot measurements. For instance, in an experiment to compare the effects of treatments on plant dry weight, you may have five plots per treatment. If you just measure one plant from each plot, you will have five replications per treatment, but the plant dry weights within each treatment may not be normally distributed. If you sample 10 plants from each plot, you should use the five sample means as the five replications per treatment. These means are likely to be almost exactly normally distributed. Do not use the 50 individual weights as 50 replicates per treatment.

There are situations where the measurements are markedly non-normal. Examples are counts of insects on a plant, counts of weeds in a plot, number of seeds germinating, scores on a scale of one to five. For one treatment, there may be five plots, and the number of weeds of a particular species may be, 6, 8, 18, 125, 410. In addition, this type of data usually violates other assumptions.

14.1.3 Homogeneity of Variance

In Chapter 6, we discussed the independent samples t-test. You will recall that one assumption of this test is that the two sets of data come from populations with the same variance. If this assumption is justified, the two sample variances can be pooled and the test completed.

In Chapter 9 we discussed the CRD and one-way analysis of variance. We pooled the variance within each treatment to get a pooled residual mean square (RMS). This was used to calculate an SED in order to compare treatment means. It was also used to calculate a pooled standard error of a treatment mean (SEM) which in turn could be used to calculate confidence intervals for population means.

CHECKING THE ASSUMPTIONS AND TRANSFORMATION OF DATA

This pooling is only valid if the within-treatment variation is similar for each treatment, and does not depend markedly on the treatment applied. In most experiments where the differences in treatment effects are small, homogeneity of variance can be assumed and the usual analysis is valid. However, if you propose to carry out an analysis of variance on counts, you should check the assumption of homogeneity of variance. If this assumption is violated a transformation is called for (Section 14.2). Alternatively you could use a generalised linear model (not in this book).

In experiments where the treatment means are very different, the variation within treatments with large means is often much larger than the variation within treatments with small means. Failure to check the assumptions in these cases can lead to serious errors of interpretation of results. We illustrate this problem in the next example.

Example 14.1
Three different herbicides (A, B, C) were compared using a CRD with six plots per herbicide. Included in the design were six untreated plots (D). The number of weeds per plot were counted. The results are shown in Table 14.1. In the following discussion we refer to A, B, C and D as treatments.

Many students would simply proceed with the analysis of variance, yet it is obvious there is something strange about these data. The variance of the treatment D values is 73 times as large as that for treatment A. What happens if we proceed with the usual analysis?

The pooled variance estimate is the average of the separate variances because of equal replication. It is 62.675, the RMS value in the ANOVA table (Table 14.2). The VR of 17.9 is very highly significant: it is much greater than the 0.1% F-table value of 8.098 on (3, 20) df.

The pooled SED is
$$\sqrt{\frac{2 \times RMS}{r}} = \sqrt{\frac{2 \times 62.675}{6}} = 4.571$$

and the 5% LSD value is
$$LSD_{5\%} = t_{(20, 2.5\%)} \times SED = 2.086 \times 4.571 = 9.53$$

Table 14.1. Data from a CRD on weed counts showing non-homogeneity of variance

	A	B	C	D
	4	8	25	33
	5	11	28	21
	2	9	20	48
	5	12	15	18
	4	7	14	53
	1	7	30	31
Mean	3.5	9	22	34
s^2	2.7	4.4	45.2	198.4

Table 14.2. Analysis of variance for Example 14.1

Source	DF	SS	MS	VR
Treatments	3	3361.125	1120.375	17.9***
Residual	20	1253.510	62.675	
Total	23	4614.625		

***Significant at 0.1% level.

The difference between the A and B means is 9.0 − 3.5 = 5.5. This difference is less than 9.53 so we conclude that these two means are not significantly different. This is clearly a misleading conclusion because, on inspection of the data, there is no overlap in the values for these two treatments. The difference between the means of treatments C and D is 34.0 − 22.0 = 12.0. This difference is significant based on the LSD test. The pooling of the separate variances has resulted in biased t-tests. The test of the difference between the A and B means is based on a within-treatment variance vastly greater than it should be because it includes the larger variances from C and D (Table 14.1).

The reader should now:

- Carry out a t-test to compare the A and B means based only on the data from A and B. Confirm that the SED value is 1.088, the 5% LSD value is 2.42 and conclude that the difference in means of 5.5 is now significant ($t = 5.5/1.088 = 5.06$ on 10 df). This is the opposite conclusion to that obtained earlier.
- Repeat these calculations for C and D. The SED value is 6.372 and the 5% LSD value is 14.20. The difference in means of 12.0 is now not significant ($t = 1.88$ on 10 df).

When the pooled variance of all four treatments was used, the small variation within A and B made the SED smaller than it should have been for comparing C and D. **When variances are wrongly pooled, confidence intervals for individual means become meaningless and often nonsensical.**

From the ANOVA (Table 14.2), the pooled standard error of a treatment mean is

$$SEM = \sqrt{\frac{RMS}{r}} = \sqrt{\frac{62.675}{6}} = 3.232$$

A 95% confidence interval for the mean of treatment mean A, based on this pooled SEM, is

$$\bar{x}_A \pm t_{(20, 2.5\%)} \times SEM = 3.5 \pm (2.086 \times 3.232) = 3.5 \pm 6.7 = (-3.2, 10.2)$$

This is clearly meaningless as you cannot have a negative yield. However, a 95% confidence interval for the mean of treatment A, based only on data for A, is

$$\bar{x}_A \pm t_{(5, 2.5\%)} \times \sqrt{s^2/n} = 3.5 \pm (2.571 \times 0.671) = 3.5 \pm 1.7 = (1.8, 5.2)$$

This is a more sensible interval, but it is only based on six values (Table 14.1). You obtain more precision by pooling, **but only if there is homogeneity of variance.**

With data such as these it is often safer just to present individual means and standard errors as follows:

Treatment	A	B	C	D
Mean	3.5	9	22	34
SE	0.67	0.86	2.75	5.75

With this information, and knowing the number of replications, you have the option of carrying out further tests. Another option is to carry out a transformation (Section 14.2). If the transformed data satisfy the assumptions (Section 14.1), they can be used to perform significance tests.

Example 14.1 illustrates the dangers of misinterpretation if you proceed with analysis of variance without first checking the assumptions. It shows that sometimes it is better not to pool the data, but to do individual t-tests. For this example, two separate t-tests could be carried out. The first would compare A and B using only the data for A and B, the second would compare C and D using only the data for C and D. If the data from one treatment is very different from of all the others in the experiment, or its yields are almost all zero, it is best to omit that treatment from the analysis of variance.

A quick check of the assumption of homogeneity of variance in the CRD is to divide the largest s^2 by the smallest and carry out an F-test (Section 6.5). For other designs, the best way to assess this assumption is to obtain a graph of the residuals versus the fitted values. This graph should show a random pattern of residuals on either side of 0. If the residuals increase with the fits this is evidence of heterogeneity of variance.

Note: If Example 14.1 had used a RBD, the variation within treatments after allowing for blocks could have been investigated after removing the block effects. These are removed by subtracting (corresponding block mean minus grand mean) from each yield. This does not affect the treatment means and each residual can be obtained by subtracting the appropriate treatment mean from the adjusted yield.

14.1.4 Additivity

Validity of the analysis is based on the assumption that the yields are made of components which are added together. For example, when carrying out an independent samples t-test you assume the two population variances are equal. This is a reasonable assumption, if effects are additive. Suppose one treatment (A) is expected to give higher yields than the other (B). Treatment A should have the effect of raising the yield by an equal amount on all the plots to which it is applied, over and above the yield which would have been obtained from B. If the plots receiving each treatment are selected at random, the additivity assumption predicts that the variation in the yields of the two sets of plots should be the same.

For the CRD, each yield value is assumed to be made up of an overall mean value plus a treatment effect plus a random component. This means that the replicate values for a given treatment differ only in their random components. These random components are estimated by the residuals, which are the differences between the yields and the corresponding treatment means.

For the RBD each yield is assumed to be made up of an overall mean plus a block effect plus a treatment effect plus a random component. For example, in a low-fertility block, the yields for treatments A and B may be 20 and 30 respectively. In a high-fertility block, the yield gap is still expected to be 10; if the yield of A is 40, the yield of B is expected to be 50. The treatment and block effects are **additive**. Treatment B is expected to outyield treatment A by 10 units in all conditions. The high fertility has raised both yields by 20 units.

If treatment B is expected to outyield B by a constant percentage in all conditions, the block and treatment effects are **multiplicative**. In the low-fertility block the yield of B is 50% greater than the yield of A. If the effects are multiplicative, and the high fertility doubles the yields, A is raised from 20 to 40 units and B's yield is expected to double from 30 to 60 units. The difference in the low-fertility block is 10 units (20 to 30), and the difference in the high-fertility block is 20 units (40 to 60). B outyields A by a constant 50%. These additive and multiplicative effects are summarised in Table 14.3.

The effects of non-additivity are similar to the effects of non-homogeneity of variance on the analysis of variance. Often the non-homogeneity of variance is due to non-additivity so a transformation that corrects for one corrects for the other. For example, insecticide treatments are likely to be non-additive. If one insecticide halves the number of insects, a plot that had 20 insects before spraying should have 10 afterwards. A plot that had 1000 insects should have 500 after spraying. In the first plot, the spraying should reduce the number of insects by 10; in the second, it should reduce the number by 500.

Before analysing multiplicative data you should take logarithms. The logarithms conform to an additive model. For example, the logarithms (to base e) of 20 and 10 are 3.00 and 2.31 respectively; a reduction of 0.69. The logarithms of 1000 and 500 are 6.91 and 6.22; again a reduction of 0.69. Also the variances of the transformed data are more likely to be homogeneous. Hence, the analysis of variance of the logarithms is likely to be valid (Section 14.2.1).

Note: You cannot take the logarithm of zero or a negative number. If your data are counts and include zeros, add 1 (or a number between 0 and 1) to every count before taking logarithms.

Table 14.3. Illustration of additive and multiplicative effects

	Additive model				Multiplicative model		
	A	B	B−A		A	B	B−A
Low	20	30	10	Low	20	30	10
High	40	50	10	High	40	60	20
High−Low	20	20		High−Low	20	30	

14.2 TRANSFORMATIONS

If your data do not conform to the assumptions which validate the analysis of variance and t-test procedures, it may be possible to transform them to a new scale conforming to the assumptions. In this case, analysis of variance and t-tests carried out on the transformed data will be valid. Presentation of results should include the untransformed and transformed treatment means, and the SED value for comparing the transformed means.

In general, transformations should be avoided unless they are obviously necessary to avoid biased tests of significance. Now that analyses can be carried out very quickly on the computer, you are advised to compare the results with and without transformations. If alternative analyses lead to the same conclusions, it is best to present the results based on the untransformed data. The transformation chosen should be the one which most closely satisfies the analysis of variance assumptions; this may not be the one which gives you the results you are looking for!

Some types of data should be routinely transformed. For example, data on insect counts should be transformed by taking logarithms of the counts or the counts plus one. Data on proportions often need to be transformed using an arcsin or angular transformation. Data on area may need a square root transformation.

Before making a decision to use transformations you should always use the computer to carry out an analysis of the residuals in order to check the assumptions of analysis of variance.

The checks most used are a histogram of the residuals, a normal plot of the residuals and graph of the residuals versus the fitted values. In regression models plots of the residuals versus the X-values are also useful.

If the homogeneity of variance assumption is true, a scatter graph of the residuals against the fitted values will have a random distribution. Any tendency of the variation in the residuals to increase or decrease with the fitted values is evidence of non-homogeneity of variance associated with non-additivity and/or non-normality. This scatter graph will also identify **outliers**. These are extreme observations which have a large influence on the outcome of the analysis. Outliers should be checked to find out if they are due to recording errors. If they are, they should be removed from the analysis or taken as missing values.

Because it is very easy to obtain scatter graphs displayed on a computer monitor, the checking for failure of assumptions should be done as a matter of course. Figure 14.1 shows the residual plots produced by Minitab for the data of Example 14.1

The normal plot is far from a straight line, indicating a violation of the normality assumption. The histogram is peaked showing a large number of very small residuals, and a few large positive and negative residuals. The individual chart shows the residuals in data order from treatment one to treatment four. Some of the largest residuals are almost three standard deviations from the mean of zero. The graph of residuals against fitted values (Fits) shows that, as the fitted values increase, so the variation in the residuals increases. This suggests a log or square root transformation is required. The residuals are set along vertical lines because **in one-way analysis of variance the fitted values are the treatment means**.

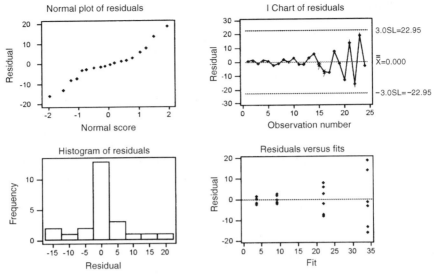

Figure 14.1. Residual plots for Example 14.1 — no transformation

The transformation that best gives rise to a random scatter graph of residuals against fitted values and to a symmetrical histogram of residuals is the best one to use for the analysis of variance and *t*-tests. The three most common transformations are the **log**, **square root**, and **arcsin**.

14.2.1 The Logarithmic Transformation

The log transformation is often required for counts such as the number of diseased plants per plot and number of insects per plant. These data are counts that frequently cover a wide range from single-digit numbers to numbers in hundreds or thousands. If the data contain zeros, add one to each number before taking logs. Figure 14.2 shows the residual plots produced by Minitab for the data of Example 14.1 after a log transformation to natural logarithms (log to the base e, denoted by $\log_e$ or ln). The normal plot is more like a straight line but the histogram is rather skewed. However, the graph of residuals versus fitted values indicates that the homogeneity of variance assumption may be valid for the log transformed data. Note that a transformation that corrects for one violation of assumption may introduce another, thus the distribution of residuals is no longer symmetric.

14.2.2 The Square-root Transformation

Data needing a square-root transformation are typically number of events per unit of time or space when the event is considered rare, for example numbers of insects caught in a trap. Some researchers recommend the square root transformation for leaf area because it is measured in square units. Small increases in length when the

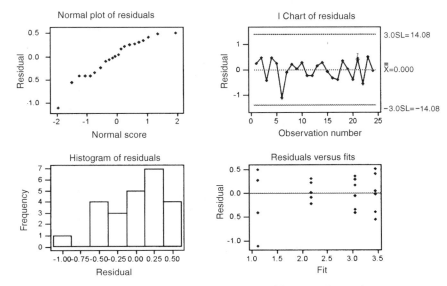

Figure 14.2. Residual plots for Example 14.1 — natural log transformation

leaf length is small are associated with small increases in area. Small increases in length when the length is large are associated with large increases in area. This transformation is also recommended for percentage data in the range 0–30% or 70–100%. Figure 14.3 shows the residual plots produced by Minitab for the data of Example 14.1 after a square root transformation.

The normal plot is a quite a good fit to a straight line and the histogram is fairly symmetrical. However, the graph of residuals versus fitted values shows non-homogeneity of variance.

14.2.3 Deciding between a Log and a Square Root Transformation.

Violation of the homogeneity assumption is more serious than a moderate violation of the normality assumption. A comparison of Figures 14.2 and 14.3 suggests that a log transformation would be better than a square root transformation.

Before it was possible to carry out residual diagnostic tests as a matter of course by computer, other methods were and are still used to decide an appropriate transformation. These involve investigating possible relationships between the treatment means and the within-treatment standard deviations and variances.

If the ratio of standard deviation to treatment mean is constant, for data from a CRD so that treatments with large means have large s values, a log transformation is recommended. Check by making a graph of standard deviation against treatment mean. If the points lie approximately on a straight line, a log transformation is appropriate. However, if it is the variance and not the standard deviation which is proportional to the treatment mean, the square root transformation is recommended.

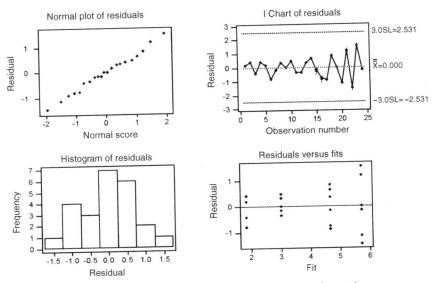

Figure 14.3. Residual plots for Example 14.1 — square root transformation

Example 14.2
The ratios of mean to standard deviation and mean to variance for the data of Example 14.1 are shown in Table 14.4. The values of $s/\bar{x}$ are much more stable than the values of $s^2/\bar{x}$ and this suggests that a log transformation would be much better than a square-root transformation. If the log transformed counts satisfy the assumptions, the analysis of variance should be carried out using them. Genstat output 14.1 shows the analysis of variance of the natural logs of the counts. If you use the log transformed SED value (0.2505) for comparing A and B, you will find these log transformed means (1.114 and 2.175) are significantly different. If the SED for the untransformed data (4.57) is used, the A and B means of 3.5 and 9.0 (Table 14.1) are not significantly different. However, this SED should not be used, as the conditions for pooling are not valid.

In addition, the conclusion based on the log transformation, is that treatment C does not produce significantly different counts from the untreated plots. These conclusions support the idea that a small difference in mean counts when counts are small is more important than a similar difference in mean counts when counts are large. A check is made on the stability of the ranges in the log scale. The log transformed data are given in Table 14.5. After transformation the log ranges are fairly stable, confirming that a log transformation is suitable.

Output 14.1 ANOVA table after transfoming the counts of Example 14.1 into natural logarithms

```
***** Analysis of variance *****
Variate: lweed

Source of variation  d.f.    s.s.      m.s.     v.r.    F pr.
herbicide             3    19.3285    6.4428   34.22   <0.001
Residual             20     3.7660    0.1883

Total                23    23.0945

***** Tables of means *****
Variate: lweed

Grand mean           2.447

herbicide              A        B        C      None
                     1.114    2.175    3.049   3.451

*** Standard errors of differences of means ***
Table             herbicide
rep.                 6
s.e.d.             0.2505
l.s.d.(5% level)   0.5226
```

Table 14.4. Relationships between mean, standard deviation and variance for Example 14.1

Treatment	A	B	C	D (Untreated)
Mean, $\bar{x}$	3.5	9	22	34
Standard deviation, s	1.643	2.098	6.723	14.09
Variance, s^2	2.7	4.4	45.2	198.4
$s/\bar{x}$	0.47	0.23	0.31	0.41
$s^2/\bar{x}$	0.77	0.49	2.05	5.84

14.2.4 The Arcsin Transformation

This transformation, also called the *angular* transformation, is suitable for percentages based on counts. Each percentage is divided by 100 to convert it to a proportion. The angle whose sine is the square root of the proportion is used as the transformed value. Thus each percentage value is converted to an angle between 0 and 90 degrees. Some statistical packages require you to input the data in the form of

Table 14.5. Log transformed data of Example 14.1

Treatment	Log counts of weeds per plot						Range
A	1.39	1.61	0.69	1.61	1.39	0	1.61
B	2.08	2.4	2.2	2.48	1.95	1.95	0.53
C	3.22	3.33	3	2.71	2.64	3.4	0.76
Untreated	3.5	3.04	3.87	2.89	3.97	3.43	1.08

proportions and give the transformed results in radians (to convert radians to degrees multiply by 57.296).

Consider an experiment on rooting to compare several growth hormones. Each hormone is applied to five groups of 20 apple cuttings (five replicates). For each group, the number of cuttings out of 20 which root is recorded. These are converted into five percentages which are the five replicate values. The possible percentages are 0%, 5%, 10%, 15%, . . ., 95%, 100%. For treatments that give

- A low incidence of rooting, the variation in the percentages will be small, say 0–10%.
- A high incidence of rooting, the variation in the percentages is also likely to be small, say 90–100%.
- A moderate incidence of rooting, the variation in the percentages is likely to be large, say 30–70%.

This is clearly an example of heterogeneity of variance. However, unlike data which require a log or square-root transformation, the within-treatment variation at first increases as the treatment mean increases from near zero up to 50%, and then decreases as the treatment mean approaches 100%. If you do not carry out a transformation you are likely to obtain confidence intervals where the lower limit is less than 0% or the upper limit is greater than 100%. The arcsin transformation makes this less likely.

This transformation has little effect on percentages within the range 30–70%, so it may not be necessary to use it if all your percentages are in this range. If the range of percentages in your data is greater than 40%, the transformation should be carried out. If you have a treatment consisting entirely of 0% values or entirely of 100% values you should remove it from the analysis. If your percentages vary between 0% and 30%, or between 70% and 100% a square root transformation may be more suitable.

The arcsin transformation is valid when all the percentages are based on the same n-value. For example, if samples each of 50 seeds are planted, and the numbers germinating per sample are recorded, the percentages germinating can be validly transformed because, in this case, n is constant at 50. If the values of n vary slightly, this transformation can still be used. However, if there are widely differing values of n, a weighted analysis or a generalised linear model should be used (not discussed in this book).

In an experiment where each treatment is applied to a single group of 50 seeds, there is effectively only one replication per treatment. For each treatment the one measurement is the number out of 50 which germinate. A chi-square analysis may be appropriate (Chapter 18). See Example 18.13.

Other types of percentages

Some researchers use the angular transformation for percentages based on ratios of measurements, for example when the dry weight of leaves is expressed as a percentage of the total plant dry weight. As these percentages arise from continuous data, the angular transformation should not normally be used. A graph of residuals against fitted values should indicate whether a transformation is needed. Various transformations can be tried and the one which best stabilises the variance should be adopted.

14.2.5 An Alternative to Transformations

When the assumptions are not valid you should consider using a non-parametric test as an alternative to carrying out a transformation (Chapter 19). A better alternative is to consider using **generalised linear models** (not discussed in this book). These are suitable if the residuals follow one of a number of distributions other than a normal distribution. See McConway, Jones and Taylor (1999) for more details and Dobson (1990) for a more mathematical introduction. For a general discussion on transformations see Fernandez (1992).

Chapter 15

Missing Values and Incomplete Blocks

15.1 INTRODUCTION

When an experiment is planned, it is expected that data will be collected from every experimental unit. In practice, data for some experimental units may be missing or illogical for some of the following reasons:

- A plot may be flooded, grazed, or vandalised.
- Plants may be destroyed by pests and diseases.
- Some measurements may be lost.
- Illogical data may be present because an experimental unit accidentally received the wrong treatment.

A distinction must be made between a zero yield and a missing value. For example, in a pot experiment in which each pot contains a single plant, the yield is zero if plant death occurs due to the application of a chosen treatment. Otherwise, it is classified as a missing value. If in an experiment designed to compare the effects of insecticides on crop yield, a control plot is destroyed by insects, the yield is zero and not a missing value. However, if in another trial seed yield is zero due to treatment, dependent characters such as 100-grain weight should be treated as missing. Other measurements, such as plant height, should be taken at different growth stages as they may be useful in explaining the zero yield. Illogical data should not be treated as missing unless it can be established that values are due to errors of measurement.

15.2 MISSING VALUES IN A COMPLETELY RANDOMISED DESIGN

If the experimental design is a CRD, the effect of data loss on the analysis minimal. If there are an unequal number of replications per treatment there is no common *SED* for comparing any two treatments.

If one treatment has n_1 replications and another has n_2 replications, the SED value used to compare these two treatment means is

$$SED_{(1,2)} = \sqrt{RMS\left(\frac{1}{n_1} + \frac{1}{n_2}\right)}$$

See Section 9.8 for details on how to proceed with the analysis.

15.2.1 Using a Computer

If your only measurement is yield, data should be entered into a spreadsheet with the treatment codes in one column and the yields in another. The number of rows should be the number of non-missing values. Ask for a one-way analysis of variance.

If you have measured a number of response variables on each experimental unit not all variables may have the same missing value pattern. It is best to enter the data with one column for the treatment codes and one for each of the variables, so that there is one row for each unit. Missing values should be denoted by a special character such as an asterisk (*) or a full stop (.).

Statistical packages have different ways of handling missing values. If you give an instruction to analyse all the variables simultaneously, all units that have at least one missing value may be deleted from the analysis. If you do not want this to happen, you should analyse each variable separately. This should result in the appropriate units being deleted for each variable. However, some packages may estimate the missing values and give yet a different analysis.

Example 15.1
Table 15.1 shows data from a completely randomised design with four treatments and three replications per treatment. Measurements were made on two variables (Y_1 and Y_2) and there were five missing values (*). If the computer is instructed to carry out a simultaneous analysis of Y_1 and Y_2, units 2, 3, 8 and 10 may be deleted. Table 15.2 shows the resulting analysis of variance table and means for Y_1, Table 15.3 the corresponding analysis when Y_1 is analysed separately after deleting just units 2 and 10 and Table 15.4 the results of analysing Y_1 after estimating the two missing values. The estimates are the means of the non-missing values. Thus the estimate for unit 2 is $(6.4 + 5.6)/2 = 6.0$ and that for unit 10 is 8.7. The treatment and residual sums of squares are calculated after inserting the estimates but the corresponding degrees of freedom have been reduced by the number of missing values (2). The total sum of squares is calculated from the ten non-missing values.

This example illustrates the dangers of blindly submitting data for analysis by computer. You should find out how missing values are handled and select the required option. If there are missing values for all the variables of a given unit, the row for that unit should be deleted from the spreadsheet. The method described in the first paragraph of this section can always be used to analyse each variable separately.

Table 15.1. Data on two variables (Y_1 and Y_2) and four treatments with missing values (*)

Unit	Treatment	Y_1	Y_2
1	1	6.4	12.2
2	1	*	13.4
3	1	5.6	*
4	2	9.8	15.5
5	2	8.7	16.3
6	2	7.2	17.8
7	3	7.3	10.4
8	3	6.1	*
9	3	6.4	10.6
10	4	*	*
11	4	8.0	16.8
12	4	9.4	17.6

Table 15.2. Analysis of variance of Y_1 after deleting units 2, 3, 8 and 10

Source	DF	SS	MS	VR	P
Treatment	3	7.068	2.356	1.96	0.261
Residual	4	4.792	1.198		
Total	7	11.860			

Means T_1: 6.400 T_2: 8.567 T_3: 6.850 T_4: 8.700.

Table 15.3. Analysis of variance of Y_1 after deleting units 2 and 10

Source	DF	SS	MS	VR	P
Treatment	3	13.222	4.407	4.82	0.049
Residual	6	5.487	0.914		
Total	9	18.709			

Means T_1: 6.000 T_2: 8.567 T_3: 6.600 T_4: 8.700.

Table 15.4. Analysis of variance of Y_1 after estimating missing values for units 2 and 10

Source	DF	SS	MS	VR	P
Treatment	3	16.900	5.6333	6.16	0.029
Residual	6 (2)	5.4867	0.9144		
Total	9 (2)	18.709			

Means T_1: 6.000 T_2: 8.567 T_3: 6.600 T_4: 8.700.

15.3 MISSING VALUES IN A RANDOMISED BLOCK DESIGN

When observations are missing in a RBD, the treatment and block effects become mixed up. For example, the mean of treatment one may be based on three blocks, whereas that for treatment two may be based on four, which may or may not include the three blocks containing treatment one. Hence, the difference between two treatment means is partly due to block differences. There are two ways of dealing with this problem: (1) estimate the missing values and carry out the analysis (with minor modifications) using the estimates, or (2) use only the available data and fit models.

If all the yields for a particular treatment are missing or zero, do not include this treatment in your analysis. Similarly, if all the yields in a block are zero or missing, this block should be excluded.

15.3.1 Estimation of Missing Values

Do not use estimation if more than 10% of the data values are missing. There are formulae for estimating missing values. These do not supply any extra information, they just allow the experimenter to apply the usual balanced analysis of variance. **Data which are missing cannot be retrieved by statistical analysis**.

The residual degrees of freedom are reduced by the number of missing values estimated. The estimates are chosen so as to minimise the residual sum of squares. You do not have to learn how to estimate missing values because the computer will do it for you. When entering your data into a computer spreadsheet insert an asterisk (*) to represent a missing yield. The advantage of using asterisks is that when you want to analyse several characters simultaneously you can have a separate column for each. Characters such as yield may have missing values, whereas characters such as plant height may have no missing values, but be aware that the computer may regard heights as missing for the units with missing values for yield.

Example 15.2

Consider the data of Example 10.1 in which a randomised block experiment is used to compare four varieties (treatments) in three blocks. Suppose that two values (Block 1, V_1) and (Block 3, V_3) are missing. Table 15.5 shows the revised spreadsheet.

If you attempt to analyse these data using the standard balanced randomised blocks program in Minitab you will get the error message 'Unbalanced design'. Genstat will automatically find estimates for the missing values. These are called **least squares estimates**. They are the values which, if used in place of the missing values, minimise the residual sum of squares. If any other two values are used, a larger residual sum of squares is obtained. Output 15.1 shows the results produced by Genstat.

The estimates of the missing values are 8.34 and 4.94 respectively. The blocks, variety and residual sums of squares (15.3417, 8.0911, 1.1229) are obtained after including the missing value estimates. The total sum of squares (18.016) is calculated using only the ten non-missing values. This is why it is not the sum of the other three sums of squares. The residual degrees of freedom and the total degrees of freedom have both been reduced by 2 (the number of missing values). They would have been 6 and 11 respectively if no values had been missing. The residual mean square (0.2807) is calculated by dividing the residual sum of squares (1.1229) by 4. The rest of the ANOVA table is calculated in the usual way.

The means for varieties 1 and 3 printed by Genstat (6.81 and 6.11) are calculated using the missing value estimates. They are called the **least squares means (LS means)**. Their use enables variety comparisons to be made on a more equal footing than if the means of the actual data (6.05 and 6.70) had been used. The raw means for V_1 and V_2 should not be compared because they contain different block effects. For this reason, when presenting the results of an experiment containing missing values, the least squares means should be quoted. They are the same as the raw means when there are no missing values.

The SED is not adjusted for missing values. It is calculated as

$$SED = \sqrt{\frac{2 \times 0.2807}{3}} = 0.433$$

where 0.2807 is the *ResidSS* divided by 4 (the adjusted residual degrees of freedom).

The corresponding 5% LSD value is $t_{(4, 2.5\%)} \times SED = 2.776 \times 0.433 = 1.201$ and may be used to compare any two least square means. However, it should be used with caution because it is based on estimates of missing values which may be far from the true values. For example, V_1 and V_3 should be compared using a larger SED than that used to compare V_1 and V_2 because V_3 has less real data than V_2.

The advantage of this approach is that the balance of the design is maintained. However, if too many values are missing, the results become unreliable. The unadjusted SED becomes unsuitable for making comparisons and the general linear model method should be used.

Note: SAS calculates an SED using the harmonic mean (Section 2.3.11) of the number of replications per treatment. For this example, the harmonic mean is 2.4, so the corresponding SED is 0.484 and the LSD is 1.343.

15.3.2 The Exact Analysis of a RBD with Missing Values

When values are missing, the design becomes unbalanced. The analysis proceeds using the *general linear model* (GLM) approach.

First, a model is fitted with only blocks in it. This is equivalent to a one-way ANOVA with blocks as the only factor. The residual sum of squares is noted. Then a model is fitted in which treatments are added. The reduction in the residual sum of squares is that due to treatments after fitting blocks. An ANOVA table is

Table 15.5. Data of Example 10.1 with two missing values

Variety	Block	Yield
1	1	*
1	2	6.5
1	3	5.6
2	1	9.8
2	2	6.8
2	3	6.2
3	1	7.3
3	2	6.1
3	3	*
4	1	9.5
4	2	8.0
4	3	7.4

formed in which the effect of treatments after allowing for block differences can be tested for.

The procedure is illustrated, using the same data as in the previous example. The results using the general linear model procedure in Minitab are shown in Output 15.2. Similar results can be obtained by using PROC GLM in SAS.

Notice that the least square means and the error sum of squares (1.1229) are the same as those produced by Genstat after estimating the missing values. In the sequential sum of squares (Seq SS) column the sum of squares due to fitting blocks ignoring variety is 10.5993. The extra sum of squares due to variety after allowing for blocks is 6.2938. These are called Type I sums of squares in the SAS system. They depend on the order of fitting the model. If the order of model terms had been variety block, the Seq SS column would have shown 7.1110 for variety and 9.7821 for block.

The adjusted sum of squares column (Adj SS) shows that the sum of squares due to blocks after allowing for variety is 9.7821 and the sum of squares for variety after allowing for blocks is 6.2938. These are called Type III sums of squares in the SAS system. These two types of sums of squares are identical when there are no missing values.

The F-tests are carried out using the adjusted sums of squares. There are significant differences in the variety means after allowing for block effects because the P-value is 0.041 (<0.05).

15.3.3 Comparing the Least Square Means

The SED value in Output 15.1 was used to compare the LS means but it was not adjusted for missing values. The exact analysis requires that each comparison has its own SED. Output 15.2 includes all pairwise comparisons. Note that only the SED

Output 15.1 Genstat output for data of Example 10.1 with two missing values

```
**** Analysis of variance *****

Variate: Yield

Source of variation    d.f.(m.v.)    s.s.      m.s.      v.r.     F pr.

Block stratum                    2  15.3417   7.6708    27.33

Block.*Units* stratum
Variety                          3   8.0911   2.6970     9.61    0.027
Residual                      4(2)   1.1229   0.2807

Total                         9(2)  18.0160

***** Tables of means *****

Variate: Yield

Grand mean   7.21

    Variety      1      2      3      4
              6.81   7.60   6.11   8.30

*** Standard errors of differences of means ***
Table                Variety
rep.                    3
d.f.                    4
s.e.d.                  0.433

(Not adjusted for missing values)

*** Least significant differences of means (5% level) ***

Table                Variety
rep.                    3
d.f.                    4
l.s.d.                  1.201

(Not adjusted for missing values)

***** Missing values *****

Variate: Yield
    Unit     estimate
     1          8.34
     9          4.94
```

for comparing varieties 2 and 4 (0.4326) is the same as that given in Output 15.1. The SED for comparing V_1 and V_3 (0.5804) is larger than that for comparing V_1 and V_2 (0.5013) because V_3 contains less real data than V_2.

The LS mean for V_2 exceeds that of V_1 by 0.7857. The SED value for this comparison is 0.5013 so the corresponding t-value is $0.7857/0.5013 = 1.567$. There are 4 degrees of freedom for this t-test (the error degrees of freedom from the ANOVA table). The corresponding P-value using Tukey's procedure is 0.4832 so this comparison is not significant. Similar conclusions can be made for the other

Output 15.2 Minitab General Linear Model output for data of Example 10.1 with two missing values

```
Analysis of Variance for Yield, using Adjusted SS for Tests

Source     DF    Seq SS    Adj SS    Adj MS     F        P
Block       2   10.5993    9.7821    4.8911   17.42    0.011
Variety     3    6.2938    6.2938    2.0979    7.47    0.041
Error       4    1.1229    1.1229    0.2807
Total       9   18.0160

Least Squares Means for Yield

Variety   Mean     StDev
1         6.814    0.3972
2         7.600    0.3059
3         6.114    0.3972
4         8.300    0.3059

Tukey Simultaneous Tests
Response Variable Yield
All Pairwise Comparisons among Levels of Variety

Variety=1 subtracted from:

Level    Difference    SE of                   Adjusted
Variety  of Means      Difference   T-Value    P-Value
2         0.7857       0.5013        1.567     0.4832
3        -0.7000       0.5804       -1.206     0.6551
4         1.4857       0.5013        2.964     0.1286

Variety=2 subtracted from:

Level    Difference    SE of                   Adjusted
Variety  of Means      Difference   T-Value    P-Value
3        -1.486        0.5013       -2.964     0.1286
4         0.700        0.4326        1.618     0.4615

Variety=3 subtracted from:

Level    Difference    SE of                   Adjusted
Variety  of Means      Difference   T-Value    P-Value
4         2.186        0.5013        4.360     0.0400
```

comparisons. The only significant difference is that between the LS means for V_3 and V_4.

To make these comparisons using two-tailed t-tests (at the 5% level) you should compare the t-values with the table value $t_{(4, 2.5\%)} = 2.776$. Using this method the comparisons between V_1 and V_4 and between V_2 and V_3 are also significant. However using t-tests to make all possible comparisons is equivalent to the LSD method discussed in Chapter 13 where the dangers of its use were pointed out. The P-values corresponding to each of these t-tests can be obtained using one of the options of PROC GLM in SAS. Output 15.3 shows that the P-value for comparing the V_1 and V_2 LS means corresponding to the t-value of 1.567 is 0.1921. The

Output 15.3 Comparison of least square means for Example 15.1 using PROC GLM in SAS

```
General Linear Models Procedure
                    Least Squares Means
   VARIETY       YIELD    Pr >|T| H0:   LSMEAN(i) = LSMEAN(j)
                 LSMEAN   i/j       1        2        3        4

      1       6.81428571    1       .     0.1921   0.2942   0.0414
      2       7.60000000    2     0.1921     .     0.0414   0.1809
      3       6.11428571    3     0.2942  0.0414      .     0.0121
      4       8.30000000    4     0.0414  0.1809   0.0121      .

NOTE:  To ensure overall protection level, only probabilities associated with
       pre-planned comparisons should be used.
```

comparison of V_1 and V_4 is significant at the 5% level because the *P*-value is 0.0414 (<0.05).

15.4 OTHER TYPES OF EXPERIMENT

When missing values occur in more complicated experiments such as Latin square and factorial experiments the same principles apply:

- The easiest option is to carry out the usual analysis with asterisks representing missing values. The least squares means will be given, but individual comparisons will be biased if the SED values not adjusted for missing values are used. The balance of the design is thus maintained.
- If more than 10% of values are missing or if most missing values occur in one or two treatments, the exact analysis should be carried out.

Note: In survey work data are rarely balanced. There are unlikely to be equal subclass numbers in a factorial structure and no attempt should be made to estimate missing values. The GLM method should be used when there is a lack of balance.

15.5 INCOMPLETE BLOCK DESIGNS

When the number of treatments is large, it is often not possible to have a complete set in each block of a RBD. For example, in a greenhouse experiment the width of the bench may restrict the number of pots which can form a block. In a field experiment, it may not be possible to find uniform areas of land large enough to accommodate all the treatments. In these cases incomplete blocks can be formed which contain less than the complete set of treatments. The analysis can be carried out using the GLM method (Section 15.3) as a special case of missing values in a RBD. The missing values are by design, but unless the design is chosen according to special rules valid treatment comparisons cannot be made. If a particular pair of

MISSING VALUES AND INCOMPLETE BLOCKS   235

treatments does not occur together in any block, any differences in yields may be due to blocks. The design should be balanced such that each pair of treatments occur together the same number of times within blocks.

15.5.1 Balanced Incomplete Blocks

A balanced incomplete block design (BIB) with t treatments and b blocks has k ($k < t$) experimental units per block such that each treatment occurs in r blocks, no treatment occurs more than once in a block and each pair of treatments occurs together in λ blocks. In this way all possible pairs of treatments can be compared with equal precision.

The symbols t, b, k, r and λ are called the parameters of the design but only certain combinations of t, b, k and r will work as λ must be an integer and $t \times r = b \times k$. No design with $b < t$ is possible. Also it can be shown that $\lambda(t-1) = r(k-1)$.

Construction

It is always possible to construct a design (an unreduced design) for any combination of t and k although in practice when t is large, the number of blocks required is unmanageable. The number of blocks is the number of combinations of all possible sets of t treatments taken k at a time ($k < t$). The formula is

$$\text{Number of blocks} = b = {}^tC_k = \frac{t!}{k!(t-k)!}$$

For example, for five treatments with three per block ($t = 5$, $k = 3$), the number of blocks is

$$b = \frac{5!}{3!(5-3)!} = \frac{5!}{3! \times 2!} = \frac{5 \times 4}{2 \times 1} = 10$$

The design is shown in Table 15.6. As $b \times k = t \times r$ the number of replications per treatment is $r = (b \times k)/t = (10 \times 3)/5 = 6$. Also, as $r(k-1) = \lambda(t-1)$,

$$\lambda = \frac{r(k-1)}{t-1} = \frac{6 \times 2}{4} = 3$$

Table 15.6. A BIB unreduced design with five treatments, three per block and ten blocks

Block	Treatments	Block	Treatments
1	1, 2, 3	6	1, 4, 5
2	1, 2, 4	7	2, 3, 4
3	1, 2, 5	8	2, 3, 5
4	1, 3, 4	9	2, 4, 5
5	1, 3, 5	10	3, 4, 5

Table 15.7. A BIB design with nine treatments, three per block and twelve blocks

Block	Treatments	Block	Treatments
1	1, 2, 3	7	1, 6, 8
2	4, 5, 6	8	2, 4, 9
3	7, 8, 9	9	3, 5, 7
4	1, 4, 7	10	1, 5, 9
5	2, 5, 8	11	2, 6, 7
6	3, 6, 9	12	3, 4, 8

Table 15.8. Design and mean plant height per pot for Example 15.3

Block	Treatments and mean height (cm)		
1	T_1 14.0	T_3 12.0	T_2 13.0
2	T_4 7.3	T_2 12.5	T_1 15.5
3	T_2 12.8	T_3 15.5	T_4 11.0
4	T_3 12.5	T_1 18.3	T_4 9.3

so each possible pair of treatments occurs together in three blocks. For example, the pair 2,5 occurs in blocks 3, 8 and 9.

In general, if t is large, r and b will be large. A BIB design with nine treatments, three per block, would require 84 blocks and have 28 replications per treatment if constructed using the above method. However, a balanced design may exist which consists of fewer than the ^tC_k blocks and hence fewer replications than required for the unreduced design. Such a design does exist with $t = 9, b = 12, k = 3, r = 4, \lambda = 1$ as shown in Table 15.7. The total number of experimental units in the design is $t \times r = b \times k = 9 \times 4 = 12 \times 3 = 36$. See Cochran and Cox (1957) for plans of other possible designs.

Randomisation

Once you have found a design it is important to randomise the order of treatments within each block and to randomly allocate the blocks of the design to the physical blocks of the experiment. Some computer packages will find a design and carry out the randomisation.

Example 15.3

Finally, we present a small numerical example with four treatments to illustrate the method of analysis by computer. A greenhouse experiment was conducted to compare four different fertiliser treatments on the growth of forage rape. There were four blocks, each of three pots. Each pot contained four plants and the mean heights in cm of the plants within the pots at the end of the experiment were used in the analysis to compare the treatments. The design and mean heights are given in Table 15.8 and the results are shown in Output 15.4.

In this example, $t = 4$, $k = 3$, $b = 4$, $r = 3$ and $\lambda = 2$ (each pair of treatments occurs in two blocks). Each block has a different one of the four treatments missing. This balance enables each pair of treatments to be compared with equal precision. Output 15.4 gives 1.267 as the common SED value; it is not $\sqrt{(2 \times 2.139)/3} = 1.194$ because it has been adjusted for the missing values. The means given are the least square means (16.43, 12.97, 13.01, 8.82). The raw means (15.93, 12.77, 13.33, 9.20) should not be quoted as they are based on different blocks. For example, 15.93 has no contribution from block 3 and 12.77 has no contribution from block 4 so a big difference between these blocks would bias the comparison of treatments 1 and 2. The least square means are adjusted for the missing values. In the analysis of variance table the sum of squares due to blocks ignoring treatments is 4.556 and the sum of squares due to treatments after allowing for blocks is 77.619. The residual sum of squares is 10.694 on 5 df. The least squares means are significantly different ($VR = 12.10$, $P = 0.010$).

Note: This example can also be analysed using the general linear model procedure in Minitab or SAS (Section 15.3.2). An output similar to Output 15.2 would be obtained using Minitab.

Output 15.4 Analysis of variance for data of Example 15.3 (Genstat output)

```
Variate: height
Source of variation   d.f.    s.s.       m.s.      v.r.    F pr.

block stratum
treat                  3      4.556     1.519

block.*Units* stratum
treat                  3     77.619    25.873    12.10    0.010
Residual               5     10.694     2.139

Total                 11     92.869

***** Tables of means *****

Variate: height

Grand mean     12.81

    treat            1        2        3        4
                  16.43    12.97    13.01     8.82

*** Standard errors of differences of means ***

Table      treat
rep.         3
d.f.         5
s.e.d.     1.267
```

Chapter 16

Split Plot Designs

16.1 INTRODUCTION

Split plot designs can take many forms. The most common is a randomised block design for a two-factor experiment with restricted randomisation of the treatment combinations within each block. The plots do not have to be plots of land so these designs are often called split unit designs.

Let a represent the number of levels of factor A and b the number of levels of factor B. There are thus $a \times b$ treatment combinations. Each block is divided into a large plots, called **main plots** or whole plots and each main plot is divided into b small plots, called **subplots**. Within each block, the levels of factor A are randomly assigned to the main plots. Within each main plot, the levels of factor B are randomly assigned to the subplots. A typical layout with three blocks, three levels of factor A and four levels of factor B is shown in Figure 16.1.

16.2 USES OF THIS DESIGN

(1) A split plot design is used in factorial experiments where the levels of one factor must be applied to large plots, while the levels of the other can be applied to small plots. Examples for large plots are ploughing and irrigation treatments, while fertilisers and varieties can be tested using smaller plots. A factorial experiment to compare several different ploughing depths on the yields of several varieties could be carried out as a split plot experiment, with ploughing as the main plot factor (A), and variety as the subplot factor (B).

(2) A split plot design is used in factorial experiments where one factor (A) is more important than the other (B). Each level of factor A occurs in r main plots, one in each block (r = number of blocks). Each level of factor B occurs in $a \times r$ subplots (a = number of main plots in each block). For example, for the design in Figure 16.1, each level of A occurs three times, and each level of B occurs $3 \times 3 = 9$ times. The subplot error is generally smaller than the main plot error. In other words, the variation in conditions between subplots within main plots

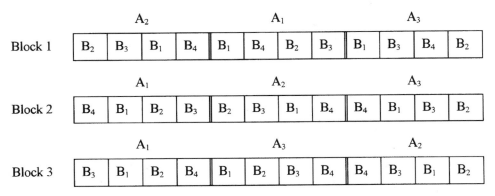

Figure 16.1. A typical layout for split plot design

is generally less than the variation in conditions between main plots within blocks. For these reasons the main effects of the levels of the main plot factor are generally estimated with less precision than the main effects of the subplot factor and the interaction effects.

(3) A split plot design can be used when one experiment is superimposed on another that is already in progress. Consider the layout shown in Figure 16.1. The nine main plots could be the blocks in a randomised block experiment designed to compare four varieties (the four levels of factor B). While this experiment is in progress a decision could be made to compare three types of insecticide. The nine original blocks could be divided into three groups of adjacent blocks, thus creating a split plot experiment. Each group is a new block, and the original blocks are main plots. Alternatively, the layout of Figure 16.1 could represent a randomised block experiment with three blocks to compare three varieties. The main plots within each block would be sown with a different variety. At a later date it may be decided to compare the effects of four levels of a fungicide on the three varieties. Each main plot is then divided into four subplots and a different level of fungicide applied at random to each subplot.

(4) A split plot design can be used when plots that have to be treated in the same way need to be grouped together. For example, if factor A is planting date or harvest date, then all the plots within each main plot are planted or harvested at the same time. In growth cabinet experiments the whole cabinet is a main plot for a factor such as temperature while fertiliser treatments can be applied to subplots within cabinets (Example 16.3).

The major disadvantage of a split plot design is that the main effects of the main plot factor are estimated with lower precision than the interaction and main effects of the subplot factor. For this reason a split plot experiment should not be carried out if equal precision is required on both factors. Another disadvantage is that the analysis is more complicated than that for a factorial experiment in randomised blocks. Even

Table 16.1. Main plot skeleton ANOVA table for design given in Figure 16.1

Source	DF
Blocks	$r - 1 = 2$
A	$a - 1 = 2$
Main plot residual	$(r - 1)(a - 1) = 4$
Main plot total	$(ar - 1) = 8$

Table 16.2. Subplot skeleton ANOVA table for design given in Figure 16.1

Source	DF
B	$b - 1 = 3$
A × B	$(a - 1)(b - 1) = 6$
Subplot residual	18
Subplot total	27

though computer programs replace the tedium of carrying out the calculations by hand, the interpretation of the output is not straightforward.

16.3 THE SKELETON ANALYSIS OF VARIANCE TABLES

When designing a split plot experiment you should work out the degrees of freedom in the ANOVA table. This table is in two parts. The first deals with the **main plot analysis** and the second with the **subplot analysis**. The main plot skeleton ANOVA table for the design of Figure 16.1 is given in Table 16.1.

Once the analysis of the data has been carried out, the main effects of A are compared using the main plot residual mean square which is based on only 4 degrees of freedom. Hence the main effects of A are compared with very low precision. The grand total degrees of freedom are $(abr - 1) = 35$. The subplot total degrees of freedom are (grand total df − main plot total df) $= 35 - 8 = 27$. The subplot skeleton ANOVA table is given in Table 16.2. The subplot residual degrees of freedom are found by subtraction and are $27 - 3 - 6 = 18$.

Example 16.1
Consider a field experiment testing all combinations of three depths of ploughing (P), and four drilling techniques (D). The twelve treatment combinations are to be replicated four times using four blocks. In practice, it is necessary to use large plot areas (main plots) for the different ploughings. Each main plot is divided into four subplots, one for each drilling technique. A typical layout of one block could be as follows:

	P₂				P₃				P₁		
D₁	D₄	D₂	D₃	D₄	D₁	D₃	D₂	D₂	D₃	D₁	D₄

Note that within each block the ploughing treatments are assigned at random to the main plots, and within each main plot the drilling techniques are assigned at random to the subplots.

If this experiment had been carried out using a randomised blocks design, each of the 12 treatment combinations would have been assigned at random to 12 plots within each block. In this case different ploughing treatments would have occurred in adjacent plots as can be seen in the following table which represents one block. As a result the plot sizes would have to be considerably increased, so making the blocks very large.

P₁	P₃	P₂	P₁	P₃	P₂	P₃	P₂	P₃	P₁	P₂	P₁
D₂	D₁	D₁	D₄	D₄	D₂	D₃	D₄	D₂	D₁	D₃	D₃

It is informative to compare the skeleton analyses of variance for these twelve treatment combinations when using randomised block factorial and split plot designs each having four replications (blocks). The skeleton ANOVA tables for the split plot design are shown in Table 16.3.

There are 12 main plots and three of them in a given block are assigned a different ploughing treatment. The total number of subplots altogether is 48 and so the grand total degrees of freedom are 47. The subplot total degrees of freedom are obtained by subtracting the main plot total degrees of freedom from the grand total degrees of freedom. In this example the subplot total degrees of freedom are $47 - 11 = 36$. Note that the ploughing treatments are compared with low precision as the main plot residual degrees of freedom are only 6. The subplot treatments, namely the drilling techniques, are compared with much greater precision as the subplot residual degrees of freedom are 27.

In the randomised blocks factorial design where there is unrestricted randomisation within each block, the main and interaction effects are compared with equal precision because the same pooled residual mean square is used in all significance tests. Thus, if the ploughing effect is as important as the drilling effect, the split plot design should be avoided. The skeleton ANOVA table is given in Table 16.4.

Table 16.3. Skeleton ANOVA tables for Example 16.1

Main plot table

Source	DF
Blocks	3
Ploughing	2
Main plot residual	6
Main plot total	11

Subplot table

Source	DF
Drilling	3
Ploughing × Drilling	6
Subplot residual	27
Subplot total	36

Table 16.4. Skeleton ANOVA table assuming a randomised blocks factorial design for Example 16.1

Source	DF
Blocks	3
Ploughing	2
Drilling	3
Ploughing × Drilling	6
Residual	33
Total	47

16.4 AN EXAMPLE WITH INTERPRETATION OF COMPUTER OUTPUT

Example 16.2
An experiment was carried out to find the effect of spraying an insecticide to control aphids on the yield of three varieties of field beans. The design was a split plot in four randomised blocks (replicates). Factor A was spraying at two levels; the main plot treatments were A_1 = no insecticide (plots sprayed with water), A_2 = + insecticide in the same volume of water. Factor B was variety; the subplot treatments were B_1 = variety 1, B_2 = variety 2, B_3 = variety 3.

Each block contained two equal-sized main plots which were divided into three equal subplots. The varieties were assigned at random within each main plot so that for factor B (variety) a randomised block design with $4 \times 2 = 8$ blocks was used. Within each original block the spray treatment (A_2) was assigned at random to one of the two main plots; the other main plot received only water. The spraying can be regarded as being a separate experiment superimposed on an experiment originally designed to compare three varieties in eight blocks. Each area sprayed consisted of three adjacent subplots each containing a different variety. The design and the yields of beans in kg/subplot are given in Table 16.5:

r = number of blocks = 4 $\qquad a$ = number of levels of factor A = 2
b = number of levels of factor B = 3 $\quad t$ = number of treatment combinations = 6

The experimental units are the subplots of which there are $2 \times 3 \times 4 = 24$, so $N = 24$. Comparisons of the A effects are expected to be less precise than comparisons of the B effects and A × B interactions. This is because (1) main plots are further apart in space than subplots and this implies greater residual variation, and (2) main plot treatments have less replication. For these reasons two separate ANOVA tables are formed; one for the analysis of the main plot treatments and the other for the subplot treatments and the interaction. There are two residual mean squares. The first is called the main plot error and is denoted by E_a, the second is the subplot error and is denoted by E_b. E_a is used to find the variance ratio for A and to calculate the SED for comparing the spraying (factor A) means.

E_b is used to find the variance ratios for B and A × B and to calculate the *SED* values for comparing variety (factor B) means and the A × B means. E_b should be less than E_a. If it is not, this is probably because the residual variation **between** main plots is less than that **within** main plots possibly due to not having the blocks at right angles to any site gradient.

16.4.1 The Analysis

Using a computer you should make a spreadsheet with 24 rows, one for each subplot. There should be four columns, one for the block codes, one for the factor A (spraying) codes or labels, one for the factor B (variety) codes or labels and one for the yields. In the program you have to specify that spraying is the main plot factor and variety the subplot factor. Output 16.1 shows the results produced by Genstat for this example. Because of its complexity the interpretation is presented in two parts. In part 1 we give the main conclusions and in part 2 we explain how the entries were obtained.

16.4.2 Interpretation (Part 1)

The two-way table of means shows that spraying increased the yields for V_1 and V_2, while the yield of V_3 was high in the presence and absence of spraying. As this interaction is very highly significant ($VR = 88.13$, $P < 0.001$) it would be misleading to present the overall spraying means even though they are significantly different

Table 16.5. Design and yields (kg/subplot) for Example 16.2

Block

									Block total
A_2	25.5	24.9	25.8	A_1	26.1	18.0	21.7		142.0
	B_2	B_1	B_3		B_3	B_2	B_1		
A_1	21.1	17.9	28.9	A_2	27.6	29.4	29.3		154.2
	B_1	B_2	B_3		B_1	B_3	B_2		
A_1	28.6	19.5	23.2	A_2	29.7	30.3	29.5		160.8
	B_3	B_2	B_1		B_2	B_3	B_1		
A_2	29.3	29.2	29.8	A_1	21.3	31.0	25.8		166.4
	B_3	B_2	B_1		B_2	B_3	B_1		

Output 16.1 Results of the split plot analysis for Example 16.2 using Genstat

***** Analysis of variance *****

Variate: yield

Source of variation	d.f.	s.s.	m.s.	v.r.	F pr.
block stratum	3	55.0583	18.3528	5.16	
block.spraying stratum					
spraying	1	136.3267	136.3267	38.32	0.008
Residual	3	10.6733	3.5578	7.39	
block.spraying.variety stratum					
variety	2	98.3700	49.1850	102.23	<0.001
spraying.variety	2	84.8033	42.4017	88.13	<0.001
Residual	12	5.7733	0.4811		
Total	23	391.0050			

***** Tables of means *****

Variate: yield

Grand mean 25.97

spraying	No spray	Spray	
	23.59	28.36	

variety	V1	V2	V3
	25.45	23.80	28.67

spraying variety	V1	V2	V3
No spray	22.95	19.17	28.65
Spray	27.95	28.42	28.70

*** Standard errors of differences of means ***

Table	spraying	variety	spraying variety
rep.	12	8	4
s.e.d.	0.770	0.347	0.868
d.f.	3	12	4.76

Except when comparing means with the same level(s) of
spraying   0.490
d.f.   12

*** Least significant differences of means (5% level) ***

Table	spraying	variety	spraying variety
rep.	12	8	4
l.s.d.	2.451	0.756	2.266
d.f.	3	12	4.76

Except when comparing means with the same level(s) of
spraying   1.069
d.f.   12

***** Stratum standard errors and coefficients of variation *****

Variate: yield

Stratum	d.f.	s.e.	cv%
block	3	1.749	6.7
block.spraying	3	1.089	4.2
block.spraying.variety	12	0.694	2.7

($VR = 38.32$, $P = 0.008$). They suggest that spraying increases yields from 23.59 to 28.36 on average. The mean for variety three (28.67) is much larger than for the other two varieties because of its high yield without spray, possibly due to genetic resistance to aphids. This is why the VR for variety is extremely large (102.23). However, due to the interaction, the two-way table of means should be presented. A visual impression of the effect of spraying is shown in Figure 16.2. This includes the SED value of 0.868 which is used to compare two means at the same level of variety (we could have put the LSD value of 2.266 but this is not recommended). To compare two means at the same level of spraying you would use the SED value of 0.490 or the LSD value of 1.069 (see part 2 for a more detailed explanation).

You can conclude that spraying significantly increased yield for V_1 and V_2 but had no effect on V_3. If there had been no significant interaction the overall variety means would have been compared using the SED value of 0.347 or LSD value of 0.756. Similarly, the overall spraying means would have been compared using the SED value of 0.770 or LSD value of 2.451.

16.4.3 Interpretation (Part 2)

You will note that Output 16.1 refers to three strata. To understand this you have to consider the structure of the plots before any treatments are applied. The existence of a **block stratum** implies a background block-to-block variation. Within each block there is a background variation between the main plots, and within each main plot there is a background variation between subplots. After the treatments have been applied and the yields measured these three variations can be estimated. The **block stratum** sum of squares ($BlockSS = 55.0583$) represents the background block-to-block variation. It should not be affected by the treatments as all six treatment combinations are equally represented in each block.

In the **block.spraying** stratum, the variation between main plots within each block is pooled to give the main plot residual sum of squares ($MPResidSS = 10.6733$). This

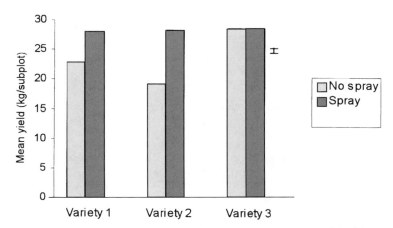

Figure 16.2. Effect of insecticide spray on the yields of three varieties of field beans

represents the background variation between main plots within blocks. It is used to test the effect of the main plot factor (spraying). In the **block.spraying.variety** stratum, the variation between subplots within each main plot is pooled to give the subplot residual sum of squares ($SPResidSS = 5.7733$). This represents the background variation between subplots within main plots within blocks. It is used to test the effect of the subplot factor (variety) and to test if there is an interaction between the main and subplot factors.

The total variation in all 24 yields is represented by the grand total sum of squares $GTotalSS = 391.0050$. This is the corrected sum of squares (the Sxx value) of all 24 yields. The corresponding degrees of freedom are $24 - 1 = 23$.

The main plot analysis

This can be carried out as a randomised blocks analysis using the main plot totals. However, all the sums of squares should be divided by 3, as each of these totals is made up of ($b = 3$) yields. Table 16.6 shows the block × spraying totals.

The main plot total sums of squares is found by first calculating the corrected sum of squares of the eight main plot totals (65.8, 76.2, . . ., 88.3). The answer (606.175) is divided by 3 because each main plot consists of three subplots. Hence $MPTotalSS = 202.0583$. This is the sum of the first three entries in the s.s. column of Output 16.1 and represents the variation in the main plot yields on a subplot basis. The corresponding degrees of freedom are 7 (one less than the total number of main plots).

Note: To calculate a corrected sum of squares using your calculator in SD or STATS mode, find the sample standard deviation of the relevant entries, square the result and multiply by one less than the number of entries.

The sums of squares for blocks ($BlockSS = 55.0583$) is obtained by calculating the corrected sum of squares of the block totals and dividing by 6 because each block total is the sum of yields from six subplots. The corresponding degrees of freedom are 3 (one less than the number of blocks).

To calculate the sums of squares for factor A (spraying), denoted by $SprayingSS$, first find the corrected sum of squares of the spraying totals (283.1 and 340.3), then

Table 16.6. Two-way table of Block × A totals for Example 16.2

	A_1 (no spray)	A_2 (spray)	Total
Block 1	65.8	76.2	142.0
Block 2	67.9	86.3	154.2
Block 3	71.3	89.5	160.8
Block 4	78.1	88.3	166.4
Total	283.1	340.3	623.4
Mean	23.59	28.36	

SPLIT PLOT DESIGNS 247

divide by 12 because each spraying total is made up of 12 subplot yields. The result is 136.3267 on 1 degree of freedom (one less than the number of main plots within each block).

The main plot residual sum of squares and degrees of freedom are found by subtraction:

$$MPResidSS = MPTotalSS - BlockSS - SprayingSS$$
$$= 202.0583 - 55.0583 - 136.3267 = 10.6733$$
$$MPResidDf = MPTotalDf - BlockDf - SprayingDf = 7 - 3 - 1 = 3$$

The VR for Blocks is found by dividing the Blocks MS (18.3528) by the Main plot residual MS (3.5578). The value of 5.16 is less than 9.28, the 5% F-table value on (3, 3) df. This shows that blocking was not particularly successful. The VR for spraying is found by dividing the spraying MS (136.3267) by the Main plot residual MS. The value of 38.32 is greater than 34.12, the 1% F-table value on (1, 3) df ($P = 0.008$ confirms this). This indicates a considerable overall spraying effect, but because of the large spraying × variety interaction, the effect of spraying should be investigated for each variety separately.

The SED for comparing the factor A (spraying) means of 23.59 (unsprayed) and 28.36 (sprayed) is

$$SED_A = \sqrt{\frac{2 \times E_a}{r \times b}} = \sqrt{\frac{2 \times 3.5578}{12}} = 0.770 \quad \text{on 3 df}$$

The corresponding $LSD\ (5\%) = t_{(3, 2.5\%)} \times SED_A = 3.182 \times 0.770 = 2.45$.

The difference between the A means is larger than 2.45 so you conclude that, averaged over the three varieties, spraying increased yield. However, there may be interaction, i.e. spraying may not have increased the yield of all varieties as some genetic resistance to aphid attack may have been introduced by plant breeding. To investigate further you must carry out the subplot analysis.

The subplot analysis

Table 16.7 shows the A × B table of totals. In this example A = spraying and B = variety. Each A × B total is made up of $r = 4$ yields. The SS for the subplot total is

$$SPTotalSS = GTotalSS - MPTotalSS = 391.0050 - 202.0583 = 188.9467$$

and the corresponding degrees of freedom are

$$SPTotalDf = GTotalDf - MPTotalDf = 23 - 7 = 16$$

The sums of squares for factor B (variety), denoted by $VarietySS$, is obtained by calculating the corrected sum of squares of the variety totals (203.6, 190.4, 229.4) and dividing by 8 because each variety total is the sum of yields from eight subplots. The result is 98.3700 and the corresponding degrees of freedom are 2 (one less than the number of varieties).

Table 16.7. The A × B table of totals for Example 16.2

	A$_1$ (no spray)	A$_2$ (spray)	Total	Mean
B$_1$ (Variety 1)	91.8	111.8	203.6	25.45
B$_2$ (Variety 2)	76.7	113.7	190.4	23.80
B$_3$ (Variety 3)	114.6	114.8	229.4	28.68

The interaction sum of squares is obtained by subtracting *SprayingSS* and *VarietySS* from the treatment sum of squares (*TreatSS*). To find *TreatSS* first calculate the corrected sum of squares of the treatment combination totals (the six numbers in the A$_1$ and A$_2$ columns of Table 16.7), then divide by 4 because each treatment total is made up of four subplot yields, one from each block. The result is 319.5000 on 5 degrees of freedom (one less than the number of treatment combinations):

$$InteractionSS = TreatSS - SprayingSS - VarietySS$$
$$= 319.5000 - 136.3267 - 98.3700 = 84.8033$$

The corresponding degrees of freedom are

$$InteractionDf = TreatDf - SprayingDf - VarietyDf = 5 - 1 - 2 = 2$$

The subplot residual SS is found as follows:

$$SPResidSS = SPTotalSS - VarietySS - InteractionSS$$
$$= 188.9467 - 98.3700 - 84.8033 = 5.7734$$

The corresponding degrees of freedom are found by subtraction, hence

$$SPResidDf = SPTotalDf - VarietyDf - InteractionDf = 16 - 2 - 2 = 12$$

The *VR* for variety is found by dividing the variety MS (49.1850) by the subplot residual MS (0.4811). The value of 102.23 is greater than 12.97, the 0.1% *F*-table value on (2, 12) df. This suggests a significant variety effect, but because of the large spraying × variety interaction, the varieties should be compared at each level of spraying.

The SED for comparing main effects of B (variety) is

$$SED_B = \sqrt{\frac{2 \times E_b}{r \times a}} = \sqrt{\frac{2 \times 0.4811}{8}} = 0.347$$

The overall variety means are 25.45, 23.80 and 28.67. The LSD for comparing these at the 5% level is

$$t_{(12, 2.5\%)} \times SED_B = 2.179 \times 0.347 = 0.756$$

All the overall variety means differ by more than this so they are all significantly different from each other at the 5% level.

The *VR* for the spraying variety interaction is found by dividing the spraying variety MS (42.4017) by the subplot residual mean square (0.4811). The value of 88.13 is greater than 12.97, the 0.1% *F*-table value on (2, 12) df. This shows that there is a considerable interaction between spraying and variety. To investigate the interaction, compare the means in the spraying variety table of means in Output 16.1.

In this type of split plot experiment there are two SED values for comparing the means in the A × B table. One is used for comparing two A × B means at the same level of A. This is likely to be fairly small as it is based on the subplot error. The other SED is used to compare two A × B means not at the same level of A. This is likely to be larger than the first SED because its calculation includes the main plot error. In this example the main plot error was found to be significantly greater than the subplot error because the *VR* for main plot residual (7.39) is greater than 5.95, the 1% *F*-table value on (3, 12) df. This *VR* is found by dividing the main plot residual MS (3.5578) by the subplot residual MS (0.4811).

The SED for comparing two spraying variety means at the same level of spraying is

$$\sqrt{\frac{2 \times E_b}{r}} = \sqrt{\frac{2 \times 0.4811}{4}} = 0.490$$

The corresponding LSD at the 5% level is $LSD\,(5\%) = t_{(12, 2.5\%)} \times SED = 2.179 \times 0.490 = 1.069$. This shows that when spray is not used, there are significant differences between the variety yields, and that when spray is used there are no significant differences between the variety yields.

The SED for comparing two spraying variety means not at the same level of spraying is

$$\sqrt{\frac{2 \times [(b-1) \times E_b + E_a]}{r \times b}} = \sqrt{\frac{2 \times [2 \times 0.4811 + 3.5578]}{12}} = 0.868$$

Note that the expression in the square root symbol is $2/r$ times a weighted average of the two residual mean squares E_a and E_b where the weights are 1 and $b-1$ respectively.

You cannot find an exact LSD for this example because the difference between two A × B means not at the same level of A, when divided by the appropriate SED, does not come from a *t*-distribution. Genstat calculates an approximate LSD (2.266) based on an effective degrees of freedom (4.76). You can therefore conclude that spraying significantly increases yield for varieties 1 and 2 but has no effect on variety 3.

The calculations of the coefficients of variation per main plot and per subplot are as follows:

$$\text{Coefficient of variation per main plot} = \frac{\sqrt{\frac{E_a}{b}}}{\bar{X}} = \frac{\sqrt{\frac{3.5578}{3}}}{25.975} = 4.19\%$$

where $\bar{X}$ = overall mean = $\frac{G}{N} = \frac{623.4}{24} = 25.975$ and N = total number of yields = 24

$$\text{Coefficient of variation per subplot} = \frac{\sqrt{E_b}}{\bar{X}} = \frac{\sqrt{0.4811}}{25.975} = 2.67\%$$

16.5 THE GROWTH CABINET PROBLEM

Treatment factors such as temperature, humidity, light and CO_2 concentration cannot be applied at different levels to individual plants within a growth cabinet. They have to be applied to whole cabinets and are thus referred to as between-cabinet factors. Consider an experiment to compare the effects of different temperatures on plant growth. Suppose one cabinet is used for each temperature. If each contains several identical replicate plants placed at random, it is tempting to analyse the experiment as a completely randomised design. However, the temperature effect is confounded with the cabinet effect. Differences in the means can only be attributed to temperature if it can be assumed that the cabinets differ only in this respect. Ideally you should have several cabinets per temperature and use the growth characteristic means per cabinet in the analysis. This may not be satisfactory if only two cabinets per treatment are available as this would result in too few residual degrees of freedom.

Example 16.3

Experiments often involve a between-cabinet and a within-cabinet factor, for example temperature and variety or humidity and nutrient concentration. Consider an experiment with five nutrient levels and three temperatures. The following design is not appropriate. Allocate one cabinet for each temperature and within each cabinet have four pots with each nutrient level randomly placed. Although this could be analysed as a factorial experiment with four replications per treatment combination, the interpretation of the results would not be valid because all the pots at a given temperature are in one cabinet. There is effectively no temperature replication so this design is unsatisfactory. You could analyse the results within each cabinet separately as three different experiments but you will not be able to compare the temperature effects statistically. A split plot design should be used. The minimum requirement is two cabinets randomly allocated to each temperature. These are the main plots. Within each cabinet, each nutrient level is randomly assigned to one of five pots and these are randomly placed. There are thus six main plots and five subplots within each main plot. The main plots are arranged in a completely randomised design. Table 16.8 gives hypothetical fresh weights per pot for such an experiment and the analysis performed by Genstat is given in Output 16.2.

Note that there is no row for blocks in the main plot table because the main plots are arranged in a completely randomised design. The analysis is similar to that for Example 16.2 but there is no sum of squares for blocks to calculate. The main plot residual mean square is 2.9457 on 3 df. The subplot residual mean square is much smaller (0.7890) and has many more degrees of freedom (12) leading to the nutrient and interaction effects being assessed with greater precision than the temperature effects. In this case there appears to be a significant temperature effect ($VR = 12.87$, $P = 0.034$) but the significant interaction ($VR = 3.24$, $P = 0.033$) indicates that you should investigate the two-way table of means.

Table 16.8. Hypothetical plant fresh weights (g per pot) in a growth cabinet experiment

Cabinet	Temp (°C)	Nutrient levels				
		1	2	3	4	5
1	20	13.0	16.7	20.4	27.7	26.9
2	20	10.7	17.2	18.5	24.3	25.3
3	25	14.2	17.9	21.5	30.4	30.8
4	25	12.7	16.8	22.0	28.7	32.4
5	30	13.8	19.0	23.3	29.8	32.0
6	30	14.3	19.2	22.9	30.5	33.8

Output 16.2 Analysis of variance for Example 16.3 using Genstat

```
Variate: freshwt

Source of variation    d.f.      s.s.       m.s.      v.r.    F pr.
temp.rep stratum
temp                    2      75.8247    37.9123    12.87    0.034
Residual                3       8.8370     2.9457     3.73

temp.rep.*Units* stratum
nutrient                4    1241.9187   310.4797   393.51   <0.001
temp.nutrient           8      20.4453     2.5557     3.24    0.033
Residual               12       9.4680     0.7890

Total                  29    1356.4937
```

***** Tables of means *****

Variate: freshwt

Grand mean   22.22

```
      temp         20.00     25.00     30.00
                   20.07     22.74     23.86

nutrient              1         2         3         4         5
                  13.12     17.80     21.43     28.57     30.20

temp nutrient         1         2         3         4         5
20.00             11.85     16.95     19.45     26.00     26.10
25.00             13.45     17.35     21.75     29.55     31.60
30.00             14.05     19.10     23.10     30.15     32.90
```

*** Standard errors of differences of means ***

```
Table           temp    nutrient    temp
                                    nutrient
rep.             10        6          2
s.e.d.          0.768    0.513      1.105
d.f.             3         12         10
Except when comparing means with the same level(s) of
   temp                              0.888
   d.f.                                12
```

Do not ignore the split plot nature of this kind of experiment when analysing the results. This could lead to the reporting of an effect which is not present (a Type I error). If you, erroneously, analyse the above as a completely randomised factorial experiment, you will obtain 15 residual df and a residual mean square of 1.220 for estimating all the effects. The temperature effect will appear to be much more significant ($VR = 31.07$, $P < 0.001$) and the interaction will not be significant ($VR = 2.09$, $P = 0.103$).

These types of experiments are often not satisfactory when few growth cabinets are available. This is because there are likely to be too few main plot residual degrees of freedom for comparing the levels of the between-cabinet factor.

16.6 OTHER TYPES OF SPLIT PLOT EXPERIMENT

The split plot principle can be used in many ways. We have described in detail a split plot experiment where the main plots are arranged in a randomised complete block design (Example 16.2). Other designs such as the completely randomised design (Example 16.3) and Latin square can also be used for the main plot factor. In addition, the treatments applied to the main plot or subplots or both could have a factorial structure.

The principle can be extended to **split split plot experiments**. Each subplot is further divided into subsubplots. The analysis is complicated as there are now main plot, subplot, and subsubplot errors to consider. Another modification of the split plot principle is the split block or criss-cross design. Within each block, the plots are arranged in rows and columns. The levels of factor A are applied to whole rows and the levels of factor B to complete columns. The analysis is complex. For further details and discussion see Mead (1988) and Steel, Torrie and Dickey (1996).

16.7 REPEATED MEASURES

The term *repeated measures* sometimes refers to experiments where each unit (often an animal) receives different treatments separated in time, and sometimes to experiments where each unit receives a given treatment but measurements are repeatedly made on it over time. We consider only the latter type of experiment here. The observations are called **serial measurements**.

Consider a simple experiment to compare several treatments in a CRD or a RBD. Suppose mean plant height per plot is measured at weekly intervals for several weeks. A separate analysis can be carried out on the data for each week. This enables comparisons between the treatments to be made for several development stages. A multiple graph of the treatment mean against time should be presented for each treatment. A separate *SED* value should be included for each time period.

How can we obtain a single analysis which takes into account the development from week to week? It is incorrect to analyse this as a factorial experiment with time as a factor, because the experimental units measured at Time 1 for Treatment 1 are the same units that are measured at Time 2 for Treatment 1. In an experiment with four treatments and five replicates per treatment, there are 20 units. If measurements are taken at six weekly intervals, there are still 20 units, not 120. The same problem occurs with destructive sampling when plant subsamples are repeatedly taken from the same plot for dry weight measurements. However, it is valid to use a factorial analysis with time as a factor if for each treatment time combination, different replicate plots have been allocated in advance using an appropriate experimental design.

The results of an experiment with repeated measurements on each experimental unit are often analysed as if they came from a split plot experiment with time as the subplot factor. The units are considered to be main plots with treatment as the main plot factor, and time as the subplot factor. The several time measurements for a given plot are assumed to be measurements from several subplots. The main criticism of this analysis is that the heterogeneity assumption is very unlikely to be valid. At early times the residual variation is likely to be small and this is expected to increase rapidly with time. Another criticism is that the time measurements do not come from actual time subplots which have been randomly assigned different times. Time cannot be randomised. In reality the time measurements for a given plot are likely to be positively correlated. There is positive correlation between the time 1 and time 2 measurements if plots with above-average time 1 measurements have above-average time 2 measurements. This relationship is likely to decrease for times further apart. A **multivariate analysis** of variance takes into account these different correlations, but it is not often carried out due to the complexities of interpretation.

To take into account the time effect in a single analysis of variance without using time as a subplot factor you could use a derived variable. If the variable, y, is measured on each unit at each time point, a simple derived variable is the last minus the first y-value; another is the average of the y-values. You could also

(1) Obtain a graph of y against time for each unit, fit a smooth curve and estimate one or more of (a) the area under the curve, (b) the maximum y attained and (c) the time taken to reach the maximum y; then analyse one or more of these derived variables.
(2) Fit a response curve or straight line for each unit. Then analyse the slopes or other estimated parameters.

Exercise 16.1
An experiment was carried out to compare the effects of three concentrations of a chemical seed dressing with a control on the yield of oats. Three varieties were used in the trial because it was suspected that the response to the seed treatments would depend on the variety used. A split plot design was laid out in five randomised blocks. Varieties were assigned at random to the main plots within each block, and the seed treatments and control were assigned at random to the subplots within each main plot. The design and yields in kg per subplot were as follows:

A_1 = Variety 1 A_2 = Variety 2 A_3 = Variety 3
B_1 = Control B_2 = Conc 1 B_3 = Conc 2 B_4 = Conc 3

	Block 1	Block 2	Block 3	Block 4	Block 5
	B_1 42.9	B_2 67.3	B_3 41.4	B_2 46.3	B_3 54.1
A_1	B_4 44.4 A_3	B_3 65.3 A_2	B_4 44.1 A_1	B_4 34.7 A_3	B_1 56.5
	B_3 49.5	B_1 65.6	B_2 42.4	B_3 39.4	B_2 60.2
	B_2 53.8	B_4 69.4	B_1 45.4	B_1 30.8	B_4 57.2
	B_3 59.8	B_2 69.6	B_1 28.9	B_2 58.5	B_4 34.0
A_2	B_1 53.3 A_2	B_3 65.8 A_1	B_3 40.7 A_3	B_3 51.0 A_1	B_2 41.2
	B_2 57.6	B_4 57.4	B_4 28.3	B_4 47.4	B_1 35.1
	B_4 64.1	B_1 69.6	B_2 43.9	B_1 52.7	B_3 37.4
	B_3 68.8	B_3 53.8	B_1 54.0	B_3 45.4	B_4 51.8
A_3	B_1 75.4 A_1	B_2 58.5 A_3	B_4 56.6 A_2	B_4 51.6 A_2	B_3 50.3
	B_2 70.3	B_1 41.6	B_2 57.6	B_1 35.1	B_2 52.7
	B_4 71.6	B_4 41.8	B_3 45.6	B_2 51.9	B_1 48.3

Analyse these data as fully as possible and state your conclusions as to the effect of the concentration of seed dressing on yield for each variety. See Example 16.2 for details of the calculations.

Answers to Exercise 16.1

The ANOVA table and the two-way table of means are as follows:

Source	DF	SS	MS	VR
Blocks	4	2931.27	732.82	24.53
Variety	2	3631.54	1815.77	60.77
Main plot residual	8	239.02	29.88	
Main plot total	14	6801.83		
Conc.	3	350.20	116.73	8.32
Conc. × Variety	6	368.28	61.38	4.37
Subplot residual	36	505.36	14.04	
Subplot total	45	1223.84		
Grand total	59	8025.67		

	B_1	B_2	B_3	B_4
A_1	35.86	48.74	44.16	36.64
A_2	50.34	54.84	52.54	53.80
A_3	60.84	62.78	56.96	60.44

The varieties differ significantly as 60.77 is greater than $F_{(2,8,0.1\%)} = 18.49$. The SED for comparing two variety means = 1.728 (the means are 41.35, 52.88, 60.26). The main plot coefficient of variation is 5.31%. However, the differences between the variety means depend on the concentration as indicated by the significant interaction (4.37 is greater than $F_{(6,36,1\%)} = 3.35$). The concentrations differ significantly as 8.32 is greater than $F_{(3,36,0.1\%)} = 6.74$. Due to the significant interaction it is not appropriate to use the SED = 1.368 for comparing two concentrations averaged over all varieties.

The A_1B_1 mean is 35.86 and the A_1B_2 mean is 48.74. The SED for comparing these two concentration means for the same variety is 2.370 and the corresponding $LSD_{5\%}$ is $2.370 \times t_{(36,2.5\%)} = 2.370 \times 2.028 = 4.806$. The difference in means is $48.74 - 35.86 = 12.88$. Hence, variety 1, concentration 2, produces much higher yields than the control. However, for variety 2 the corresponding difference in means is $54.84 - 50.34 = 4.50$ which is not significant at the 5% level.

The A_2B_1 mean is 50.34. To compare this with the A_1B_1 mean of 35.86 (two AB means not at the same level of A) the appropriate SED is 2.683. To compare the A_1B_1 and A_2B_2 means the same SED value applies. There is no exact LSD but Genstat gives an approximate value of 5.464 based on 32.22 df.

Plot a graph of mean yield against concentration for each variety separately and mark on it the SED for comparing two variety means at the same level of concentration, namely 2.683. The graph shows that for varieties 2 and 3 the chemical has little effect and their yields remain well above those for variety 1. For variety 1 its yield is greatly increased if the chemical is applied at level 1 or level 2.

Chapter 17

Comparison of Regression Lines and Analysis of Covariance

17.1 INTRODUCTION

In Chapter 7 we discussed simple linear regression which involves fitting a straight line to a number of points using the method of least squares. In this chapter we show how to compare two or more regression lines. We also show how this technique can be used with analysis of variance to control experimental error, a method called analysis of covariance.

17.2 COMPARISON OF TWO REGRESSION LINES

In this section we describe how to test whether regression lines fitted to two independent sets of data are parallel. Using the two independent estimates of slope $\hat{\beta}_1$, $\hat{\beta}_2$ based on n_1 and n_2 observations respectively, you wish to test the hypothesis that $\beta_1 = \beta_2$. If you can assume parallel lines you can go on to test whether the intercepts are equal. If they are you can fit an overall line to the data.

Example 17.1
A greenhouse experiment was carried out to investigate the response of forage rape to two types of fertiliser (F_1 and F_2). Four plants were grown in each pot and twelve pots were treated at random with each fertiliser. At harvest time various measurements were made on each pot, but for this example we are only interested in the total plant dry weight (X) and the total leaf area (Y). The results are shown in Table 17.1.

A straight line fitted to all 24 points gives the equation $Y = -9.51 + 468.74X$. You may want to use this equation to estimate leaf area from plant dry weight. However, the regression lines may be different for the two fertilisers. Figure 17.1

shows the two fitted lines. The dotted line is fitted to the data of F_1 and the full line to the data of F_2.

The fitted equations are $Y_1 = -36.50 + 528.54 X_1$ and $Y_2 = 2.49 + 446.00 X_2$. You can also calculate the following from Table 17.1:

$\bar{X}_1 = 0.3650 \quad \bar{Y}_1 = 156.4167 \quad \bar{X}_2 = 0.3008 \quad \bar{Y}_2 = 136.6667$

$Sxx_1 = 0.1053 \quad Syy_1 = 34\,220.92 \quad Sxy_1 = 55.6550 \quad ResidSS_1 = 4805.161$

$Sxx_2 = 0.1021 \quad Syy_2 = 27\,254.67 \quad Sxy_2 = 45.5333 \quad ResidSS_2 = 6946.599$

Table 17.1. Measurements of plant dry weight (X) and leaf area (Y) for two fertilisers (F_1, F_2)

F_1	X_1	0.29	0.43	0.21	0.53	0.27	0.33	0.47	0.40	0.48	0.30	0.37	0.30
	Y_1	144	180	60	226	105	111	217	221	218	137	153	105
F_2	X_2	0.27	0.37	0.42	0.19	0.30	0.25	0.35	0.48	0.22	0.30	0.14	0.32
	Y_2	129	206	172	80	124	89	134	220	138	105	62	181

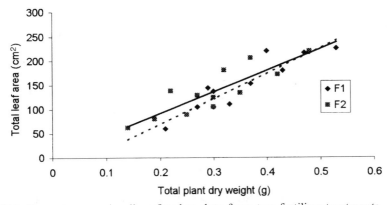

Figure 17.1. Separate regression lines fitted to data from two fertiliser treatments

17.2.1 Testing for Equality of Slopes

To test the hypothesis that $\beta_1 = \beta_2$ carry out a t-test by calculating $t = (\hat{\beta}_1 - \hat{\beta}_2)/SED$ and compare it with t-tables on $n_1 + n_2 - 4$ df where SED is the standard error of the difference between the two fitted slopes. Its formula is

$$SED = \sqrt{s_p^2 \left\{ \frac{1}{Sxx_1} + \frac{1}{Sxx_2} \right\}}$$

where

$$s_p^2 = \frac{ResidSS_1 + ResidSS_2}{(n_1 - 2) + (n_2 - 2)}$$

is an estimate of the assumed common population residual variance. When $n_1 = n_2$ this is the average of the two residual mean squares RMS_1 and RMS_2.

To test the assumption of common population variance you divide the larger RMS by the smaller and compare with F-tables. In this example the residual degrees of freedom for both groups are $12 - 2 = 10$ so the corresponding RMS values are $RMS_1 = 480.5$ and $RMS_2 = 694.7$. The F-value is $694.7/480.5 = 1.45$. This is less than $F_{(10, 10, 5\%)} = 2.98$ so we may justifiably obtain a common estimate of variance:

$$s_p^2 = \frac{ResidSS_1 + ResidSS_2}{(n_1 - 2) + (n_2 - 2)} = \frac{4805 + 6947}{(12 - 2) + (12 - 2)} = \frac{11\,752}{20} = 587.6$$

From this you obtain

$$SED = \sqrt{587.6 \left\{ \frac{1}{0.1053} + \frac{1}{0.1021} \right\}} = 106.47$$

Hence to test the hypothesis that the population slopes are equal find

$$t = \frac{528.54 - 446.00}{106.47} = \frac{82.54}{106.47} = 0.78$$

This should be compared with the t-distribution on $n_1 + n_2 - 4 = 20$ degrees of freedom. It is not significant as it is much less than $t_{(20, 2.5\%)} = 2.086$. The data are thus in agreement with the hypothesis of a common slope. The estimate of the common slope is

$$\hat{\beta} = \frac{Sxy_1 + Sxy_2}{Sxx_1 + Sxx_2} = \frac{55.655 + 45.534}{0.1053 + 0.1021} = \frac{101.189}{0.2074} = 487.89$$

You can now assume that the lines have a common slope. This does not necessarily imply they are identical, only that they are parallel. You can only use the overall fitted line if you can also show that the intercepts are not significantly different. Assuming a common slope the two intercepts are estimated as

$$\hat{\alpha}_1 = \bar{Y}_1 - \hat{\beta}\bar{X}_1 = 156.4167 - (487.89 \times 0.3650) = -21.6632$$
$$\hat{\alpha}_2 = \bar{Y}_2 - \hat{\beta}\bar{X}_2 = 136.6667 - (487.89 \times 0.3008) = -10.0906$$

These two estimates are used in the following test.

17.2.2 Testing for Equality of Intercepts

It only makes sense to compare the intercepts if the lines are parallel. When they are parallel, a test of equality of intercepts is a test of the difference in elevations of the two lines.

To test the hypothesis that $\alpha_1 = \alpha_2$ carry out a t-test by calculating $t = (\hat{\alpha}_1 - \hat{\alpha}_2)/SED$ and compare it with t-tables on $n_1 + n_2 - 3$ df where SED is the standard error of the difference between the two fitted intercepts. Its formula is

$$SED = \sqrt{s_{y.x}^2 \left\{ \frac{1}{n_1} + \frac{1}{n_2} + \frac{(\bar{X}_1 - \bar{X}_2)^2}{Sxx_1 + Sxx_2} \right\}}$$

where

$$s_{y.x}^2 = \frac{Syy_1 + Syy_2 - (Sxy_1 + Sxy_2)^2/(Sxx_1 + Sxx_2)}{n_1 + n_2 - 3}$$

For this example

$$s_{y.x}^2 = \frac{34\,220.92 + 27\,254.67 - (55.6550 + 45.5333)^2/(0.1053 + 0.1021)}{12 + 12 - 3} = 576.518$$

and

$$SED = \sqrt{576.518 \left\{ \frac{1}{12} + \frac{1}{12} + \frac{(0.3650 - 0.3008)^2}{0.1053 + 0.1021} \right\}} = 10.3703$$

Hence to test the hypothesis that the population intercepts are equal assuming equal slopes find

$$t = \frac{-21.66 - (-10.09)}{10.3703} = \frac{-11.57}{10.3703} = -1.116$$

This should be compared with the t-distribution on $n_1 + n_2 - 3 = 21$ degrees of freedom. It is not significant as it is less than $t_{(21, 2.5\%)} = 2.080$. Thus you can assume equal intercepts. Three degrees of freedom are subtracted because three parameters have been estimated, the two intercepts (-21.66, -10.09) and the common slope (487.89). As the data are in agreement with the hypothesis of a common slope and a common intercept you can use the overall line $Y = -9.51 + 468.74X$.

17.2.3 The Adjusted Y means

Assuming equal slopes, the difference between the intercepts is the same as the difference between the fitted Y-values when $X = \bar{X}$, the overall X mean. We show this as follows. The fitted equation of line one is

$$Y_1 = \hat{\alpha}_1 + \hat{\beta}X = (\bar{Y}_1 - \hat{\beta}\bar{X}_1) + \hat{\beta}X = \bar{Y}_1 - \hat{\beta}(\bar{X}_1 - X)$$

Thus when $X = \bar{X}$, the fitted Y-value is

$$\bar{Y}_1 - \hat{\beta}(\bar{X}_1 - \bar{X}) = 156.42 - 487.89(0.3650 - 0.3329) = 156.42 - 15.66 = 140.76$$

This is the adjusted Y mean for the first line. Similarly, the adjusted Y mean for the second line is

$$\bar{Y}_2 - \hat{\beta}(\bar{X}_2 - \bar{X}) = 136.67 - 487.89(0.3008 - 0.3329) = 137.67 + 15.66 = 152.33$$

The difference is $140.76 - 152.33 = -11.57$, which is the difference between the intercepts.

Thus a test of this difference is a test of the difference between the adjusted Y means. The calculations involved are part of the analysis of covariance which is dealt with in the next section and the method enables several regression lines to be compared.

17.3 ANALYSIS OF COVARIANCE

The technique of analysis of covariance can be applied to all the common experimental designs to increase the precision of treatment comparisons. A covariate, X, is measured on the same experimental units as the yield, Y. The treatment means are adjusted to what it is estimated they would have been if all the treatments had the same mean X-value, and the adjusted treatment means are compared. The amount of the adjustment depends on the form of relationship between the Y- and X-values. It is most commonly assumed that this relationship is linear and the slope is the same for each treatment. It is further assumed that the X-values are not affected by the treatments. This is likely to be true if the X-values are measured before the treatments are applied. If the X-values are measured at the same time as the yields it is possible that they are affected by the treatments. The analysis of covariance can still be carried out but the interpretation is different. One might still be interested to know to what extent differences in the Y-values can be explained by differences in the X-values.

An analysis of variance carried out on the Y-values may show no significant difference between the treatment means. However, one of the treatments may have much lower than average X-values and so would have given a higher yield if it had had average X-values assuming a positive linear relationship between Y and X.

Analysis of covariance is often used in animal feeding trials where the mean weight gains of animals fed on different diets are adjusted for initial weight. In a crop experiment yields may be adjusted by a measure of fertility as shown by last year's yields from the same plots. Other covariates could be frost or pest damage or plant establishment if these are not affected by the treatments. Thus the yields could be adjusted for several covariates but we will be considering only one covariate.

17.4 ANALYSIS OF COVARIANCE APPLIED TO A COMPLETELY RANDOMISED DESIGN

This technique can be used as an alternative to blocking to control experimental error in field experiments or it could be used in addition to blocking. Together with each plot yield, Y, a suitable covariate, X, unaffected by the treatment is measured. In addition to the usual assumptions of analysis of variance and regression we assume that for each treatment there is a linear regression between Y and X with a common slope, β. For every unit increase in X we expect Y to increase by β units within a given treatment. For example, for treatment 1 the difference between $\bar{X}_1$ and

COMPARISON OF REGRESSION LINES AND ANALYSIS OF COVARIANCE

the overall X mean is $(\bar{X}_1 - \bar{X})$. So the Y mean should be adjusted by $\beta(\bar{X}_1 - \bar{X})$ to give an estimate of what it would have been if each treatment had the same X mean. Hence the adjusted treatment mean for treatment 1 is $\bar{Y}_1 - \beta(\bar{X}_1 - \bar{X})$. A similar argument applies to the other treatments. The analysis involves estimating β, the assumed common slope, and testing for equality of intercepts (equality of adjusted treatment means). We illustrate this with an example and later test the assumption of equality of slopes.

Example 17.2

An experiment on wheat was laid out in a completely randomised design with three treatments and five plots per treatment. The variable of interest was crop yield (Y kg/plot) and the covariate used was the yield, X, from the same plots in the previous year when no treatments were applied. The data are given in Table 17.2 and Figure 17.2 shows separate fitted regression lines.

To carry out the analysis by computer you should enter these data in three columns. The first column should contain the treatment codes, the second the X-values and the third the Y-values. Separate analysis of variance tables for Y and X are shown in Tables 17.3(a) and 17.3(b).

The small VR and large P-value in the Y table indicate that unadjusted treatment means (22.80, 19.28 and 24.48) are not significantly different. The X table shows that the X means do not differ significantly between treatments ($P = 0.085$).

We now test whether the treatment means after adjusting for a common X value are significantly different. Most computer packages allow you to specify a covariate when carrying out an analysis of variance. In SAS and Minitab you should use the General Linear Model procedure. The Minitab output for this example is given in Output 17.1.

17.4.1 Interpretation of Computer Output

The inclusion of the covariate has reduced the residual sum of squares from 235.156 to 53.291, a reduction of 181.865. This change is significant ($F = 37.54$, $P = 0.000$)

Table 17.2. Present wheat yields (Y) with previous yields (X) used as a covariate

	Treatment 1		Treatment 2		Treatment 3	
	X_1	Y_1	X_2	Y_2	X_3	Y_3
	29.5	16.5	24	17.2	22.4	18.3
	30.4	20.3	26.4	21.4	29.7	25.6
	29.8	22.2	30.7	22.9	30.5	30
	36	29	31.1	22	28.1	24
	32.5	26	23.9	12.9	27.1	24.5
Total	158.2	114	136.1	96.4	137.8	122.4
Mean	31.64	22.8	27.22	19.28	27.56	24.48

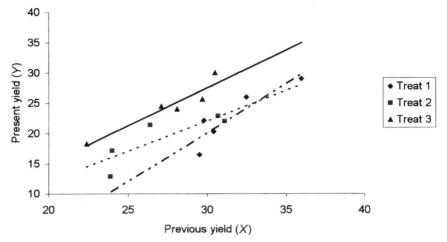

Figure 17.2. Separate regression lines fitted to data of Example 17.2

Table 17.3(a). ANOVA table for Y

Source	df	SS	MS	VR	P
Treatment	2	70.4213	35.2107	1.80	0.21
Residual	12	235.1560	19.5963		
Total	14	305.5773			

Table 17.3(b). ANOVA table for X

SS	MS	VR	P
60.4973	30.2487	3.05	0.085
118.8320	9.9027		
179.3293			

which indicates that the covariate was useful in increasing the precision of the treatment comparisons. It shows that 1.2371, the estimate of the assumed common slope, is significantly different from zero. The corresponding standard error is 0.2019 and the t-value of $1.2371/0.2019 = 6.13$ (the square root of 37.54) provides an equivalent test.

The regression sum of squares obtained by fitting a regression line to all 15 points is 150.826 and after fitting parallel lines the regression sum of squares increases by 101.460. This increase is significant ($F = 10.47$, $P = 0.003$) indicating that a model with three parallel lines fits better than a model with a single line. This shows that the intercepts are significantly different assuming a common slope and hence that the adjusted treatment means (19.29, 21.24 and 26.02) are significantly different. These are calculated as:

$$\bar{Y}_1 - \hat{\beta}(\bar{X}_1 - \bar{X}) = 22.80 - 1.2371(31.64 - 28.8067) = 22.80 - 3.51 = 19.29$$

$$\bar{Y}_2 - \hat{\beta}(\bar{X}_2 - \bar{X}) = 19.28 - 1.2371(27.22 - 28.8067) = 19.28 + 1.96 = 21.24$$

$$\bar{Y}_3 - \hat{\beta}(\bar{X}_3 - \bar{X}) = 24.48 - 1.2371(27.56 - 28.8067) = 24.48 + 1.54 = 26.02$$

Output 17.1 Analysis of covariance for Example 17.2 using Minitab

```
Analysis of Variance for Y, using Adjusted SS for Tests
Source    DF     Seq SS    Adj SS    Adj MS       F        P
X          1    150.826   181.865   181.865    37.54    0.000
Treat      2    101.460   101.460    50.730    10.47    0.003
Error     11     53.291    53.291     4.845
Total     14    305.577

Term           Coef     StDev        T        P
Constant    -13.450     5.844    -2.30    0.042
X             1.2371    0.2019    6.13    0.000

Means for Covariates
Covariate    Mean     StDev
X           28.81     3.579

Least Squares Means for Y
Treat     Mean     StDev
1        19.29     1.139
2        21.24     1.035
3        26.02     1.016

Tukey Simultaneous Tests
Response Variable Y
All Pairwise Comparisons among Levels of Treat

Treat = 1 subtracted from:
Level    Difference     SE of                  Adjusted
Treat    of Means    Difference   T-Value      P-Value
2           1.948       1.654      1.178       0.4895
3           6.727       1.618      4.159       0.0042

Treat = 2 subtracted from:
Level    Difference     SE of                  Adjusted
Treat    of Means    Difference   T-Value      P-Value
3           4.779       1.394      3.429       0.0143
```

where $\bar{X} = 28.8067$ is the mean of all the 15 X values. It is rounded to 28.81 in the output and its standard deviation is given as 3.579. The standard errors of the adjusted treatment means are 1.139, 1.035 and 1.016 respectively. Their method of calculation is given in Section 17.5.6.

Output 17.1 also shows the results of tests to compare any two adjusted treatment means. For example, the adjusted mean for treatment 2 exceeds that for treatment 1 by 1.948 and the corresponding SED value is 1.654. Using a Tukey test this difference is not significant (Adjusted $P = 0.4895$). However, the other two comparisons are significant. Some computer packages, for example, Genstat give a pooled SED value which can be used to compare the adjusted means if the X means do not differ significantly. For this example its value is 1.559 and its method of calculation is given at the end of Section 17.5.6.

The method of calculating the standard error of the common slope (0.2019) is given at the end of Section 17.5.3, and that for calculating the SED values (1.654, 1.618, 1.394) for comparing the adjusted treatment means is shown in Section 17.5.6. The assumption of a common slope is tested in Section 17.5.4.

We have shown in this example that by taking into account the fertility of the plots, we have removed the effect of some treatments being applied to more fertile plots than others. The variation within treatments due to fertility has also been removed, thereby reducing the residual variation and allowing more precise treatment comparisons than if the covariate had not been used. The unadjusted residual mean square is 19.5963 (Table 17.3(a)) and the adjusted value is 4.845. The ratio (4.04) is a measure of how much the covariate has improved the precision of the experiment. This is shown in the Genstat output (Output 17.2) in the residual row and cov.ef column.

Notice that the first three numbers in the s.s. column of Output 17.2 do not add up to 305.577. This is because 101.460 (used to test whether the adjusted treatment means are equal) is the adjusted sum of squares for treatments after allowing for the covariate, and 181.865 (used to test whether the assumed common slope is zero) is the adjusted sum of squares for the covariate after allowing for treatments. These numbers are also shown in the Adj SS column of Output 17.1. The entry of 0.80 in the cov.ef. column is called the covariance efficiency factor. It is calculated as

$$cef = \frac{ResidSS_x}{ResidSS_x + TMS_x} = \frac{118.8320}{118.8320 + 30.2487} = 0.80$$

where $ResidSS_x$ and TMS_x are the residual sum of squares and treatment mean squares in the analysis of variance of the X values (Table 17.3(b)).

In general the value of cef must lie between zero and one. A value of zero would indicate that the differences between the adjusted treatment means can be completely explained by the covariate and there is no information about the treatments. A low value may mean that the values of the covariate have been affected by the treatments, or by chance some treatments have been allocated to units with high values of the covariate and others to low ones. Values in the range of 0.8 to 0.9 are usually expected.

Finally, a question that might be asked is 'how many more replicates per treatment would be needed in this experiment to give the same precision without the covariate?'

Output 17.2 Analysis of Covariance for Example 17.2 using Genstat

```
***** Analysis of variance (adjusted for covariate) *****
Variate: Y
Covariate: X
```

Source of variation	d.f.	s.s.	m.s.	v.r.	cov.ef.	F pr.
Treat	2	101.460	50.730	10.47	0.80	0.003
Covariate	1	181.865	181.865	37.54		<0.001
Residual	11	53.291	4.845		4.04	
Total	14	305.577				

COMPARISON OF REGRESSION LINES AND ANALYSIS OF COVARIANCE

The answer is provided by multiplying the covariance efficiency factor by the ratio of the unadjusted to adjusted mean squares. For this example it is $0.80 \times 4.04 = 3.23$, which indicates that more than three times as many replicates would be required to give the same precision of treatment comparisons if the covariate were not used.

17.5 COMPARING SEVERAL REGRESSION LINES

To compare several regression lines we fit three models. In Model 1, a single line is fitted and the residual sum of squares (*ResidSSM1*) noted. In Model 2, a common slope is estimated and separate parallel lines are fitted. The residual sum of squares (*ResidSSM2*) for this model is the sum of the residual sums of squares for the separate fitted parallel lines. In Model 3, separate lines are fitted and for each the residual sum of squares found. These are added to give the residual sum of squares (*ResidSSM3*) for this model. It can be shown that *ResidSSM2* is always less than *ResidSSM1* and *ResidSSM3* is always less than *ResidSSM2*. Hence a test of the hypothesis that the lines have equal slopes involves finding whether Model 3 is a significantly better fit than Model 2. If it is, you would conclude the lines are not parallel. However, if the lines can be considered of equal slope you can then test whether they have equal intercepts. This test involves testing whether Model 2 is a significantly better fit than Model 1. If it is, you would conclude the intercepts are not equal. If it is not, you can assume equal intercepts and may use the overall regression line. We now show how to carry out these tests for Example 17.2 using the notation of Chapter 7.

17.5.1 The Overall Line: Model 1

The following summary can be calculated from the data of Table 17.2:

$n = 15$ $\Sigma X = 432.1$ $\Sigma Y = 332.8$ $\bar{X} = 28.8067$ $\bar{Y} = 22.1867$
$Sxx = 179.3293$ $Sxy = 164.4613$ $Syy = 305.5773$
$RegSS = 150.8250$ on 1 df $ResidSS = 154.7513$ on $(n-2) = 13$ df

Thus *ResidSSM1* is $(305.5773 - 150.8260) = 154.7513$. As two parameters have been estimated (the intercept and the overall slope), the corresponding residual degrees of freedom are $15 - 2 = 13$. The fitted equation of the overall line is $Y = -4.2317 + 0.9171X$ and $R^2 = 49.4\%$.

17.5.2 The Separate Lines: Model 3

Table 17.4 gives a summary of the calculations involved in fitting separate lines. The fitted equations are

Line 1: $Y_1 = -28.5452 + 1.6228X_1$ $R^2 = 81.1\%$

Table 17.4. Information required to fit separate lines to data of Table 17.2

	Line 1	Line 2	Line 3
n	5	5	5
ΣX	158.2	136.1	137.8
ΣY	114.0	96.4	122.4
$\bar{X}$	31.64	27.22	27.56
$\bar{Y}$	22.80	19.28	24.48
Sxx	29.252	49.228	40.352
Sxy	47.470	49.292	50.246
$RegSS$	77.0341	49.3561	62.5659
$ResidSS$	17.9459	20.6719	7.5821
Syy	94.980	70.028	70.148

Line 2: $Y_2 = -7.9754 + 1.0013 X_2$ $R^2 = 70.5\%$
Line 3: $Y_3 = -9.8375 + 1.2452 X_3$ $R^2 = 89.2\%$

Thus $ResidSSM3 = 17.9459 + 20.6719 + 7.5821 = 46.1999$. As six parameters have been estimated (the three intercepts and the three slopes), the corresponding residual degrees of freedom are $15 - 6 = 9$.

17.5.3 The Parallel Lines: Model 2

First we find $\hat{\beta}$, the estimate of the common slope, β. This is a weighted average of the separate slopes, the weights being the Sxx values. It is the sum of the separate Sxy values divided by the sum of the separate Sxx values. Hence

$$\hat{\beta} = \frac{Sxy_1 + Sxy_2 + Sxy_3}{Sxx_1 + Sxx_2 + Sxx_3} = \frac{47.470 + 49.292 + 50.246}{29.252 + 49.228 + 40.352} = \frac{147.008}{118.832} = 1.2371$$

We now assume parallel lines, so the three intercepts are estimated to be

$\hat{\alpha}_1 = \bar{Y}_1 - \hat{\beta}\bar{X}_1 = 22.80 - (1.2371 \times 31.64) = -16.3418$
$\hat{\alpha}_2 = \bar{Y}_2 - \hat{\beta}\bar{X}_2 = 19.28 - (1.2371 \times 27.22) = -14.3939$
$\hat{\alpha}_3 = \bar{Y}_3 - \hat{\beta}\bar{X}_3 = 24.48 - (1.2371 \times 27.56) = -9.6145$

Hence the equations of the three parallel lines are estimated to be

$Y_1 = -16.3418 + 1.2371 X_1$
$Y_2 = -14.3939 + 1.2371 X_2$
$Y_3 = -9.6145 + 1.2371 X_3$

For each equation the fitted values can be found by substituting the X values from Table 17.2. The residuals are found by subtracting the fitted values from the Y-values. The three residual sums of squares are found to be 22.2975, 23.4091 and 7.5847 respectively. These add to 53.2913 which is $ResidSSM2$, the residual sum of

COMPARISON OF REGRESSION LINES AND ANALYSIS OF COVARIANCE

squares after fitting Model 2. As four parameters have been estimated (the three intercepts and the common slope), the corresponding residual degrees of freedom are $15 - 4 = 11$. Hence the mean square, denoted by $s_{y.x}^2$, is $53.2913/11 = 4.845$ (see the Error line of Output 17.1). The value of *ResidSSM2* can be found directly from the information in Table 17.4 as follows:

$ResidSSM2 = \Sigma Syy - (\Sigma Sxy)^2/\Sigma Sxx$

$\Sigma Syy = 94.980 + 70.028 + 70.148 = 235.156$ (this is the residual sum of squares in Table 17.3(a))

$\Sigma Sxy = 47.470 + 49.292 + 50.246 = 147.008$

$\Sigma Sxx = 29.252 + 49.228 + 40.352 = 118.832$ (this is the residual sum of squares in Table 17.3(b))

Hence $ResidSSM2 = 235.156 - (147.008)^2/118.832 = 235.156 - 181.865 = 53.291$.
The standard error of the estimate of common slope is calculated as

$$SE = \sqrt{\frac{s_{y.x}^2}{\Sigma Sxx}} = \sqrt{\frac{4.845}{118.832}} = 0.2019$$

17.5.4 Testing the Assumption of a Common Slope

Analysis of covariance is often carried out without first checking whether this assumption is reasonable. An X, Y plot of the data with different symbols for each treatment such as Figure 17.2 should always be obtained prior to analysis. The analysis is biased if the slopes are significantly different.

The test involves comparing Model 2 (the parallel slopes model) with Model 3 (the separate lines model). The reduction in the residual sums of squares due to fitting Model 3 after Model 2 is $ResidSSM2 - ResidSSM3 = 53.2913 - 46.1999 = 7.0914$. This reduction has $11 - 9 = 2$ df. So the change mean square is $7.0914/2 = 3.5457$ on 2 df. The residual mean square for Model 3 is $ResidSSM3/9 = 5.1333$. The F-value for testing the equality of slopes is obtained by dividing this change mean square by the mean square for Model 3. Hence

$$F = \frac{7.0914/2}{46.1999/9} = \frac{3.5457}{5.1333} = 0.69 \quad \text{on (2, 9) df}$$

This is very small, so Model 3 is not a significantly better fit than Model 2, which implies that the assumption of a common slope is valid.

This test can be performed on the data of Example 17.2 using Minitab by putting X Treat X*Treat in the model box and X in the covariate box. The resulting ANOVA table is shown in Output 17.3 where the Treat*X line gives the above F-test.

Note that in the Seq SS column of Output 17.3:

- 150.826 is the regression sum of squares after fitting Model 1 (a single overall line). Hence, the corresponding residual sum of squares is $ResidSSM1 = 305.577 - 150.826 = 154.751$.

Output 17.3 Test of a common slope for data of Example 17.2 (Minitab output)

```
Analysis of Variance for Y, using Adjusted SS for Tests

Source    DF    Seq SS    Adj SS    Adj MS      F        P
X          1   150.826   188.839   188.839    36.79   0.000
Treat      2   101.460     9.188     4.594     0.89   0.442
Treat*X    2     7.091     7.091     3.546     0.69   0.526
Error      9    46.200    46.200     5.133
Total     14   305.577


Term              Coef     StDev        T        P
Constant       -15.453     6.260    -2.47    0.036
X                1.2898    0.2126    6.07    0.000
X*Treat
  1              0.3330    0.3220    1.03    0.328
  2             -0.2885    0.2828   -1.02    0.334
```

- 101.460 is the increase in the regression sum of squares due to fitting Model 2 (parallel lines) after fitting Model 1. It is also the decrease in the residual sum of squares ($ResidSSM1 - ResidSSM2$), where $ResidSSM2 = 305.577 - (150.826 + 101.460) = 53.291$.
- 7.091 is the increase in the regression sum of squares due to fitting Model 3 (separate lines) after fitting Model 2. It is also the decrease in the residual sum of squares ($ResidSSM2 - ResidSSM3$).
- 46.200 is $ResidSSM3$, the residual sum of squares after fitting Model 3 (Section 17.5.2).

Finally, you should be aware that the above F-test requires the assumption that the residual variances about the individual fitted lines are not significantly different. They are given by the three residual mean squares (5.98, 6.89 and 2.53) which can be obtained from the $ResidSS$ line in Table 17.4 after dividing by $(n - 2) = 3$. As the largest ratio, 2.72 is much less than the 5% F-table value of 9.28 on (3, 3) df we can make the required assumption.

17.5.5 Comparing the Slopes

If the slopes are found to be significantly different you may be interested to make pairwise comparisons. The criticisms of the LSD test also apply, so you should make no more comparisons than one less than the number of lines. We have already shown that, for Example 17.2, the slopes are not significantly different. However, we use the data from this example to illustrate the method of making pairwise slope comparisons when the slopes cannot be assumed equal. In general the SED value for comparing two slopes depends on which pair are being compared. For example, the SED for comparing slopes one and two is

$$\sqrt{s_p^2 \left(\frac{1}{Sxx_1} + \frac{1}{Sxx_2} \right)} = \sqrt{5.133 \left(\frac{1}{29.252} + \frac{1}{49.228} \right)} = 0.529$$

COMPARISON OF REGRESSION LINES AND ANALYSIS OF COVARIANCE   269

where $s_p^2 = 5.133$ is the pooled residual mean square for Model 3, and can be found by adding the individual residual sums of squares (Table 17.4) and dividing by the sum of the residual degrees of freedom. Thus, $s_p^2 = (17.9459 + 20.6719 + 7.5821)/(3 + 3 + 3) = 46.1999/9 = ResidSSM3/9$. This calculation is shown in the error row of Output 17.3.

Output 17.4 is obtained from Genstat using 'simple linear regression with groups'. It shows the estimates after fitting Model 3. The estimate of slope 1 is 1.623. The estimate of slope 2 minus slope 1 is -0.621 and the corresponding SED value is 0.529; this difference is not significant ($P = 0.270$). Similarly -0.378 is the estimate of slope 3 minus slope 1. It has an SED of 0.550 and a P-value of 0.510. The other entries in Output 17.4 concern the separate intercepts of Model 3 and do not concern us; there is no point in comparing intercepts when the slopes are not equal. In Output 17.3, the value of -15.453 is the average of the intercepts for Model 3, and 1.2898 is the average of the separate slopes. The value of 0.3330 is slope 1 minus the average slope, and -0.2885 is slope 2 minus the average slope.

17.5.6 Testing the Assumption of Equality of Intercepts

If the slopes can be assumed equal as in this example the intercepts can be compared. This is done by comparing Model 2 (the parallel slopes model) with Model 1 (the single line model).

The reduction in the residual sums of squares due to fitting Model 2 after Model 1 is $ResidSSM1 - ResidSSM2 = 154.751 - 53.291 = 101.460$. This reduction has $13 - 11 = 2$ df. So the change mean square is $101.460/2 = 50.730$ on 2 df. The residual mean square for Model 2 is $ResidSSM2/11 = s_{y.x}^2 = 4.845$.

The F-value for testing the equality of intercepts is obtained by dividing this change mean square by the mean square for Model 2. Hence

$$F = \frac{101.460/2}{53.291/11} = \frac{50.730}{4.845} = 10.47 \quad \text{on (2, 11) df}$$

This is the same as the F-test for comparing the adjusted treatment means (Outputs 17.1 and 17.2).

The standard error (1.139) of the adjusted treatment 1 mean (19.29) shown in Output 17.1 is

Output 17.4 Comparison of slopes and intercepts for data of Example 17.2 (Genstat output)

	estimate	s.e.	t(9)	t pr.
Constant	-28.5	13.3	-2.15	0.060
X	1.623	0.419	3.87	0.004
Treat 2	20.6	16.0	1.29	0.230
Treat 3	18.7	16.6	1.13	0.288
X.Treat 2	-0.621	0.529	-1.18	0.270
X.Treat 3	-0.378	0.550	-0.69	0.510

$$SE_1 = \sqrt{s_{y.x}^2 \left\{ \frac{1}{n_1} + \frac{(\bar{X}_1 - \bar{X})^2}{\Sigma Sxx} \right\}} = \sqrt{4.845 \left\{ \frac{1}{5} + \frac{(31.64 - 28.8067)^2}{118.832} \right\}} = 1.139$$

The standard errors (1.035 and 1.016) for the other two adjusted means are found using the corresponding values of the X means (27.22 and 27.56).

The SED values used to compare two intercepts or adjusted treatment means depends on which two means are being compared. For example, the SED for comparing treatments 1 and 2 is

$$SED_{1,2} = \sqrt{s_{y.x}^2 \left\{ \frac{1}{n_1} + \frac{1}{n_2} + \frac{(\bar{X}_1 - \bar{X}_2)^2}{\Sigma Sxx} \right\}} = \sqrt{4.845 \left\{ \frac{1}{5} + \frac{1}{5} + \frac{(31.64 - 27.22)^2}{118.832} \right\}} = 1.654$$

Similarly you can calculate $SED_{1,3} = 1.618$ and $SED_{2,3} = 1.394$ (Output 17.1).

When there are a large number of treatments, each with the same number of replicates, n, a pooled standard error of the difference between two adjusted means is often quoted. For this example it is calculated as

$$SED = \sqrt{\frac{2}{n} \times s_{y.x}^2 \left\{ 1 + \frac{TMS_x}{\Sigma Sxx} \right\}} = \sqrt{\frac{2}{5} \times 4.845 \left\{ 1 + \frac{30.2487}{118.832} \right\}} = 1.559$$

where TMS_x is the treatment mean square in the analysis of variance of the X-values (Table 17.3(b)).

17.5.7 Common X-values

Many of the calculations involved in comparing regression lines are simplified if each line has the same X-values. In this case, the pooled slope (Model 2) is the average of the separate slopes (Model 3) and is also equal to the overall slope (Model 1). There is then a common SED value for comparing any two slopes in Model 3 and for comparing any two intercepts (elevations) in Model 2. Their formulas are

$$\sqrt{\frac{2 \times s_p^2}{Sxx}} \quad \text{and} \quad \sqrt{\frac{2 \times s_{y.x}^2}{n}}$$

and respectively, where n is the number of points for each line, and Sxx is the common corrected sum of squares of the X-values for each line.

17.6 CONCLUSION

We have shown how the analysis of covariance in a completely randomised design and the comparison of several regression lines are related. The information required for the various tests can be obtained by fitting the three models described in Section 17.5. These models are compared automatically by requesting 'simple linear regression with groups' in Genstat.

Analysis of covariance can be applied to other experimental designs such as randomised blocks. Its main use is to reduce the residual (error) mean square so increasing precision or to allow less replication than would be needed without a covariate, and there can be more than one covariate. The interested reader may consult Steel, Torrie and Dickey (1996) or Snedecor and Cochran (1989).

Chapter 18

Analysis of Counts

18.1 INTRODUCTION

In this chapter we discuss the analysis of categorical data, having shown in previous chapters how to analyse data measured on a continuous scale such as heights, weights, etc. Suppose the experimental unit is a plant and ten plants receive a particular treatment. If plant height is measured you are almost certain to obtain ten different values. However, if the disease state is measured, there are only two possibilities: healthy and diseased. You may be interested in the proportion of plants which are infected. In germination tests you may wish to compare the proportions of seeds germinating for several treatments. In a genetics experiment you may want to compare the proportions of offspring of various phenotypes. We start by considering attributes which can only have two possible values; for example a seed may or may not germinate or a cutting may or may not root, and how to find confidence intervals for and test hypotheses about a population proportion. These techniques are based on the theory of the binomial and normal distributions. In later sections the chi-square distribution is used to test goodness of fit and independence in two-way contingency tables.

18.2 THE BINOMIAL DISTRIBUTION

Consider a series of n independent *trials*. Assume that there are only two possible outcomes (success or failure) at each *trial*, and that p is the constant probability of a success. This implies the probability of a failure at each *trial* is $1 - p$ (denoted by q). Hence $p + q = 1$. In this context a *trial* could be the planting of a seed or the injection of a viral strain into a plant. Success would then be germination, or the incidence of disease. Independence implies that the outcome of any one *trial* does not influence the outcome of any other.

Let X represent the number of successes in n *trials*. It is a random variable which could take any value between 0 and n. The probability that X is equal to k is denoted by $\Pr(X = k)$, abbreviated to P_k.

The probability of obtaining exactly k successes in the n trials is given by the formula

$$P_k = \frac{n!}{k!(n-k)!} p^k q^{(n-k)}$$

When using this formula you need to know the meaning of $n!$ (n factorial). We illustrate the meaning with 4 factorial: $4! = 4 \times 3 \times 2 \times 1 = 24$. Similarly $5! = 120$. You also need to know that $0! = 1$ and $p^0 = 1$; in fact any integer raised to the power $0 = 1$.

Example 18.1
Four identical seeds are planted and treated in the same way. Suppose each has a probability of 0.8 of germinating. Find the probability distribution of the number of seeds germinating. In other words, for each value of k from 0 to 4 find P_k.

You can use the above formula to find P_0, P_1, P_2, P_3 and P_4. Alternatively you can consult tables of the binomial distribution or use a computer package. Confirm that the distribution is

k	0	1	2	3	4	Total
P_k	0.0016	0.0256	0.1536	0.4096	0.4096	1

Thus the probability of exactly two seeds germinating is 0.1536 and the probability of two or less germinating is given by $P_0 + P_1 + P_2 = 0.0016 + 0.0256 + 0.1536 = 0.1808$. The probability of at least three germinating is $P_3 + P_4 = 0.4096 + 0.4096 = 0.8192$. This is also 1 minus the probability of two or less germinating ($0.8192 = 1 - 0.1808$). The distribution also implies that if you planted 100 pots each containing four seeds, you would expect none to have no seeds germinating, three to have one, 15 to have two, 41 to have three and 41 to have all four seeds germinating.

Example 18.2
In general the binomial distribution is skew. Table 18.1 show the cases with $n = 6$ and $p = 0.2, 0.3, 0.4, 0.5$ and 0.8 respectively. Figure 18.1 shows the last two of these distributions diagrammatically.

Note that the probabilities for $p = 0.2$ are identical to those for $p = 0.8$ but in reverse order. This is a special case of the general rule that for a given n, $\Pr(X = k)$ when $p = p'$ is the same as $\Pr(X = n - k)$ when $p = 1 - p'$. When $p = 0.8$ and $n = 6$ there is a negligible chance of obtaining either zero or one success.

18.2.1 Mean and Variance

It can be shown that the mean of the binomial distribution is np and the variance is npq. For example, if you carry out six *trials* and the probability of success at each *trial* is 0.2 you would expect 1.2 successes. Obviously you cannot have 1.2

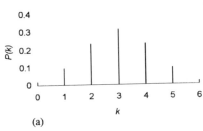 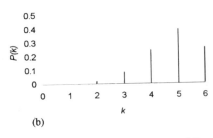

Figure 18.1. The binomial distribution (a) with $n = 6$, $p = 0.5$ and (b) with $n = 6$, $p = 0.8$

Table 18.1. The binomial distribution for $n = 6$ and various values for p

k	$p = 0.2$	$p = 0.3$	$p = 0.4$	$p = 0.5$	$p = 0.8$
0	0.262144	0.117649	0.046656	0.015625	0.000064
1	0.393216	0.302526	0.186624	0.093750	0.001536
2	0.245760	0.324135	0.311040	0.234375	0.015360
3	0.081920	0.185220	0.276480	0.312500	0.081920
4	0.015360	0.059535	0.138240	0.234375	0.245760
5	0.001536	0.010206	0.036864	0.093750	0.393216
6	0.000064	0.000729	0.004096	0.015625	0.262144

successes in a single *trial*. However, if you carry out the six *trials* a large number of times and note the number of successes at each *trial*, you should find the average number of successes per *trial* is close to 1.2 with a standard deviation of $\sqrt{6 \times 0.2 \times 0.8} = 0.98$.

18.2.2 The Normal Approximation to the Binomial Distribution

Provided the number of trials, n, is large and the probability of success at each trial is not too near zero or one, probabilities associated with the binomial distribution can be approximated by assuming a normal distribution with mean np and variance npq. In practice, the approximation is acceptable if neither np nor nq is less than five.

The probability of k or fewer successes in n independent trials is $\Pr(X \leq k) = P_0 + P_1 + \ldots + P_k$. This can be approximated by finding the area under the normal distribution curve to the left of $k + \frac{1}{2}$. The $\frac{1}{2}$ is the continuity correction reflecting the fact that a discrete distribution is being approximated by a continuous one. The corresponding z-value is

$$z = \frac{(k + 0.5) - np}{\sqrt{npq}}$$

Tables of the standard normal distribution (Appendix 1) may be consulted to find the probability of a smaller z. To find $\Pr(X \geq k)$ use the equation $\Pr(X \geq k) = 1 - \Pr(X \leq k - 1)$.

ANALYSIS OF COUNTS 275

Example 18.3
Find an approximation to the probability of seven or fewer successes in 15 trials with $p = 0.4$. The z-value is
$$z = \frac{(7 + 0.5) - (15 \times 0.4)}{\sqrt{15 \times 0.4 \times 0.6}} = \frac{7.5 - 6}{1.897} = 0.79$$
The corresponding table value is 0.7852 (Appendix 1) so this is an approximation to the required probability. The exact answer is 0.7869. This is also 1 minus the probability of eight of more successes.

18.3 CONFIDENCE INTERVALS FOR A PROPORTION

Suppose you have a large batch of seed and wish to know the proportion which will germinate. You could take a random sample of seeds, carry out a germination test to find the proportion germinating and use this as an estimate for the whole batch. The reliability of this estimate depends on the sample size and can be assessed by calculating a confidence interval based on the binomial distribution. For large samples the normal approximation to the binomial distribution can be used provided certain assumptions are made. In the following sections it is assumed that the population is large compared to the sample size.

18.3.1 The Exact Method

Let p be the proportion of the population possessing a certain attribute. For example, a plant may or may not be diseased, an animal may be male or female, a cutting may root or not, etc. In a random sample of size n, let k be the observed number of individuals possessing the attribute. The best estimate of p is $\hat{p} = k/n$. In another sample, it is likely a different value for k would be obtained. Assuming that the possession of the attribute by one individual (success) does not affect that of another, k can be considered to be a value from a binomial distribution with n trials and probability of success at each trial of p. A 95% confidence interval for p can be obtained as follows.

Let p_L and p_U be the required lower and upper confidence limits. p_U is the value of p such that the probability of k or fewer successes in the n trials is 0.025 and p_L is that value of p such that the probability of k or more successes in the n trials is 0.025. The following example should clarify the method.

Example 18.4
A random sample of 20 seeds is taken and four germinate. Find a 95% confidence interval for the population germination rate.
    The point estimate is $4/20 = 0.2$ (20%). To find p_U you need to find that value of p for a binomial distribution with $n = 20$ such that the probability of getting four or fewer successes is 0.025. In other words you have to solve the equation

$0.025 = P_0 + P_1 + P_2 + P_3 + P_4$. This can be done with the aid of tables of the cumulative probabilities in binomial distributions such as Lindley and Scott (1995) which shows that p_U is just under 0.44. To find p_L you need to find that value of p such that the probability of getting four or more successes when $n = 20$ is 0.025. This is 1 minus the probability of getting three or fewer successes so you need p such that $0.975 = P_0 + P_1 + P_2 + P_3$. From tables the answer for p_L is just under 0.06. Minitab gives the exact probabilities as 0.057334 and 0.436614. Thus you can be 95% confident that between 5.7% and 43.7% of seeds will germinate.

Note: This confidence interval is not symmetrical about 0.2 (if the observed proportion had been 0.5 it would have been symmetrical about 0.5; Minitab gives 0.272 to 0.728). The interval is not unique. Another 95% confidence interval is obtained by replacing 0.025 and 0.025 in the calculations of p_U and p_L by 0.01 and 0.04 to give $p_L = 0.07$ and $p_U = 0.48$. Any combination could be used provided they add to 0.05, but it is traditional to use 0.025 and 0.025.

18.3.2 The Approximate Method

The exact method is rather difficult to understand and use. Fortunately there is a much easier method which gives almost the same results provided n is large and the observed proportion, $\hat{p}$, is not too close to 0 or 1. In practice if neither $n\hat{p}$ nor $n(1-\hat{p})$ is less than 5 you can use the normal approximation (Chapter 4) to the binomial distribution (Section 18.2.2). Provided the population is large compared to the sample you can assume that k follows (approx.) a normal distribution with mean np and variance npq. This implies that k/n follows a normal distribution with mean $np/n = p$ and a variance $npq/n^2 = pq/n$ where $q = 1 - p$.

Approximate confidence intervals for a population proportion can be obtained by assuming that $\hat{p}$ comes from a normal distribution with a mean of p and a standard deviation $\sqrt{pq/n}$. As p is unknown a further approximation is obtained by using the sample estimate, $\hat{p}$. Hence an approximate 95% confidence interval for p is $\hat{p} \pm 1.96\sqrt{\hat{p}\hat{q}/n}$.

An approximate 99% confidence interval is obtained by replacing 1.96 by 2.58 in the above formula. The value of $\sqrt{\hat{p}\hat{q}/n}$ is called the standard error of the proportion.

Example 18.5

In a random sample of 200 seeds taken from a large batch, 120 were found to germinate. $\hat{p} = 120/200 = 0.6$ and $\hat{q} = 1 - \hat{p} = 0.4$. Hence a 95% confidence interval for p is

$$0.6 \pm 1.96\sqrt{\frac{0.6 \times 0.4}{200}} = 0.6 \pm (1.96 \times 0.035) = 0.6 \pm 0.068 = (0.532, 0.668)$$

You are therefore 95% confident that the true percentage of seeds germinating in the batch is between 53.2% and 66.8%. Your best estimate is 60% with a standard error of 3.5%.

ANALYSIS OF COUNTS 277

Caution: Do not use the normal approximation when it is not valid. If you use this method for the data of Example 18.4, you will obtain (0.025, 0.375) whereas the correct interval is (0.057, 0.437). If the observed proportion was 0.2 based on a sample of size 10, the approximate method gives (−0.048, 0.448) which is clearly nonsense as a proportion cannot be negative. The exact method gives (0.025, 0.556). This is a very wide interval and so may not be much use in practice. A larger sample is needed to give a smaller interval.

18.3.3 Estimation of Sample Size

Suppose you want to ensure that, whatever the true population proportion, a 95% confidence interval extends no further than δ on each side of the sample proportion. How many sample observations should you take?

Using the normal approximation you require n such that

$$1.96 \times \sqrt{\frac{pq}{n}} < \delta$$

This implies that n should be at least $(1.96^2 \times pq)/\delta^2$. Although p is unknown, pq has a maximum value of 0.25 when $p = 0.5$. For example, if $p = 0.5$ and you set δ to 0.04, you will need a sample size of at least $(3.8416 \times 0.25)/0.0016 = 600$. If p is any other value the required sample size will be smaller. Check that if $p = 0.2$ the required size is 385. The method will not work if the true p is very close to 0 or 1 because the normal approximation to the binomial distribution does not apply.

18.4 HYPOTHESIS TEST OF A PROPORTION

To test whether a population proportion is a particular value, say p_0, you could find a 95% confidence interval for the true proportion, p. If p_0 is not in the interval you would reject the hypothesis at the 5% level on a two-tailed test. Alternatively you could proceed as follows.

18.4.1 Large-sample Test

Set up the null hypothesis that the population proportion, p, is p_0 and decide on a level of significance, say 5%. Also decide on a one- or two-tailed alternative. Take a large random sample of size n and count the number of individuals, k, with the chosen attribute. The observed proportion is thus $\hat{p} = k/n$. This is approximately normally distributed with a mean p_0 and a standard deviation $\sqrt{(p_0 q_0)/n}$ if the null hypothesis is true, and assuming that the conditions for the normal approximation to the binomial distribution hold (Section 18.2.2). Hence the test statistic is

$$z = \frac{\hat{p} - p_0}{\sqrt{(p_0 q_0)/n}}$$

If the test is two-tailed at the 5% level, reject the null hypothesis if the magnitude of z is greater than 1.96. If it is one-tailed with the alternative that p is greater than p_0 reject at the 5% level if z is greater than 1.65. If you use the computer, reject if the P-value is less than 0.05. An alternative but equivalent method is described in Example 18.9.

Example 18.6

A plant breeder wishes to know whether a single gene is responsible for the stem colour of a tomato plant. Genetic theory predicts that a cross of two parents should result in seedlings which are purple- and green-stemmed in the ratio 3:1. The offspring consisted of 112 purple- and 46 green-stemmed seedlings. Hence the observed proportion of purple plants was $112/158 = 0.709$. The null hypothesis is that $p = \frac{3}{4} = 0.75$ and the test is two-tailed as it is not known in advance whether the ratio will be greater or less than that expected. The test statistic is

$$z = \frac{0.709 - 0.750}{\sqrt{(0.75 \times 0.25)/158}} = \frac{-0.041}{0.0344} = -1.19$$

This is less than 1.96 in magnitude so there is no reason to doubt the 3:1 ratio. The Minitab output for this problem is

```
Test of p = 0.75 vs p not = 0.75
Sample    X     N    Sample p      95.0 % CI           Z-Value  P-Value
1        112   158   0.708861   (0.638025, 0.779696)   -1.19    0.232
```

The P-value of 0.232 is not less than 0.05, confirming that the data are consistent with the null hypothesis. This conclusion is also supported by the fact that 0.75 is included in the confidence interval. See Example 18.9 for an alternative method.

18.4.2 Small-sample Test

The use of the normal approximation may lead to false conclusions when small samples are involved. The exact test makes use of the binomial distribution.

Example 18.7

A manufacturer claims that a plant growth hormone will cause at least 90% of cuttings to root. However, when ten identical cuttings are treated only seven root. Should you reject the claim based on this evidence?

The null hypothesis is that $p = 0.9$ versus the alternative that p is less than 0.9. You carry out a one-tailed test because you will only refute the claim if a significantly smaller proportion than 0.9 root. The sample proportion is 0.7 so you need to find the probability of obtaining a proportion as low as this or lower when the population proportion is 0.9. This is the same as observing seven or fewer successes in a binomial distribution with $n = 10$ and $p = 0.9$. Using the formula in Section 18.2, the computer or tables of the binomial distribution,

$P_0 + P_1 + \ldots + P_7 = 0.070$. As this P-value is not less than 0.05 you do not have enough evidence to reject the claim. The Minitab output is

```
Test of p = 0.9 vs p < 0.9
                                                    Exact
Sample   X    N    Sample p    90.0 % CI           P-Value
1        7    10   0.700000    (0.393376, 0.912736) 0.070
```

The 90% CI is used because the test is one-tailed. The hypothesised proportion (0.9) is in this interval which confirms the conclusion reached based on the P-value. The large-sample method gives a z-value of -2.11, and a P-value of 0.018 so if this method were used the claim would have been rejected; a false conclusion.

Note: With a sample size of 20, and 14 cuttings rooting (observed p still 0.7) the exact P-value becomes 0.011 which would provide evidence against the claim. The moral is: small samples don't tell you much. Even large samples give little information when the population proportion is very small or very close to 1; in 100 trials you may not observe any successes or you may observe 100 successes.

18.5 COMPARING TWO PROPORTIONS

A random sample of size n_1 is taken from population 1 and the number of successes k_1 is noted. An independent sample of size n_2 is taken from population 2 and the number of successes k_2 is counted. The unknown population proportions are p_1 and p_2 and the sample proportions are $\hat{p}_1 = k_1/n_1$ and $\hat{p}_2 = k_2/n_2$. An approximate 95% confidence interval for $p_1 - p_2$ is given by

$$(\hat{p}_1 - \hat{p}_2) \pm 1.96 \sqrt{\frac{\hat{p}_1 \hat{q}_1}{n_1} + \frac{\hat{p}_2 \hat{q}_2}{n_2}} \quad \text{where } \hat{q}_1 = 1 - \hat{p}_1 \text{ and } \hat{q}_2 = 1 - \hat{p}_2$$

A 99% interval is obtained by replacing the 1.96 by 2.58. The approximation is not good if the observed proportions are near zero or one or if the sample sizes are small.

Example 18.8
An experiment is conducted to compare two treatments (seed dressings) on the viability of wheat seeds. One hundred seeds of a particular variety are randomly divided into two groups of 50 each. One group is given treatment one and the other treatment two. Subsequently, all the seeds are left to germinate on a moistened filter paper. It is found that 80% of the seeds given treatment one and 66% of those given treatment two germinate.

An approximate 95% confidence interval for the difference in proportions germinating is

$$(0.80 - 0.66) \pm 1.96 \sqrt{\frac{0.80 \times 0.20}{50} + \frac{0.66 \times 0.34}{50}} = 0.14 \pm 0.17 = (-0.03, 0.31)$$

You are therefore 95% confident that the proportion germinating lies between 31% in favour of treatment one to 3% in favour of treatment two. As this interval includes zero you cannot reject, at the 5% level, the null hypothesis that these two treatments give equal germination rates. Check that if the same sample proportions had been based on sample sizes of 80 each, the confidence interval would have been (0.004, 0.276). Thus a difference of 14% based on sample sizes of 50 is not significant, but when based on sample sizes of 80 it is. This emphasises that large samples are required to detect small differences in observed proportions.

When comparing two treatments on a measured variable such as plant height, a small number, say ten replicates per treatment, may be sufficient. You will obtain ten different readings per treatment. However, if you measure an attribute such as plant survival, you only have one reading per treatment; the proportion surviving. As your individual measurements are zeros or ones (success or failure) you will need a large number of plants per treatment to detect a real difference in survival rates.

Note: While the above method can be used to carry out hypothesis tests to compare two population proportions, the contingency table method (Section 18.9) is more commonly used.

18.6 THE CHI-SQUARE GOODNESS OF FIT TEST

This test is used to compare observed frequencies (O) in several categories (cells) corresponding with expected frequencies (E). The expected frequencies are calculated assuming a particular null hypothesis (H_0) to be true. Under the null hypothesis you would expect the Os and Es to be close together. To test H_0 the quantity $(O - E)^2/E$ is calculated for each O and the results added. Thus the test statistic is

$$\chi^2 = \sum \frac{(O - E)^2}{E}$$

This is the calculated chi-square which, if the null hypothesis is true can be assumed to come approximately, from a χ^2 distribution. The degrees of freedom, ν, is the number of cells minus the number of sample constants used to calculate the expected frequencies, which add to the sum of the observed frequencies. The null hypothesis is rejected if the value of χ^2 is sufficiently large, reflecting a significant discrepancy between observed and expected frequencies, hence the test is one-tailed. To carry out the test, tables of the χ^2 distribution may be consulted (Appendix 7). If the calculated value of χ^2 is greater than the table value in the row corresponding to ν and the column headed 5% you reject the hypothesis at the 5% level. However, if the calculated chi-square is less than the degrees of freedom there is no need to consult tables. This is because, under the null hypothesis, the expected value of the calculated chi-square is the degrees of freedom and only much higher values would lead to rejection of the null hypothesis.

Note: This approximate test is not reliable when some expected frequencies are less than 5. Some statisticians do not recommend using this test if more than 20% of the

ANALYSIS OF COUNTS 281

expected frequencies are less than 5, especially if these cells make a large contribution to the total chi-square value. The test should not be used if any cells have expected frequencies less than 1. However, in some situations cells with low expected frequencies may be combined.

18.6.1 Goodness of Fit with Two Categories

The simplest application of the chi-square goodness of fit test is the test of the null hypothesis that a population proportion is a specified value. It is equivalent to the test described in Section 18.4.1 and is only valid if the conditions for the normal approximation to the binomial distribution apply.

Example 18.9

We apply this method to Example 18.6. A 3:1 ratio of purple- to green-stemmed seedlings is expected. The observed frequencies (O) were 112 and 46 respectively giving a total of 158. Hence the corresponding expected frequencies (E) are $3 \times 158/4 = 118.5$ and $1 \times 158/4 = 39.5$. Thus

$$\chi^2 = \sum \frac{(O-E)^2}{E} = \frac{(112-118.5)^2}{118.5} + \frac{(46-39.5)^2}{39.5} = 0.356 + 1.070 = 1.426$$

This test has one degree of freedom because there are two categories and the sample total was used to calculate the expected frequencies. This ensures that the sum of the observed is equal to the sum of the expected frequencies. As 1.426 is less than the 5% chi-square table value of 3.84 the null hypothesis is not rejected. Note that the square root of 1.426 is 1.19, the z-value in Example 18.6, and the square root of 3.84 is 1.96 showing that the two tests are equivalent.

18.6.2 Continuity Correction

The chi-square distribution is continuous whereas calculated values are discrete. If the observed frequencies in the previous example had changed by 1 to 113 and 45, the calculated chi-square would have changed from 1.426 to 1.021, a discrete jump. With a total of 158 plants and expected frequencies of 118.5 and 39.5 it is not possible to obtain a calculated chi-square between these values. When the degrees of freedom $\nu = 1$, a continuity correction is recommended where, for each cell, the magnitude of the difference between the O and E values is reduced by 0.5 before squaring. This reduces the value of chi-square and so makes the test more conservative. For Example 18.9 the adjusted calculated chi-square is

$$\chi^2_{adj} = \sum \frac{(|O-E|-0.5)^2}{E} = \frac{(|112-118.5|-0.5)^2}{118.5} + \frac{(|46-39.5|-0.5)^2}{39.5}$$

$$= \frac{6.0^2}{118.5} + \frac{6.0^2}{39.5} = 0.304 + 0.911 = 1.215$$

and the conclusion is the same. In another example the unadjusted chi-square maybe more than 3.84 and the adjusted value below leading to two different conclusions. If the expected frequencies are large the adjustment should not make much difference. However, it should be routinely applied when $\nu = 1$.

18.6.3 Goodness of Fit with More than Two Categories

The same chi-square test is used as for two categories. The degrees of freedom are the number of cells (categories) minus the number of data restrictions used to calculate the expected frequencies.

Example 18.10
A parental cross is expected to produce progeny in categories A, B, C, D with a ratio 9:3:3:1. The observed frequencies of 250 progeny were 150, 42, 50, 8. Under the null hypothesis of a 9:3:3:1 ratio in the population, the expected frequencies are 9/16, 3/16, 3/16 and 1/16 of 250 = 140.625, 46.875, 46.875 and 15.625. These are given to three decimal places to prevent rounding errors in subsequent calculations. Table 18.2 summarises the calculations. The calculated χ^2 is 5.061.

Only one sample constant (250) was used to calculate the expected frequencies so the degrees of freedom are ν = number of categories − 1 = 3. This can also be found by noting that there is one restriction; that the sum of the observed frequencies is equal to the sum of the expected frequencies. Thus, if the first three values of E are calculated the fourth can be found by subtraction from 250.

From Appendix 7, $\chi^2_{(3, 5\%)} = 7.815$. As 5.061 is less, the hypothesis is not rejected. The P-value is 0.1674 as found by computer. This is a measure of the strength of evidence against the null hypothesis; it is the probability of obtaining a calculated chi-square greater than 5.061 if the hypothesis is true. As P is not less than 0.05 you do not reject the hypothesis.

Example 18.11 Testing the fit to a binomial distribution
Consider an experiment in which 100 pots are each planted with six seeds of one barley variety, and the growing conditions are uniform. If the probability of any given seed germinating is 0.8 and the seeds germinate independently you would expect the number of seeds germinating in any particular pot to follow a binomial distribution with probabilities given in the last column of Table 18.1. By multiplying these probabilities by 100 the expected numbers of pots with 0, 1, 2, 3, 4, 5 or 6 are found. Now suppose the observed frequencies are 0, 2, 3, 12, 19, 43, 21. The calculations for the test of the null hypothesis that the seeds germinate independently with a probability of 0.8 are summarised in Table 18.3.

In this table O represents the observed frequency of pots in which k seeds germinate. For example, there are 12 pots each with three germinations. The calculated χ^2 is 27.945 and the degrees of freedom are $\nu = 7 - 1 = 6$. From tables, $\chi^2_{(6, 0.1\%)} = 22.46$. As 27.945 is greater than this ($P < 0.001$) you may be tempted to conclude there is very strong evidence against the null hypothesis.

Caution: There are three expected frequencies less than 5, of which two are less than 1, so the test is unreliable. When an expected frequency is less than 1 a change in O of 1 can have a large effect on χ^2. In this example if the O corresponding to $k = 1$ had been 1 instead of 2, its contribution to χ^2 would have been 4.648 instead of 22.128 and a non-significant result probably obtained. A more reliable test is to combine the cells with expected frequencies less than 5 to obtain $O = 0 + 2 + 3 = 5$ and $E = 0.006 + 0.154 + 1.536 = 1.696$. The corresponding $(O - E)^2/E$ is 6.437. On adding this to the values for $k = 3$ to 6 the calculated χ^2 is 10.853. The degrees of freedom are reduced to 4 (one less than the new number of cells). The critical table value is $\chi^2_{(4, 5\%)} = 9.488$ so the result is still significant ($P < 0.05$) and provides evidence against the null hypothesis. However, the expected frequency of 1.696 is less than 5 and some authors would recommend a further combination. The rules about when to combine cells are open to debate. Some statisticians would allow one expected frequency less than 5 provided it was greater than 1. If the cell with $k = 3$ is also combined, the degrees of freedom are reduced to 3 and the 5% table value becomes 7.815. You should check that the resulting new total χ^2 value is 7.76, which is a non-significant result.

Note: If a value for p cannot be assumed, it may be estimated from the data by dividing the total number of seeds germinating by the number planted:

$$\hat{p} = \frac{\Sigma(k \times O)}{n \times \Sigma O}$$

$$= \frac{(0 \times 0) + (1 \times 2) + (2 \times 3) + (3 \times 12) + (4 \times 19) + (5 \times 43) + (6 \times 21)}{6 \times 100}$$

$$= \frac{461}{600} = 0.768$$

The modified null hypothesis is that the seeds germinate independently with a constant but unknown probability. The calculations would be repeated using this value for p but the degrees of freedom would be reduced by one to take into account that p was estimated from the data. Try this as an exercise. The calculated chi-square is 13.22 on 5 df which is significant at the 5% level as $\chi^2_{(5, 5\%)} = 11.07$. If the first three cells are combined it is 5.44 on 3 df which is not significant as $\chi^2_{(3, 5\%)} = 7.815$. This is a more realistic conclusion as the effects of small expected frequencies make the test very unreliable.

Table 18.2. Chi-square calculations for Example 18.10

Category	A	B	C	D	Total
O	150	42	50	8	250
E	140.625	46.875	46.875	15.625	250
O − E	9.375	−4.875	3.125	−7.625	0
$(O - E)^2/E$	0.625	0.507	0.208	3.721	5.061

Table 18.3. Chi-square calculations for Example 18.11

k	0	1	2	3	4	5	6	Total
O	0	2	3	12	19	43	21	100
E	0.006	0.154	1.536	8.192	24.576	39.322	26.214	100
O − E	−0.006	1.846	1.464	3.808	−5.576	3.678	−5.214	0
$(O - E)^2/E$	0.006	22.128	1.395	1.770	1.265	0.344	1.037	27.945

18.7 $r \times c$ CONTINGENCY TABLES

Each member of a sample is classified according to two criteria. In general, the first criterion has r classes and the second has c classes. The observed frequencies (counts) can then be summarised in a contingency table having r rows and c columns. The data may arise as the result of a survey or an experiment, and the chi-square goodness of fit test can be used to test whether the two criteria of classification are independent or associated. In some cases the row totals or column totals are fixed in advance and sometimes just the grand total is fixed. If any row or column consists entirely of zeros it should be deleted. Misleading conclusions may also be made if the table contains some cells with very small counts. If the table has two rows or two columns the test can be used to compare two or more proportions.

Important note: The data in a contingency table must be in the form of counts for the chi-square test to be valid. Do not use proportions or percentages in the formula for chi-square.

Table 18.4. Contingency table with observed and expected frequencies for Example 18.12

Petal colour	Seed colour				Total
	Dark brown	Brown	Light brown	White	
Purple	21 (12.60)	32 (27.02)	13 (27.30)	4 (3.08)	70
Mauve	55 (42.30)	102 (90.71)	70 (91.65)	8 (10.34)	235
White	14 (35.10)	59 (75.27)	112 (76.05)	10 (8.58)	195
Total	90	193	195	22	500

Example 18.12
You wish to know whether seed colour and petal colour in field beans are associated. If they are independent (not associated) then knowledge of seed colour would give no information about petal colour. You carry out a survey in which you obtain a random sample of 500 bean plants from the population of interest and classify each plant according to seed colour and petal colour. Suppose you obtain the results which are summarised in Table 18.4.

The numbers in parentheses are the expected frequencies assuming independence and their method of calculation is explained as follows. The estimate of the proportion of plants with purple petals is 70/500. If there is no association, you would expect this proportion of those with dark-brown seeds to have purple petals. Thus the expected frequency of plants with dark-brown seeds and purple petals is $(70/500) \times 90 = 12.60$. The expected frequencies in the other cells are calculated in a similar way. The general formula is

$$\text{Expected frequency} = \frac{(\text{Row total}) \times (\text{Column total})}{\text{Grand total}}$$

The expected frequencies are not independent because for any row or column the sum of the expected frequencies is equal to the sum of the observed frequencies. Thus, in each of the first two rows there are $(4 - 1) = 3$ degrees of freedom because after the first three expected frequencies are calculated using the above formula, the fourth can be found by subtraction from the corresponding row total. There are no degrees of freedom left for the third row as the expected frequencies can be found by subtraction from the column totals. The degrees of freedom for the table are thus $2 \times 3 = 6$. This is the same as the number of cells left after crossing out the last row and last column. In general the degrees of freedom for a table with r rows and c columns is $(r - 1) \times (c - 1)$.

The formal test of association is to carry out a chi-square goodness of fit test on 6 df. The $(O - E)^2/E$ value is calculated for each of the 12 cells in the table and the results added. The following Minitab output shows the individual contributions:

```
Chi-Sq= 5.600 + 0.918 + 7.490 + 0.275 +
        3.813 + 1.405 + 5.114 + 0.530 +
        12.684 + 3.517 + 16.994 + 0.235 = 58.575
DF = 6, P-Value = 0.000

1 cells with expected counts less than 5.0
```

If you perform the calculations by hand you need to compare the calculated chi-square of 58.575 with the table value $\chi^2_{(6, 5\%)} = 12.59$. This shows that the P-value is less than 0.05. In fact it is less than 0.0005 as $\chi^2_{(6, 0.05\%)} = 24.10$. The result of the test confirms what is obvious by looking at the table of observed frequencies, that is, that there is a high degree of association between seed colour and petal colour. The next step is to look at the contributions to chi-square. These are highest for the dark-brown seed, white petal and light-brown seed, white petal combinations. For those with dark-brown seeds, the observed proportion with white petals is $14/90 = 0.16$ whereas of those with light-brown seeds, the proportion with white petals is $112/195 = 0.57$. Further tests can be carried out on subsections of the original table. There are many subtables with sizes 3×3, 2×4, 3×2, 2×3, and 2×2. You should be cautious about applying tests suggested by the data and some of these comparisons will not be independent. Finally, it should be noted that there is not much data on the white-seeded plants and the expected frequency for white seeds and purple petals is 3.08 which is less than 5. As there is only one such cell in the table and the E is not less than 1 the problem is not too serious. You may, however, wish to repeat the test after eliminating the column for white seeds. In another example you may want to combine two or more rows or columns.

18.8 $2 \times c$ CONTINGENCY TABLES: COMPARISON OF SEVERAL PROPORTIONS

The method of analysing data from a $2 \times c$ table is the same as that for an $r \times c$ table. The observed frequencies may result from a survey or an experiment. In some situations only the grand total is fixed by the investigator. In other cases either the row totals or the column totals are fixed. In all cases the analysis is the same and the null hypothesis is that there is no association between the two criteria of classification or equivalently that the c population proportions are equal.

Note: When reporting proportions or percentages you should also state the sample sizes on which they are based.

Table 18.5. Contingency table with observed and expected frequencies for Example 18.13

Storage method	A	B	C	D	E	Total
No. germinating	112 (105.30)	76 (76.43)	88 (101.90)	43 (42.46)	92 (84.92)	411
No. not germinating	12 (18.70)	14 (13.57)	32 (18.10)	7 (7.54)	8 (15.08)	73
Total	124	90	120	50	100	484
Percentage germinating	90.3	84.4	73.3	86.0	92.0	

Example 18.13

Five storage methods were tested to compare their effects on the viability of a variety of pea seed. Table 18.5 shows the observed frequencies with the expected frequencies in parentheses.

It is assumed that the column totals were fixed by the experimenter. For example, 124 seeds treated by method A were planted and 112 germinated (90.3%). The null hypothesis is that the population germination rates are equal versus the alternative that they are not. Assuming the null hypothesis to be true the expected frequencies for each cell are calculated by multiplying the row total by the column total and dividing by the grand total. The degrees of freedom are $(2-1) \times (5-1) = 4$. The following Minitab output shows the contributions to chi-square and the results of the test:

```
Chi-Sq = 0.427 + 0.002 + 1.896 + 0.007 + 0.591 +
         2.402 + 0.013 + 10.676 + 0.039 + 3.326 = 19.379
DF = 4, P-Value = 0.001
```

The P-value indicates that there is a very small chance of obtaining a calculated chi-square greater than or equal to 19.379 if the null hypothesis is true. Thus the data provide strong evidence against this hypothesis. The conclusion of this test is that the germination rates are significantly different. The largest contribution to chi-square (10.676) is due to storage method C having a much smaller germination rate (73.3%) than the rest.

Although you should not carry out tests of hypotheses suggested by the data after examining them, you may be tempted to ignore the data for method C and carry out a chi-square test on the remaining 2 × 4 table. You should confirm that the calculated chi-square is 3.411 on 3 df. From Appendix 7, $\chi^2_{(3, 5\%)} = 7.815$. This result suggests that methods A, B, D and E give similar population germination rates. You could also form a 2 × 2 table which has C for one column and the combined observed frequencies from the other methods as the other column. Comparisons of all possible pairs of methods could be made by forming all possible pairs of 2 × 2 tables, but they would not be independent. If significant differences are found they should be a guide to developing further hypotheses to be tested with a new set of data.

Note: When presenting proportions or percentages you should also give counts on which the proportions are based. For example, just to say that method A (90%) is better than method C (73%) is misleading. If these percentages were based on 10 counts each they would not be significantly different, whereas if they were based on 100 they would be.

18.9 2 × 2 CONTINGENCY TABLES: COMPARISON OF TWO PROPORTIONS

This is a special case of the $r \times c$ table (Section 18.7). There is only one degree of freedom because $(r - 1) \times (c - 1) = (2 - 1) \times (2 - 1) = 1$. Only one expected frequency needs to be calculated by the formula $E = $ (row total × column total)/grand total. The other three can then be found by subtraction from row and column totals. The chi-square test can be used to test the null hypothesis that two population proportions are equal versus the alternative that they are not (a two-tailed test of a difference in the proportions). The chi-square test is one-tailed as a large discrepancy in the observed proportions in either direction would result in a large calculated chi-square. The chi-square 5% table value on 1 df is 3.84 so a larger calculated value would result in rejection of the null hypothesis at the 5% level. To carry out a one-tailed test on the proportions at the 5% level you need to compare the calculated chi-square with the 10% table value which is 2.71. Alternatively, the P-value obtained by computer for the chi-square test is halved to give the appropriate P-value for the one-tailed test on the proportions. You should try to arrange for the totals on which the proportions are based to be large and equal if this is under your control.

Note: The chi-square test is not very reliable if the frequencies are small. If the grand total (N) is less than 20, or if any expected frequency is less than 1, the test should

Table 18.6. Contingency table for Example 18.14

	Hormone A	Hormone B	Total
Number rooting	25 (29.778)	42 (37.222)	67
Number not rooting	15 (10.222)	8 (12.778)	23
Total	40	50	90
Proportion rooting	0.625	0.840	

not be used. If any expected frequency is less than 5 and N is less than 40, some statisticians warn against its use. Fisher's Exact Test can be used for small frequencies. Its application is described in Table 26 of Lindley and Scott (1995) for $N \leqslant 17$.

Example 18.14
From a source of apple cuttings, 40 are chosen at random and treated with growth hormone A. Of these, 25 root after planting. Another 50 are chosen and treated with hormone B. Of these, 42 take root. We wish to test the null hypothesis that the proportion of cuttings which root is the same for both hormones, versus the alternative that they are not (a two-tailed test). Table 18.6 is the contingency table of observed frequencies with expected frequencies in parenthesess calculated using the formula

$$\text{Expected frequency } (E) = \frac{\text{Row total} \times \text{Column total}}{\text{Grand total}}$$

We test whether the two observed proportions (0.625 and 0.84) are significantly different, by using the chi-square test. Notice that for each row and for each column, the sum of the observed frequencies is the same as the sum of the expected frequencies. In this case, only one expected frequency needs to be calculated using the formula; the others can be found by subtraction. There is one degree of freedom and the calculated chi-square value is

$$\chi^2 = \sum \frac{(O-E)^2}{E} = \frac{(25-29.778)^2}{29.778} + \frac{(42-37.222)^2}{37.222} + \frac{(15-10.222)^2}{10.222} + \frac{(8-12.778)^2}{12.778}$$

$$= 0.767 + 0.613 + 2.233 + 1.787$$

$$= 5.40$$

This is greater than 3.84, the 5% point of the χ^2 distribution on 1 df. Hence we reject the null hypothesis that the two hormones are equally effective in stimulating rooting. You will find 3.84 in row 1 and the 5% column of the chi-square table in Appendix 7. This is a one-tailed test, as we only reject the null hypothesis if the O and E values are sufficiently different. It is equivalent to a two-tailed test of the difference between the two proportions. To perform a one-tailed test, at the 5% level, of the difference in the proportions compare the calculated chi-square with 2.71, the 10% point of the χ^2 distribution on 1 df.

18.9.1 Yates' Correction

If the observed frequencies are fairly small, Yates' correction should be used. The magnitude of the difference between O and E is reduced by 0.5 before squaring. For Example 18.14, the difference between O and E is 4.778 for each cell. After reduction by 0.5 we get 4.278, which, for each cell is squared and divided by the corresponding E. The adjusted value of chi-square is

$$\sum \frac{(|O - E| - 0.5)^2}{E} = \frac{4.278^2}{29.778} + \frac{4.278^2}{37.222} + \frac{4.278^2}{10.222} + \frac{4.278^2}{12.778} = 4.33$$

so the result is still significant.

The application of Yates' correction always reduces the values of chi-square, so that a significant result is less likely. However, if all the expected frequencies are large the reduction is slight.

Example 18.15

This example illustrates the fact that when data result from a survey, cause and effect cannot be assumed. In a survey of 200 smallholders selected at random the adoption of a new variety of rice was investigated. Of these, 75 had been visited by an advisor within the last two years to encourage the adoption, and 52 of them had grown the new variety. Of the 125 not visited, 54 had also adopted it. These data are summarised in Table 18.7.

Confirm that the calculated chi-square is 12.852. This indicates a very highly significant difference between the two sample proportions. The null hypothesis that the adoption rate is not affected by a visit therefore rejected. However, the difference in the proportions may be due to reasons other than the advisory visits.

18.10 ASSOCIATION OF PLANT SPECIES

An ecologist may be interested in whether two ranunculus species tend to occur together or whether the presence of one tends to preclude the presence of the other. A survey can be carried out by placing quadrats at random in the area of interest and observing the presence and absence of the two species. A contingency table can be

Table 18.7. Contingency table for Example 18.15

	Visited	Not visited	Total
Adopted	52	54	106
Not adopted	23	71	94
Total	75	125	$N = 200$
Proportion adopting	0.693	0.432	

Table 18.8. Contingency table for Example 18.16

		Species A +	Species A −	Total
Species B	+	50	40	90
	−	30	80	110
Total		80	120	N = 200

constructed and a chi-square test performed to test the null hypothesis that the occurrences of the two species are not associated. A non-significant result would imply that the chance of one of the species being present is not affected by the presence of the other.

Example 18.16
Two hundred quadrats were randomly placed in an area of meadowland to investigate the degree of association of two plant species A and B. In 50 of the quadrats both species were present and in 80 of them both were absent. Thirty contained A only, and 40 contained B only. The observed frequencies are presented in Table 18.8.

Of the 80 quadrats which contain A, the proportion containing B is $50/80 = 0.625$. Of the 120 quadrats which do not contain A, the proportion containing B is $40/120 = 0.333$. Under the null hypothesis of no association these proportions are expected to be equal. Confirm that the chi-square is 16.50 on 1 df. This is large enough to cause rejection of the null hypothesis at the 0.1% level. After applying Yates' correction the adjusted chi-square is 15.34 leading to the same conclusion. The association is positive because if A is present, B is more likely to be present than if A is absent ($0.625 > 0333$). That the association is positive can also be concluded by noting the observed frequency in the (+, +) cell (50) is greater than expected frequency ($90 \times 80/200 = 36$).

18.11 HETEROGENEITY CHI-SQUARE

The sum of individual chi-squared values is itself a chi-square value with degrees of freedom equal to the sum of the individual degrees of freedom. This fact can be used to test whether pooling results from several chi-square analyses is valid. You would like to pool the data and carry out a single chi-square test but first you should test whether the individual results are homogeneous or heterogeneous. If they are homogeneous the pooled analysis is valid. The procedure is to add the individual chi-square values and degrees of freedom to obtain a total of chi-squares. Then find a chi-square based on the totals arrived at by pooling the data. Subtract this from the total of chi-squares to find the heterogeneity chi-square with degrees of freedom also found by subtraction. If the heterogeneity chi-square is not significant, the pooling is

ANALYSIS OF COUNTS

valid and you can use the chi-square based on the totals. Examples 18.17 and 18.18 should clarify the procedures.

Example 18.17

Consider Example 18.9 in which a 3:1 ratio of purple- to green-stemmed seedlings was tested. You may have carried out several such experiments and found the individual chi-square values and their sum. Table 18.9 shows a hypothetical set of results from five experiments.

Each of the five individual chi-square tests supports the null hypothesis of a 3:1 ratio so it is tempting to pool the data and test the 3:1 ratio on the totals (155, 59). This gives a pooled chi-square of 0.754 on 1 df which is clearly not significant, again supporting a 3:1 ratio. To test if the pooling is justified find

Heterogeneity chi-square = Total of chi-squares − pooled chi-square
= 7.020 − 0.754 = 6.266 on (5 − 1) = 4 df

This is not significant as $\chi^2_{(4, 5\%)} = 9.488$ so the five sets of results can be considered homogeneous and the pooling is valid.

If the heterogeneity test had been significant you would have suspected that the data came from two or more populations. Inspection of the data suggest that samples 1, 4 and 5 came from one population with a ratio less than 3:1 and samples 2 and 3 came from another with a ratio greater than 3:1, but in this case the evidence to support this theory is not quite strong enough. With more data the individual chi-squares may not have supported a 3:1 ratio but the pooled chi-square may have.

Note: Do not use the continuity correction when computing the heterogeneity chi-square. However, if you can justify the pooling the correction can be applied to the pooled chi-square.

Table 18.9. Hypothetical data for heterogeneity chi-square analysis

Experiment	Purple	Green	Total	Prop. purple	Chi-square	df
1	40 (45.0)	20 (15.0)	60	0.67	2.222	1
2	42 (39.0)	10 (13.0)	52	0.81	0.923	1
3	28 (25.5)	6 (8.5)	34	0.82	0.980	1
4	19 (21.0)	9 (7.0)	28	0.68	0.762	1
5	26 (30.0)	14 (10.0)	40	0.65	2.133	1
Total	155	59	214	0.72	7.020	5

Example 18.18

In this example the heterogeneity test is applied to 2 × 2 contingency tables. Suppose a rooting experiment similar to that described in Example 18.14 is repeated three times. Assume a single source of cuttings. In each experiment some cuttings are treated with hormone A and others with B. The numbers of cuttings planted, rooting and not rooting and the individual chi-squares are shown in Table 18.10. The pooled data are shown in Table 18.11.

For each experiment, you would not reject the null hypothesis that the proportion rooting is the same for each hormone. This may be because there is no difference between the hormones with respect to their rooting ability. If this is your conclusion you may be committing a Type II error. There may be a real difference, but the amount of data in each experiment is not sufficient to detect the difference. The data from the three experiments may be pooled if the separate results are consistent (homogeneous).

From Table 18.11 the pooled chi-square is 7.307 on 1 df (calculations not shown). From Table 18.10 the sum of the individual chi-squares is $3.360 + 1.553 + 2.982 = 7.895$ on 3 df. The difference is the heterogeneity chi-square; it is 0.588 on $(3 - 1) = 2$ df. This is not significant as it is less than $\chi^2_{(2, 5\%)} = 5.991$, so the pooling is valid and the pooled result significant at the 5% level as 7.307 is greater than 3.841. Application of the continuity correction reduces the pooled chi-square to 6.491 and so does not affect the conclusion.

If the numbers for A and B in experiment 1 had been reversed, the calculated chi-square would still have been 3.360 but the observed proportion would have been in favour of A, and not consistent with the results from experiments 2 and 3. The pooled chi-square would have been only 0.300 and the heterogeneity chi-square would have been 7.595 on 2 df. This significant result would have meant pooling was not valid.

Table 18.10. Hypothetical data for heterogeneity chi-square analysis of 2×2 contingency tables

	Experiment 1			Experiment 2			Experiment 3		
	A	B	Total	A	B	Total	A	B	Total
Number rooting	18	32	50	14	23	37	25	30	55
Number not rooting	12	8	20	11	9	20	13	6	19
Total planted	30	40	70	25	32	57	38	36	74
Proportion rooting	0.60	0.80		0.56	0.72		0.66	0.83	
Chi-square		3.360			1.553			2.982	

Table 18.11. Pooled contingency table for Example 18.18

	Hormone A	Hormone B	Total
Number rooting	57	85	142
Number not rooting	36	23	59
Total planted	93	108	201
Proportion rooting	0.613	0.787	

Chapter 19

Some Non-parametric Methods

19.1 INTRODUCTION

The procedures used in previous chapters for carrying out significance tests and the calculation of confidence intervals rely for their validity on certain assumptions being true. In most cases we assumed that populations are normally distributed, population variances are equal, and that samples are independent. In some cases we recommended carrying out a transformation to make the transformed data satisfy these assumptions. Such tests are called **parametric** because they state hypotheses about population parameters and usually require the estimation of parameters such as population variances. They are robust in the sense that only serious departures from the assumptions lead to incorrect conclusions. When there are serious departures, conclusions based on non-parametric tests are likely to be less misleading.

In non-parametric methods no assumptions are made about specific population distributions or their parameters such as means and variances. These methods are most useful when the assumptions of the parametric methods are not valid. However, they are usually less powerful than the traditional methods in the sense that they give wider confidence intervals and are less likely to declare significant differences. They also do not use all the available information in the data. Most non-parametric methods involve replacing raw data values by ranks. For example, the following four plant heights in cm (13.5, 15.2, 23.6, 54.8) have ranks (1, 2, 3, 4). Hence, calculations based on ranks assume that the difference between the first two heights is the same as the difference between the last two. These tests are often applied to data such as subjective assessments of colour or degree of damage. However, as the data may be in the form of scores on a scale 0 to 5, for example, there are likely to be many tied ranks which should be adjusted for.

The following non-parametric tests are described in this chapter; the corresponding parametric test is shown in parentheses:

- The Sign test (one-sample t-test or paired samples t-test)

- Wilcoxon signed-ranks test (one-sample *t*-test)
- Wilcoxon matched pairs test (paired samples *t*-test or randomised blocks with two treatments)
- Mann–Whitney test (independent samples *t*-test or one-way analysis of variance with two treatments)
- Kruskal–Wallis test (*F*-test in one-way analysis of variance)
- Friedman test (*F*-test in a randomised blocks analysis of variance)

We do not discuss the theory behind these tests. For more details of these and other non-parametric tests the reader should consult Siegel and Castellan (1988).

19.2 THE SIGN TEST

This is used to test whether the median of a population is a specified value. No assumptions are made about the shape of the distribution so it can be used instead of the one-sample *t*-test (Section 5.2) if the data are very skew. Suppose a random sample of n values is obtained from the population. Under the null hypothesis that M is the true value of the median, you would expect 50% of the sample values to be greater than M, and 50% less than M. Attach a $+$ to any value greater than M and a $-$ to any value less than M. Ignore any sample value equal to M, reducing n accordingly. Hence, under this hypothesis, the number of plus signs $(n+)$ and the number of minus signs $(n-)$ follow a binomial distribution (Section 18.2) with $n = (n+) + (n-)$ and $p = 0.5$.

For a two-tailed test you would reject the null hypothesis if $n+$ and $n-$ were sufficiently different from $n/2$. If $n+$ is greater than $n-$, the *P*-value is the probability of $n+$ or more 'successes' in this binomial distribution plus the probability of $n-$ or fewer successes. Due to the symmetry of the binomial distribution with $p = 0.5$, this is double the probability of $n-$ or fewer 'successes'. If $n+$ is less than $n-$, the *P*-value is the probability of $n+$ or fewer 'successes' plus the probability of $n-$ or more 'successes'. Due to symmetry this is double the probability of $n+$ or fewer 'successes'.

To carry out the two-tailed test, find S, the smaller of $n+$ and $n-$. The *P*-value is twice the probability of S or fewer 'successes' in the binomial distribution with n 'trials' and $p = 0.5$. This probability can be found in Appendix 8 for values of n less than or equal to 20. If n is greater than 20, use the normal approximation (Section 18.2.2). Reject at the 5% level if *P* is less than 0.05.

For a one-tailed test with H_0: population median $= M$ versus H_1: population median $> M$. If $n+$ is greater than $n-$, the *P*-value is the probability of $n+$ or more successes which, by symmetry, is equal to the probability of $n-$ or fewer 'successes'. If $n+$ is less than $n-$ you would certainly not reject H_0 as the *P*-value would be greater than 0.50.

For a one-tailed test with H_0: population median $= M$ versus H_1: population median $< M$. If $n-$ is greater than $n+$, the *P*-value is the probability of $n-$ or more 'successes' which, by symmetry, is equal to the probability of $n+$ or fewer 'successes'. If $n-$ is less than $n+$ you would certainly not reject H_0 as the *P*-value would be greater than 0.50.

SOME NON-PARAMETRIC METHODS 295

To summarise: In the one-tailed (greater than) test, the P-value is the probability of n− or fewer 'successes', whereas in the one-tailed (less than) test, the P-value is the probability of n+ or fewer 'successes'

Example 19.1

When the amount of data is small, the sign test is not very powerful. When n is larger the calculations involved in the binomial distribution are tedious, but the cumulative probabilities for $p = 0.5$ can be obtained from Appendix 8. Consider the following sample of 16 observations: 12, 14, 25, 31, 11, 16, 52, 19, 9, 25, 15, 21, 18, 12, 15, 40. Test the null hypothesis that $M = 25$ versus the alternative that it is not. By assigning + and − signs as follows

12	14	25	31	11	16	52	19	9	25	15	21	18	12	15	40
−	−	+	+	−	−	+	−	−		−	−	−	−	−	+

verify that $n+ = 3$ and $n− = 11$. This gives $n = 14$ ignoring the two 25s which are equal to the hypothesised median. The smaller of $n+$ and $n−$ is $S = 3$, so the P-value is twice the probability of three or fewer successes in the binomial distribution with $n = 14$ and $p = 0.5$. From Appendix 8 this is $2 \times 0.0287 = 0.0574$. Thus the sample median (17.0) is not quite significantly different from 25 at the 5% level.

Note that if this had been a one-tailed test with the alternative that $M < 25$, the appropriate P-value is the probability of three or fewer successes = 0.0287 because $n+ = 3$.

The Minitab output for the two-tailed test is

```
Sign test of median = 25.00 versus not = 25.00

      N     Below    Equal    Above        P       Median
      16      11        2        3      0.0574      17.00
```

Tables of the binomial distribution do not usually give probabilities for n greater than 20. However, because $p = 0.5$, the binomial distribution is symmetrical, and even for relatively small n the normal approximation with a continuity correction can be used (Section 18.2.2). In the last example $n = 14$ so an approximation to the probability of three or fewer 'successes' is obtained by first calculating the z-value as

$$z = \frac{(S + 0.5) - np}{\sqrt{npq}} = \frac{3.5 - 14 \times 0.5}{\sqrt{14 \times 0.5 \times 0.5}} = \frac{-3.5}{1.8708} = -1.87$$

The required P-value is the probability of a z value larger than 1.87 in magnitude. From normal tables this is $2(1 - \Phi(1.87)) = 2(1 - 0.9693) = 0.0614$. This is quite close to the exact value of 0.0574. You should also realise that this is not significant at the 5% level because the calculated z (1.87) is not greater than 1.96 in magnitude.

19.3 THE WILCOXON SINGLE-SAMPLE TEST

This is also known as the Wilcoxon signed-rank test. Like the sign test it can also be used to test whether the median of a population is a specified value. However, it makes the assumption that the observations come from a population with a symmetric, but not necessarily normal, distribution. If the data are very skew a transformation may be applied before carrying out the test if this makes the distribution of the data more symmetrical. This is a more powerful test than the sign test if the assumption is valid because it takes account of the magnitude of the observations. The method is as follows:

(1) Ignore any observations equal to the hypothesised median (M), reducing n accordingly.
(2) Calculate the difference between each observation and M.
(3) Rank the differences in order of magnitude (smallest to largest) ignoring their signs. When ranks are tied (equal differences) an average rank is assigned.
(4) Calculate the sum of the ranks of the positive differences ($T+$) and of the negative differences ($T-$) and denote the smaller of these by T.
(5) Check that $(T+) + (T-) = n(n+1)/2$. If H_0 is true you would expect both $(T+)$ and $(T-)$ and hence T to be equal to $n(n+1)/4$.
(6) If n is small (40 or less) consult Appendix 9. For a two-tailed test, if T is less than or equal to the table value for the row corresponding to n and the column corresponding to the level of significance, the null hypothesis (H_0) of no difference in the medians is rejected. In the one-tailed (greater than) test, reject H_0 if $T-$ is less than or equal to the table value, and for the one-tailed (less than) test reject H_0 if $T+$ is less than or equal to the table value. For n larger than 40 use the normal approximation.

For larger samples use is made of the fact that under the null hypothesis, T is approximately normally distributed with a mean of $n(n+1)/4$ and a variance of $n(n+1)(2n+1)/24$. Hence, to test the null hypothesis calculate

$$z = \frac{|T - n(n+1)/4| - 0.5}{\sqrt{n(n+1)(2n+1)/24}}$$

The 0.5 is the continuity correction to account for the fact that T can only take discrete values whereas the normal distribution is continuous. Reject H_0 at the 5% level (two-tailed test) if the magnitude of z is greater than 1.96. For a one-tailed test at the 5% level, reject if z is greater than 1.65 in magnitude.

Example 19.2
Suppose that the mean yield of a standard (recommended) linseed variety when grown in South-east England is 2.0 t/ha. The yields obtained from a random sample of 10 plots sown with a new variety in the same region were 2.6, 2.2, 2.4, 2.4, 1.9, 2.3, 1.7, 2.0, 2.5, and 2.7 t/ha. Do these results provide sufficient evidence to conclude that the yield of the new variety is different from the standard?

The data are fairly symmetrical but we may not be able to assume a normal population so we apply the Wilcoxon test with the null hypothesis (H_0) that the population median (not the mean) is 2.0 versus the alternative that it is not.

Yield (y)	2.6	2.2	2.4	2.4	1.9	2.3	1.7	2.0	2.5	2.7
$y - M$	0.6	0.2	0.4	0.4	−0.1	0.3	−0.3		0.5	0.7
$\|y - M\|$	0.6	0.2	0.4	0.4	0.1	0.3	0.3		0.5	0.7
Rank	8	2	5.5	5.5	1	3.5	3.5		7	9

The yield of 2.0 is not used as it is equal to M, so reducing n to 9. The yields of 2.3 and 1.7 have the same absolute difference ($|y - M| = 0.3$) from 2.0, and share the ranks of 3 and 4 so they are both given the average rank of 3.5. Similarly, the two yields of 2.4 share the ranks of 5 and 6 and are given the average rank of 5.5.

The sum of the negative ranks is $(T-) = 1 + 3.5 = 4.5$ and the sum of the positive ranks is $(T+) = 8 + 2 + 5.5 + 5.5 + 3.5 + 7 + 9 = 40.5$. Check that $(T+) + (T-) = n(n+1)/2 = (9 \times 10)/2 = 45$. The critical value in Appendix 9 corresponding to a two-tailed test at the 5% level with $n = 9$ is 5. As the smaller of $(T+)$ and $(T-)$ is $T = 4.5$ and is less than 5, the null hypothesis is rejected and there is evidence the population median is not equal to 2.0. Confirm the sample median is 2.35.

Note: If the Sign test (Section 19.2) is applied to these data the P-value is 0.18. This shows that the Wilcoxon test is more powerful provided the assumptions for its use are valid. If the one-sample t-test (Section 5.2) is used, the P-value is 0.026. This is even more powerful provided you can assume a normally distributed population. Thus different tests applied to the same data give different answers. You should use the most powerful test only if the underlying assumptions are met and not because it is the only one that gives you a significant result.

19.4 THE WILCOXON MATCHED PAIRS TEST

This test compares the medians of two population distributions which are of the same shape but not necessarily normal. It is the non-parametric equivalent to the paired samples t-test (Section 5.8) or randomised block design with two treatments. In general n pairs of measurements or scores are obtained. These are denoted by $(x_1, y_1), (x_2, y_2), \ldots, (x_n, y_n)$. The differences $d_i = (x_i - y_i)$ are found and their magnitudes ranked. Differences of zero are ignored when finding the ranks and n is reduced accordingly. When ranks are tied (equal differences) an average rank is assigned. The sum of the ranks for the positive differences is denoted by $T+$ and of the negative differences by $T-$. The smaller of these is denoted by T. If n is small (40 or less) consult Appendix 9. If T is less than or equal to the table value for the row corresponding to n, and the column corresponding to the level of significance, the null hypothesis (H_0) of no difference in the medians is rejected.

Note: This test is the same as the Wilcoxon single-sample test (Section 19.3) applied to the differences with the null hypothesis that the median of the differences is zero. An assumption of the Wilcoxon matched pairs test is that the differences come from a symmetrical population. If you cannot make this assumption the Sign test (Section 19.2) could be used to test if the median of the differences is zero, but it is considerably less powerful when the assumption is met because it takes into account only the signs of the differences.

Table 19.1. Numbers of fungal colonies on tomato leaves after applying fungal strains A and B

Plant	A	B	Difference A − B	Absolute difference	Rank	Signed rank
1	11	9	2	2	3.5	3.5
2	12	15	−3	3	5	−5
3	5	9	−4	4	6	−6
4	7	6	1	1	1.5	1.5
5	5	14	−9	9	7	−7
6	5	16	−11	11	9	−9
7	8	8	0	0		
8	9	10	−1	1	1.5	−1.5
9	16	14	2	2	3.5	3.5
10	8	18	−10	10	8	−8

Example 19.3

Ten healthy tomato plants were injected with two strains of fungus. Strain A was applied to one leaf of each plant and strain B to another. After a period of time the numbers of fungal colonies on the affected leaves were counted. The results are shown in Table 19.1.

The fifth column shows the magnitude of the differences (the absolute differences). These are ranked in the sixth column after ignoring the difference of zero. Two of the absolute differences are equal to 1 so these share ranks 1 and 2 and are therefore each given the mean rank 1.5. Similarly the two differences of 2 share ranks 3 and 4 are each given rank 3.5. The ranks in the sixth column are entered in the last column after giving them the sign of the corresponding difference in the fourth column. The sum of the positive ranks is $T+ = 3.5 + 1.5 + 3.5 = 8.5$ and the sum of the negative ranks is $T- = 5 + 6 + 7 + 9 + 1.5 + 8 = 36.5$. Due to the one difference of zero, $n = 9$. As a check confirm that $(T+) + (T-) = n(n+1)/2 = (9 \times 10)/2 = 45$.

The null hypothesis (H_0) is that the medians of the numbers of colonies is the same for the two strains versus the alternative that they are different. From Appendix 9, the critical value of T for testing the null hypothesis at the 5% level is 8. As 8.5, the smaller of $T+$ and $T-$, is not less than or equal to 8, H_0 is not rejected.

Although the sample size is rather small we apply the normal approximation as an exercise:

$$z = \frac{|T - n(n+1)/4| - 0.5}{\sqrt{n(n+1)(2n+1)/24}} = \frac{|8.5 - 22.5| - 0.5}{\sqrt{9 \times 10 \times 19/24}} = \frac{14 - 0.5}{8.441} = 1.60$$

As z is less than 1.96 H_0 is not rejected. The P-value can be found from Appendix 1. The table value of 1.60 is $\Phi(1.60) = 0.945$. Hence P, the probability of z being greater than 1.60 in magnitude, is $2(1 - 0.945) = 0.110$.

Minitab uses this normal approximation and applies the one-sample Wilcoxon test to the column of differences. The Minitab output using the fourth column of Table 19.1 as input is

```
Test of median = 0.000000 versus median not = 0.000000

         N for    Wilcoxon              Estimated
    N    Test     Statistic      P      Median
    10   9        8.5            0.110  -3.500
```

Note that the median estimated by Minitab (−3.5) is not the same of the median of the differences (−2.0).

As an exercise carry out a single-sample t-test (Section 5.2) on the differences. This is a paired samples t-test of the data in columns 2 and 3 of Table 19.1 You should obtain $t = -2.07$ on 9 df ($P = 0.068$). This result is misleading as you cannot assume the differences are normally distributed. However, in cases where you can make this assumption the paired t-test is more powerful than the Wilcoxon test. If the assumptions for the t-test and the Wilcoxon test cannot be made you could apply the Sign test (Section 19.2) to the differences. If you apply it to this example you will obtain a P-value of 0.508.

19.5 THE MANN–WHITNEY U TEST

This test is used to compare two population medians. It is an alternative to the independent samples t-test (Section 6.2) when you cannot assume normally distributed populations. However, for the Mann–Whitney test to be valid you have to assume that the data are independent random samples from two populations that have the same shape and a common variance. If the populations have different shapes or different variances, a t-test without pooling of variances may be more appropriate (Section 6.6).

19.5.1 The Procedure

The smaller sample, of size n_1, is called Sample 1 and the larger, of size n_2 is called Sample 2. If the sample sizes are equal, either can be called Sample 1.

(1) Rank the data from the two samples as if they were from one sample. Give the smallest observation rank 1. If two or more observations are equal (tied) give each the average rank.
(2) Find the sum of the ranks (R_1) of Sample 1 and the sum of the ranks (R_2) of Sample 2. Check that $R_1 + R_2 = n(n+1)/2$ where $n = n_1 + n_2$.
(3) Calculate

$$U_1 = R_1 - \frac{n_1(n_1+1)}{2} \quad \text{and} \quad U_2 = R_2 - \frac{n_2(n_2+1)}{2}$$

and check that $U_1 + U_2 = n_1 n_2$.
(4) To carry out a two-tailed test of the null hypothesis that the population medians M_1 and M_2 are equal, consult Appendix 10. Reject at the 5% level if U, the smaller of U_1 and U_2 is less than or equal to the table value in the 2.5% column corresponding to n_1 and n_2.

For a one-tailed test at the 5% level, if the alternative hypothesis is that $M_1 < M_2$, reject if U_1 is less than or equal to the table value in the 5% column. If the alternative is that $M_1 > M_2$, reject if U_2 is less than or equal to the table value.

Note: The smallest value U can have is zero, which happens when there is no overlap. If all the values of sample 1 are less than the smallest value in sample 2, then R_1 is the sum of the integers 1 to n_1 which is $n_1(n_1+1)/2$, hence $U_1 = 0$. Similarly $U_2 = 0$ when all the values in sample 2 are less than the smallest value of sample 1.

If n_1 and n_2 are both large and on the assumption that the two populations are identical, both U_1 and U_2 come (approximately) from a normal distribution with mean $n_1 n_2/2$ and variance $n_1 n_2 (n_1 + n_2 + 1)/12$. Hence an approximate test is to find

$$z = \frac{|U - n_1 n_2/2| - 0.5}{\sqrt{n_1 n_2 (n_1 + n_2 + 1)/12}}$$

where U is either U_1 or U_2, and compare with normal tables. The continuity correction of 0.5 is applied to account for the fact that U comes from a discrete distribution. An adjustment (not given) should be made if there are many tied ranks.

Example 19.4
In Example 6.2 a t-test was carried out to compare a new variety of wheat with a standard. Six plots received the new variety and ten plots the standard in a completely randomised design. A Mann–Whitney test is now performed on these data. Table 19.2 shows the yields and their ranks.

There are two yields equal to 1.9 so instead of giving them ranks 4 and 5 they are both assigned the average rank of 4.5. Similarly, the three yields of 2.1, instead of being ranked 8, 9 and 10, are assigned the average rank of 9:

$R_1 + R_2 = 71 + 65 = 136$ and $n(n+1)/2 = 16 \times 17/2 = 136$. This is the sum of the integers 1 to 16.

$$U_1 = R_1 - \frac{n_1(n_1+1)}{2} = 71 - \frac{6 \times 7}{2} = 50 \text{ and}$$

$$U_2 = R_2 - \frac{n_2(n_2 + 1)}{2} = 65 - \frac{10 \times 11}{2} = 10$$

$U_1 + U_2 = 50 + 10 = 60$ and $n_1 n_2 = 6 \times 10 = 60$

For a two-tailed test at the 5% level consult the 2.5% column in Appendix 10 for $n_1 = 6$ and $n_2 = 10$. The smaller U is 10 which is less than the table value of 11, so the null hypothesis of equal population medians is rejected.

If it was claimed in advance of the experiment, that the new variety would give higher yields than the standard, the alternative hypothesis would be $M_1 > M_2$. You would carry out a one-tailed test at the 5% level by consulting the 5% column. The table value is 14 and the smaller U is less than this so the null hypothesis is rejected and the claim supported.

Although the normal approximation is not reliable for small samples we apply it to illustrate the method:

$$z = \frac{|U - n_1 n_2/2| - 0.5}{\sqrt{n_1 n_2 (n_1 + n_2 + 1)/12}} = \frac{|50 - 30| - 0.5}{\sqrt{60 \times 17/12}} = \frac{19.5}{9.220} = 2.115$$

This is greater than 1.96 in magnitude so reject the two-tailed hypothesis at the 5% level ($P = 0.034$). Compare this with $P = 0.017$ when the data were analysed using the independent samples t-test in Example 6.2.

Minitab applies this normal approximation with an adjustment for tied ranks. The unknown population medians are denoted by ETA1 and ETA2 (ETA is the Greek letter η). The Minitab output follows:

```
New           N = 6       Median = 2.3500
Standard      N = 10      Median = 2.0000
Point estimate for ETA1-ETA2 is         0.3000
95.5 Percent CI for ETA1-ETA2 is     (0.0000,0.5999)
W = 71.0
Test of ETA1 = ETA2   vs   ETA1 not = ETA2 is significant at 0.0344
The test is significant at 0.0335 (adjusted for ties)
```

The two sample medians are 2.35 and 2.00 respectively, but the estimate of the difference $\eta_1 - \eta_2$ is not 0.35. It is 0.30, which is the median of all the $6 \times 10 = 60$ pairwise difference in the yields where the first member of a pair is a yield from sample one. The sum of the ranks of sample one is $W = 71$ (we use R_1). The confidence interval is worked out by a complicated method (not given). It is not exactly a 95% interval because U is discrete. The P-value (0.0344) is the same as that found above. The small adjustment for ties makes very little difference.

Table 19.2. Yields (t/ha) from two wheat varieties and their ranks

New	2.6	2.1	2.5	2.4	1.9	2.3					
Rank	16	9	15	14	4.5	12.5					$R_1 = 71$
Standard	1.7	2.1	2	1.8	2.3	1.6	2.0	2.1	2.2	1.9	
Rank	2	9	6.5	3	12.5	1	6.5	9	11	4.5	$R_2 = 65$

19.6 THE KRUSKAL–WALLIS TEST

This test is used to compare several population medians and so is a generalisation of the Mann–Whitney test. It is a non-parametric alternative to one-way analysis of variance. The null hypothesis (H_0) is that the medians of the several populations are equal, versus the alternative hypothesis (H_1) that they are not equal. An assumption for this test is that the samples from the different populations are independent random samples from distributions having the same shape.

19.6.1 The Procedure

Let k be the number of samples of sizes $n_1, n_2, \ldots, n_k$ respectively and let $N = n_1 + n_2 + \ldots + n_k$:

(1) Rank all the N observations from the several samples as if they were from one sample. Give the smallest observation rank 1 and the largest rank N. If two or more observations are equal (tied) give each the average rank.
(2) For each sample, find the sum of the ranks and the average rank. If R_i is the sum of the ranks for the ith sample and n_i is the number of observations, then the average rank is $\bar{R}_i = R_i/n_i$.
(3) The test statistic is

$$H = \frac{12 \Sigma n_i (\bar{R}_i - \bar{R})^2}{N(N+1)}$$

where

$$\bar{R} = \frac{\Sigma R_i}{N} = (N+1)/2$$

the average of all the ranks. An equivalent but simpler version of H should be used if calculations are done by hand:

$$\text{Calculate } H = \frac{12}{N(N+1)} \Sigma \frac{R_i^2}{n_i} - 3(N+1)$$

(4) Under the null hypothesis, H comes (approximately) from a chi-square distribution with $k-1$ degrees of freedom. The approximation is reasonably accurate if no group has fewer than five observations. As an increase in the variation between the groups will increase H, the chi-square test is one-tailed and H_0 is rejected at the 5% level if H is greater than $\chi^2_{(k-1,5\%)}$. For small groups a special table should be consulted, for example Table 25 of Lindley and Scott (1995).

Note: Some authors suggest adjusting H when there are ties in the data; Minitab prints H(adj) for ties.

Example 19.5

To illustrate the method, we re-analyse the data of Example 9.1. In Chapter 9 the data were analysed using one-way analysis of variance, but using the Kruskal–Wallis test you do not have to assume normally distributed populations. Table 19.3 shows the yields and ranks.

Using the second equation for H given in step (3) above:

$$H = \frac{12}{20 \times 21}\left(\frac{27^2}{5} + \frac{82.5^2}{5} + \frac{40^2}{5} + \frac{60.5^2}{5}\right) - 3 \times 21 = \frac{12}{420} \times 2559.1 - 63$$

$$= 73.117 - 63 = 10.12$$

From Appendix 7, $\chi^2_{(3,5\%)} = 7.815$ and $\chi^2_{(3,1\%)} = 11.34$ so the P-value, the probability of a larger H than 10.12, is between 0.05 and 0.01. In fact it is 0.018, which is strong evidence against the null hypothesis. Compare this with 0.008 when the data were analysed by one-way analysis of variance (Example 9.1).

The Minitab output follows:

```
Kruskal-Wallis Test on Yield

Variety    N    Median    Ave Rank      Z
1          5    21.20        5.4      -2.23
2          5    27.40       16.5       2.62
3          5    22.60        8.0      -1.09
4          5    24.80       12.1       0.70
Overall   20                10.5

H = 10.12    DF = 3    P = 0.018
H = 10.12    DF = 3    P = 0.018 (adjusted for ties)
```

There was only one tied rank (15.5) so the adjustment was negligible. The z-values indicate how the mean rank for each group differs from the mean rank for all the N observations. The formula for z is

$$z_i = \frac{\bar{R}_i - \bar{R}}{\sqrt{(N+1)(N/n_i - 1)/12}} \quad \text{so for variety one,} \quad z_1 = \frac{5.4 - 10.5}{\sqrt{21 \times 3/12}} = \frac{-5.1}{2.291} = -2.23$$

This shows that the mean rank of variety one is considerably less than the overall average rank.

Having concluded that there is strong evidence of at least one difference between the groups you may wish to compare pairs of groups using the Mann–Whitney test. These tests are not very powerful when the groups are small and are subject to the same criticisms applied to multiple comparisons after analysis of variance.

Table 19.3. Wheat yields (kg/plot) and ranks for four varieties A, B, C and D

Variety	A		B		C		D	
	Yield	Rank	Yield	Rank	Yield	Rank	Yield	Rank
	22.2	7	24.1	11	25.9	14	23.9	10
	17.3	2	30.3	19	18.4	3	21.7	5
	21.2	4	27.4	17	23.2	9	24.8	12
	25.2	13	26.4	15.5	21.9	6	28.2	18
	16.1	1	34.8	20	22.6	8	26.4	15.5
Rank sum		27		82.5		40		60.5
Mean rank		5.4		16.5		8		12.1

19.7 FRIEDMAN'S TEST

This test performs a non-parametric analysis of a randomised block experiment. It is useful when the data do not fulfil the normality and homogeneity of variance assumptions of the parametric model. It is assumed there is exactly one observation per block treatment combination. The null hypothesis (H_0) is that there are no population differences between the treatments. Use the Wilcoxon matched pairs test (Section 19.4) if you have only two treatments.

19.7.1 The Procedure

Let k be the number of treatments and n the number of blocks:

(1) Rank the data within each block separately. Give the smallest observation rank 1 and the largest rank k. If two or more observations are equal (tied) give each the average rank.
(2) For each treatment, find the sum of the ranks. Let R_i be the sum of the ranks for the ith treatment.
(3) The test statistic is

$$S = \frac{12}{nk(k+1)} \sum [R_i - n(k+1)/2]^2$$

An equivalent but simpler version of S is given by

$$S = \frac{12}{nk(k+1)} \sum R_i^2 - 3n(k+1)$$

(4) Under the null hypothesis, S comes (approximately) from a chi-square distribution with $k-1$ degrees of freedom. The approximation is reasonably accurate if either n or k is at least five. As an increase in the variation between the groups will increase S, the chi-square test is one-tailed and H_0 is rejected at the 5% level if S is greater than $\chi^2_{(k-1, 5\%)}$. For small values of n and k a special table should be consulted, for example Table 24 of Lindley and Scott (1995).

Table 19.4. Yields (t/ha) of four wheat varieties with the within-block ranks in parentheses

	V_1	V_2	V_3	V_4
Block 1	7.4 (2)	9.8 (4)	7.3 (1)	9.5 (3)
Block 2	6.5 (2)	6.8 (3)	6.1 (1)	8.0 (4)
Block 3	5.6 (1)	6.2 (2)	6.4 (3)	7.4 (4)
Rank totals (R_i)	5	9	5	11
Mean rank	1.67	3	1.67	3.67

Example 19.6

To illustrate the method, we re-analyse the data of Example 10.1 in which three new varieties of wheat (V_2, V_3 and V_4) were compared with a standard variety (V_1). In Chapter 10 the data were analysed using a randomised blocks analysis of variance but using this method you do not have to assume normally distributed residuals. Table 19.4 shows the yields with the ranks in parentheses.

Using the second equation for S given in step (3) above:

$$S = \frac{12}{3 \times 4 \times 5}[5^2 + 9^2 + 5^2 + 11^2] - 3 \times 3 \times 5 = \frac{12}{60} \times 252 - 45 = 50.4 - 45 = 5.4$$

As there are four treatments, the degrees of freedom for the approximate chi-square test are 3. From Appendix 7, $\chi^2_{(3,5\%)} = 7.815$ so the P-value, the probability of a larger S than 5.4, is greater than 0.05. In fact it is 0.145, which is insufficient evidence to reject the null hypothesis. Compare this with $P = 0.037$ when the data were analysed by a randomised blocks analysis of variance (Example 10.1).

The Minitab output is:

```
Friedman test for Yield by Treat blocked by Block
S = 5.40    DF = 3    P = 0.145

                    Est         Sum of
T       N       Median          Ranks
1       3       6.200           5.0
2       3       6.800           9.0
3       3       6.100           5.0
4       3       8.000           11.0

Grand median =  6.775
```

Minitab adjusts S and P for ties. As there were no tied ranks in this example the values of S and P agree with those found above. The degrees of freedom (DF) are 3 (the number of treatments minus one). For each treatment Minitab calculates the estimated median (the grand median plus the treatment effect). For details of the method used refer to the Minitab manual. In the output, N is the number of blocks.

Note: In this example, the amount of data is very small so the chi-square approximation may not be good. From Table 24 of Lindley and Scott (1995) the critical value for testing at the 5% level is 7.400 (the chi-square table value is 7.815). As $S = 5.40$ is not larger than 7.4, the conclusion is not changed.

In general, an overall significant P-value does not indicate which treatments differ from which. Pairs of treatments can be compared using the Wilcoxon test, but you should realise that multiple comparison tests may not be independent and may be based on too few data to be of any value.

Appendix 1

The Normal Distribution Function

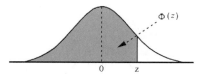

$\Phi(z)$ = proportion of total area under curve to left of z

Example: $\Phi(1.13) = 0.8708$
Example: $\Phi(-1.45) = 1 - \Phi(1.45) = 1 - 0.9265 = 0.0735$

z	0.00	0.01	0.02	0.03	0.04	0.05	0.06	0.07	0.08	0.09
0.0	0.5000	0.5040	0.5080	0.5120	0.5160	0.5199	0.5239	0.5279	0.5319	0.5359
0.1	0.5398	0.5438	0.5478	0.5517	0.5557	0.5596	0.5636	0.5675	0.5714	0.5753
0.2	0.5793	0.5832	0.5871	0.5910	0.5948	0.5987	0.6026	0.6064	0.6103	0.6141
0.3	0.6179	0.6217	0.6255	0.6293	0.6331	0.6368	0.6406	0.6443	0.6480	0.6517
0.4	0.6554	0.6591	0.6628	0.6664	0.6700	0.6736	0.6772	0.6808	0.6844	0.6879
0.5	0.6915	0.6950	0.6985	0.7019	0.7054	0.7088	0.7123	0.7157	0.7190	0.7224
0.6	0.7257	0.7291	0.7324	0.7357	0.7389	0.7422	0.7454	0.7486	0.7517	0.7549
0.7	0.7580	0.7611	0.7642	0.7673	0.7704	0.7734	0.7764	0.7794	0.7823	0.7852
0.8	0.7881	0.7910	0.7939	0.7967	0.7995	0.8023	0.8051	0.8078	0.8106	0.8133
0.9	0.8159	0.8186	0.8212	0.8238	0.8264	0.8289	0.8315	0.8340	0.8365	0.8389
1.0	0.8413	0.8438	0.8461	0.8485	0.8508	0.8531	0.8554	0.8577	0.8599	0.8621
1.1	0.8643	0.8665	0.8686	0.8708	0.8729	0.8749	0.8770	0.8790	0.8810	0.8830
1.2	0.8849	0.8869	0.8888	0.8907	0.8925	0.8944	0.8962	0.8980	0.8997	0.9015
1.3	0.9032	0.9049	0.9066	0.9082	0.9099	0.9115	0.9131	0.9147	0.9162	0.9177
1.4	0.9192	0.9207	0.9222	0.9236	0.9251	0.9265	0.9279	0.9292	0.9306	0.9319
1.5	0.9332	0.9345	0.9357	0.9370	0.9382	0.9394	0.9406	0.9418	0.9429	0.9441
1.6	0.9452	0.9463	0.9474	0.9484	0.9495	0.9505	0.9515	0.9525	0.9535	0.9545
1.7	0.9554	0.9564	0.9573	0.9582	0.9591	0.9599	0.9608	0.9616	0.9625	0.9633
1.8	0.9641	0.9649	0.9656	0.9664	0.9671	0.9678	0.9686	0.9693	0.9699	0.9706
1.9	0.9713	0.9719	0.9726	0.9732	0.9738	0.9744	0.9750	0.9756	0.9761	0.9767
2.0	0.97725	0.97778	0.97831	0.97882	0.97932	0.97982	0.98030	0.98077	0.98124	0.98169
2.1	0.98214	0.98257	0.98300	0.98341	0.98382	0.98422	0.98461	0.98500	0.98537	0.98574
2.2	0.98610	0.98645	0.98679	0.98713	0.98745	0.98778	0.98809	0.98840	0.98870	0.98899
2.3	0.98928	0.98956	0.98983	0.99010	0.99036	0.99061	0.99086	0.99111	0.99134	0.99158
2.4	0.99180	0.99202	0.99224	0.99245	0.99266	0.99286	0.99305	0.99324	0.99343	0.99361
2.5	0.99379	0.99396	0.99413	0.99430	0.99446	0.99461	0.99477	0.99492	0.99506	0.99520
2.6	0.99534	0.99547	0.99560	0.99573	0.99585	0.99598	0.99609	0.99621	0.99632	0.99643
2.7	0.99653	0.99664	0.99674	0.99683	0.99693	0.99702	0.99711	0.99720	0.99728	0.99736
2.8	0.99744	0.99752	0.99760	0.99767	0.99774	0.99781	0.99788	0.99795	0.99801	0.99807
2.9	0.99813	0.99819	0.99825	0.99831	0.99836	0.99841	0.99846	0.99851	0.99856	0.99861
3.0	0.99865	0.99869	0.99874	0.99878	0.99882	0.99886	0.99889	0.99893	0.99896	0.99900

z	3.0	3.1	3.2	3.3	3.4	3.5	3.6	3.7	3.8	3.9
$\Phi(z)$	0.99865	0.99903	0.99931	0.99952	0.99966	0.99977	0.99984	0.99989	0.99993	0.99995

Appendix 2

Percentage Points of the Normal Distribution

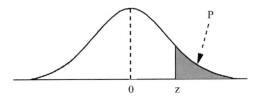

P = percentage of total area under curve to right of z

P	z	P	z	P	z
50	0.0000	2.9	1.8957	0.9	2.3656
45	0.1257	2.8	1.9110	0.8	2.4089
40	0.2533	2.7	1.9268	0.7	2.4573
35	0.3853	2.6	1.9431	0.6	2.5121
30	0.5244	2.5	1.9600	0.5	2.5758
25	0.6745	2.4	1.9774	0.4	2.6521
20	0.8416	2.3	1.9954	0.3	2.7478
15	1.0364	2.2	2.0141	0.2	2.8782
10	1.2816	2.1	2.0335	0.1	3.0902
5	1.6449	2.0	2.0537	0.09	3.1214
4.8	1.6646	1.9	2.0749	0.08	3.1559
4.6	1.6849	1.8	2.0969	0.07	3.1947
4.4	1.7060	1.7	2.1201	0.06	3.2389
4.2	1.7279	1.6	2.1444	0.05	3.2905
4.0	1.7507	1.5	2.1701	0.01	3.7190
3.8	1.7744	1.4	2.1973	0.005	3.8906
3.6	1.7991	1.3	2.2262	0.001	4.2649
3.4	1.8250	1.2	2.2571	0.0005	4.4172
3.2	1.8522	1.1	2.2904		
3.0	1.8808	1.0	2.3263		

Appendix 3

Percentage Points of the *t*-distribution

df	Percentage in top tail						
	10	5	2.5	1	0.5	0.1	0.05
1	3.078	6.314	12.71	31.82	63.66	318.3	636.6
2	1.886	2.920	4.303	6.965	9.925	22.33	31.60
3	1.638	2.353	3.182	4.541	5.841	10.21	12.92
4	1.533	2.132	2.776	3.747	4.604	7.173	8.610
5	1.476	2.015	2.571	3.365	4.032	5.894	6.869
6	1.440	1.943	2.447	3.143	3.707	5.208	5.959
7	1.415	1.895	2.365	2.998	3.499	4.785	5.408
8	1.397	1.860	2.306	2.896	3.355	4.501	5.041
9	1.383	1.833	2.262	2.821	3.250	4.297	4.781
10	1.372	1.812	2.228	2.764	3.169	4.144	4.587
11	1.363	1.796	2.201	2.718	3.106	4.025	4.437
12	1.356	1.782	2.179	2.681	3.055	3.930	4.318
13	1.350	1.771	2.160	2.650	3.012	3.852	4.221
14	1.345	1.761	2.145	2.624	2.977	3.787	4.140
15	1.341	1.753	2.131	2.602	2.947	3.733	4.073
16	1.337	1.746	2.120	2.583	2.921	3.686	4.015
17	1.333	1.740	2.110	2.567	2.898	3.646	3.965
18	1.330	1.734	2.101	2.552	2.878	3.610	3.922
19	1.328	1.729	2.093	2.539	2.861	3.579	3.883
20	1.325	1.725	2.086	2.528	2.845	3.552	3.850
21	1.323	1.721	2.080	2.518	2.831	3.527	3.819
22	1.321	1.717	2.074	2.508	2.819	3.505	3.792
23	1.319	1.714	2.069	2.500	2.807	3.485	3.768
24	1.318	1.711	2.064	2.492	2.797	3.467	3.745
25	1.316	1.708	2.060	2.485	2.787	3.450	3.725
26	1.315	1.706	2.056	2.479	2.779	3.435	3.707
27	1.314	1.703	2.052	2.473	2.771	3.421	3.689
28	1.313	1.701	2.048	2.467	2.763	3.408	3.674
29	1.311	1.699	2.045	2.462	2.756	3.396	3.660
30	1.310	1.697	2.042	2.457	2.750	3.385	3.646
35	1.306	1.690	2.030	2.438	2.724	3.340	3.591
40	1.303	1.684	2.021	2.423	2.704	3.307	3.551
50	1.299	1.676	2.009	2.403	2.678	3.261	3.496
60	1.296	1.671	2.000	2.390	2.660	3.232	3.460
80	1.292	1.664	1.990	2.374	2.639	3.195	3.416
100	1.290	1.660	1.984	2.364	2.626	3.174	3.390
120	1.289	1.658	1.980	2.358	2.617	3.160	3.373
∞	1.282	1.645	1.960	2.326	2.576	3.090	3.291

Example: $t_{(9, 2.5\%)} = 2.262$ means that the probability of a *t*-value greater than 2.262 is 2.5% for 9 df and the probability of a *t*-value outside the range −2.262 to +2.262 is 5% for 9 df.

Appendix 4a

5 per cent Points of the F-distribution

Column represents degrees of freedom (ν_1) for numerator of F-test
Row represents degrees of freedom (ν_2) for denominator of F-test

	1	2	3	4	5	6	7	8	9	10	12	24	∞
1	161.4	199.5	215.7	224.6	230.2	234.0	236.8	238.9	240.5	241.9	243.9	249.1	254.3
2	18.51	19.00	19.16	19.25	19.30	19.33	19.35	19.37	19.38	19.40	19.41	19.45	19.50
3	10.13	9.552	9.277	9.117	9.013	8.941	8.887	8.845	8.812	8.785	8.745	8.638	8.526
4	7.709	6.944	6.591	6.388	6.256	6.163	6.094	6.041	5.999	5.964	5.912	5.774	5.628
5	6.608	5.786	5.409	5.192	5.050	4.950	4.876	4.818	4.772	4.735	4.678	4.527	4.365
6	5.987	5.143	4.757	4.534	4.387	4.284	4.207	4.147	4.099	4.060	4.000	3.841	3.669
7	5.591	4.737	4.347	4.120	3.972	3.866	3.787	3.726	3.677	3.637	3.575	3.410	3.230
8	5.318	4.459	4.066	3.838	3.688	3.581	3.500	3.438	3.388	3.347	3.284	3.115	2.928
9	5.117	4.256	3.863	3.633	3.482	3.374	3.293	3.230	3.179	3.137	3.073	2.900	2.707
10	4.965	4.103	3.708	3.478	3.326	3.217	3.135	3.072	3.020	2.978	2.913	2.737	2.538
11	4.844	3.982	3.587	3.357	3.204	3.095	3.012	2.948	2.896	2.854	2.788	2.609	2.405
12	4.747	3.885	3.490	3.259	3.106	2.996	2.913	2.849	2.796	2.753	2.687	2.505	2.296
13	4.667	3.806	3.411	3.179	3.025	2.915	2.832	2.767	2.714	2.671	2.604	2.420	2.206
14	4.600	3.739	3.344	3.112	2.958	2.848	2.764	2.699	2.646	2.602	2.534	2.349	2.131
15	4.543	3.682	3.287	3.056	2.901	2.790	2.707	2.641	2.588	2.544	2.475	2.288	2.066
16	4.494	3.634	3.239	3.007	2.852	2.741	2.657	2.591	2.538	2.494	2.425	2.235	2.010
17	4.451	3.592	3.197	2.965	2.810	2.699	2.614	2.548	2.494	2.450	2.381	2.190	1.960
18	4.414	3.555	3.160	2.928	2.773	2.661	2.577	2.510	2.456	2.412	2.342	2.150	1.917
19	4.381	3.522	3.127	2.895	2.740	2.628	2.544	2.477	2.423	2.378	2.308	2.114	1.878
20	4.351	3.493	3.098	2.866	2.711	2.599	2.514	2.447	2.393	2.348	2.278	2.082	1.843
21	4.325	3.467	3.072	2.840	2.685	2.573	2.488	2.420	2.366	2.321	2.250	2.054	1.812
22	4.301	3.443	3.049	2.817	2.661	2.549	2.464	2.397	2.342	2.297	2.226	2.028	1.783
23	4.279	3.422	3.028	2.796	2.640	2.528	2.442	2.375	2.320	2.275	2.204	2.005	1.757
24	4.260	3.403	3.009	2.776	2.621	2.508	2.423	2.355	2.300	2.255	2.183	1.984	1.733
25	4.242	3.385	2.991	2.759	2.603	2.490	2.405	2.337	2.282	2.236	2.165	1.964	1.711
26	4.225	3.369	2.975	2.743	2.587	2.474	2.388	2.321	2.265	2.220	2.148	1.946	1.691
27	4.210	3.354	2.960	2.728	2.572	2.459	2.373	2.305	2.250	2.204	2.132	1.930	1.672
28	4.196	3.340	2.947	2.714	2.558	2.445	2.359	2.291	2.236	2.190	2.118	1.915	1.654
29	4.183	3.328	2.934	2.701	2.545	2.432	2.346	2.278	2.223	2.177	2.104	1.901	1.638

(*continued*)

Appendix 4a (*continued*)

	1	2	3	4	5	6	7	8	9	10	12	24	∞
30	4.171	3.316	2.922	2.690	2.534	2.421	2.334	2.266	2.211	2.165	2.092	1.887	1.622
35	4.121	3.267	2.874	2.641	2.485	2.372	2.285	2.217	2.161	2.114	2.041	1.833	1.558
40	4.085	3.232	2.839	2.606	2.449	2.336	2.249	2.180	2.124	2.077	2.003	1.793	1.509
50	4.034	3.183	2.790	2.557	2.400	2.286	2.199	2.130	2.073	2.026	1.952	1.737	1.438
60	4.001	3.150	2.758	2.525	2.368	2.254	2.167	2.097	2.040	1.993	1.917	1.700	1.389
80	3.960	3.111	2.719	2.486	2.329	2.214	2.126	2.056	1.999	1.951	1.875	1.654	1.325
100	3.936	3.087	2.696	2.463	2.305	2.191	2.103	2.032	1.975	1.927	1.850	1.627	1.283
120	3.920	3.072	2.680	2.447	2.290	2.175	2.087	2.016	1.959	1.910	1.834	1.608	1.254
∞	3.841	2.996	2.605	2.372	2.214	2.099	2.010	1.938	1.880	1.831	1.752	1.517	1.000

Example: $F_{(5,9,5\%)} = 3.482$ means that the probability of an *F*-value greater than 3.482 is 5% for (5, 9) df.

Appendix 4b

2.5 per cent Points of the F-distribution

Column represents degrees of freedom (ν_1) for numerator of F-test
Row represents degrees of freedom (ν_2) for denominator of F-test

	1	2	3	4	5	6	7	8	9	10	12	24	∞
1	647.8	799.5	864.2	899.6	921.8	937.1	948.2	956.6	963.3	968.6	976.7	997.3	1018
2	38.51	39.00	39.17	39.25	39.30	39.33	39.36	39.37	39.39	39.40	39.41	39.46	39.50
3	17.44	16.04	15.44	15.10	14.88	14.73	14.62	14.54	14.47	14.42	14.34	14.12	13.90
4	12.22	10.65	9.979	9.604	9.364	9.197	9.074	8.980	8.905	8.844	8.751	8.511	8.257
5	10.01	8.434	7.764	7.388	7.146	6.978	6.853	6.757	6.681	6.619	6.525	6.278	6.015
6	8.813	7.260	6.599	6.227	5.988	5.820	5.695	5.600	5.523	5.461	5.366	5.117	4.849
7	8.073	6.542	5.890	5.523	5.285	5.119	4.995	4.899	4.823	4.761	4.666	4.415	4.142
8	7.571	6.059	5.416	5.053	4.817	4.652	4.529	4.433	4.357	4.295	4.200	3.947	3.670
9	7.209	5.715	5.078	4.718	4.484	4.320	4.197	4.102	4.026	3.964	3.868	3.614	3.333
10	6.937	5.456	4.826	4.468	4.236	4.072	3.950	3.855	3.779	3.717	3.621	3.365	3.080
11	6.724	5.256	4.630	4.275	4.044	3.881	3.759	3.664	3.588	3.526	3.430	3.173	2.883
12	6.554	5.096	4.474	4.121	3.891	3.728	3.607	3.512	3.436	3.374	3.277	3.019	2.725
13	6.414	4.965	4.347	3.996	3.767	3.604	3.483	3.388	3.312	3.250	3.153	2.893	2.596
14	6.298	4.857	4.242	3.892	3.663	3.501	3.380	3.285	3.209	3.147	3.050	2.789	2.487
15	6.200	4.765	4.153	3.804	3.576	3.415	3.293	3.199	3.123	3.060	2.963	2.701	2.395
16	6.115	4.687	4.077	3.729	3.502	3.341	3.219	3.125	3.049	2.986	2.889	2.625	2.316
17	6.042	4.619	4.011	3.665	3.438	3.277	3.156	3.061	2.985	2.922	2.825	2.560	2.248
18	5.978	4.560	3.954	3.608	3.382	3.221	3.100	3.005	2.929	2.866	2.769	2.503	2.187
19	5.922	4.508	3.903	3.559	3.333	3.172	3.051	2.956	2.880	2.817	2.720	2.452	2.133
20	5.871	4.461	3.859	3.515	3.289	3.128	3.007	2.913	2.837	2.774	2.676	2.408	2.085
21	5.827	4.420	3.819	3.475	3.250	3.090	2.969	2.874	2.798	2.735	2.637	2.368	2.042
22	5.786	4.383	3.783	3.440	3.215	3.055	2.934	2.839	2.763	2.700	2.602	2.332	2.003
23	5.750	4.349	3.750	3.408	3.183	3.023	2.902	2.808	2.731	2.668	2.570	2.299	1.968
24	5.717	4.319	3.721	3.379	3.155	2.995	2.874	2.779	2.703	2.640	2.541	2.269	1.935
25	5.686	4.291	3.694	3.353	3.129	2.969	2.848	2.753	2.677	2.613	2.515	2.242	1.906
26	5.659	4.265	3.670	3.329	3.105	2.945	2.824	2.729	2.653	2.590	2.491	2.217	1.878
27	5.633	4.242	3.647	3.307	3.083	2.923	2.802	2.707	2.631	2.568	2.469	2.195	1.853
28	5.610	4.221	3.626	3.286	3.063	2.903	2.782	2.687	2.611	2.547	2.448	2.174	1.829
29	5.588	4.201	3.607	3.267	3.044	2.884	2.763	2.669	2.592	2.529	2.430	2.154	1.807

(*continued*)

Appendix 4b (*continued*)

	1	2	3	4	5	6	7	8	9	10	12	24	∞
30	5.568	4.182	3.589	3.250	3.026	2.867	2.746	2.651	2.575	2.511	2.412	2.136	1.787
35	5.485	4.106	3.517	3.179	2.956	2.796	2.676	2.581	2.504	2.440	2.341	2.062	1.702
40	5.424	4.051	3.463	3.126	2.904	2.744	2.624	2.529	2.452	2.388	2.288	2.007	1.637
50	5.340	3.975	3.390	3.054	2.833	2.674	2.553	2.458	2.381	2.317	2.216	1.931	1.545
60	5.286	3.925	3.343	3.008	2.786	2.627	2.507	2.412	2.334	2.270	2.169	1.882	1.482
80	5.218	3.864	3.284	2.950	2.730	2.571	2.450	2.355	2.277	2.213	2.111	1.820	1.400
100	5.179	3.828	3.250	2.917	2.696	2.537	2.417	2.321	2.244	2.179	2.077	1.784	1.347
120	5.152	3.805	3.227	2.894	2.674	2.515	2.395	2.299	2.222	2.157	2.055	1.760	1.310
∞	5.024	3.689	3.116	2.786	2.567	2.408	2.288	2.192	2.114	2.048	1.945	1.640	1.000

Example: $F_{(5,9,2.5\%)} = 4.484$ means that the probability of an F-value greater than 4.484 is 2.5% for (5, 9) df.

Use this table for checking the assumption of equal population variances prior to a t-test. Calculated F is the larger sample variance (with ν_1 df) divided by the smaller (with ν_2 df).

Appendix 4c

1 per cent Points of the *F*-distribution

Column represents degrees of freedom (ν_1) for numerator of *F*-test
Row represents degrees of freedom (ν_2) for denominator of *F*-test

	1	2	3	4	5	6	7	8	9	10	12	24	∞
1	4052	4999	5404	5624	5764	5859	5928	5981	6022	6056	6107	6234	6366
2	98.50	99.00	99.16	99.25	99.30	99.33	99.36	99.38	99.39	99.40	99.42	99.46	99.50
3	34.12	30.82	29.46	28.71	28.24	27.91	27.67	27.49	27.34	27.23	27.05	26.60	26.13
4	21.20	18.00	16.69	15.98	15.52	15.21	14.98	14.80	14.66	14.55	14.37	13.93	13.46
5	16.26	13.27	12.06	11.39	10.97	10.67	10.46	10.29	10.16	10.05	9.888	9.466	9.021
6	13.75	10.92	9.780	9.148	8.746	8.466	8.260	8.102	7.976	7.874	7.718	7.313	6.880
7	12.25	9.547	8.451	7.847	7.460	7.191	6.993	6.840	6.719	6.620	6.469	6.074	5.650
8	11.26	8.649	7.591	7.006	6.632	6.371	6.178	6.029	5.911	5.814	5.667	5.279	4.859
9	10.56	8.022	6.992	6.422	6.057	5.802	5.613	5.467	5.351	5.257	5.111	4.729	4.311
10	10.04	7.559	6.552	5.994	5.636	5.386	5.200	5.057	4.942	4.849	4.706	4.327	3.909
11	9.646	7.206	6.217	5.668	5.316	5.069	4.886	4.744	4.632	4.539	4.397	4.021	3.603
12	9.330	6.927	5.953	5.412	5.064	4.821	4.640	4.499	4.388	4.296	4.155	3.780	3.361
13	9.074	6.701	5.739	5.205	4.862	4.620	4.441	4.302	4.191	4.100	3.960	3.587	3.165
14	8.862	6.515	5.564	5.035	4.695	4.456	4.278	4.140	4.030	3.939	3.800	3.427	3.004
15	8.683	6.359	5.417	4.893	4.556	4.318	4.142	4.004	3.895	3.805	3.666	3.294	2.869
16	8.531	6.226	5.292	4.773	4.437	4.202	4.026	3.890	3.780	3.691	3.553	3.181	2.753
17	8.400	6.112	5.185	4.669	4.336	4.101	3.927	3.791	3.682	3.593	3.455	3.083	2.653
18	8.285	6.013	5.092	4.579	4.248	4.015	3.841	3.705	3.597	3.508	3.371	2.999	2.566
19	8.185	5.926	5.010	4.500	4.171	3.939	3.765	3.631	3.523	3.434	3.297	2.925	2.489
20	8.096	5.849	4.938	4.431	4.103	3.871	3.699	3.564	3.457	3.368	3.231	2.859	2.421
21	8.017	5.780	4.874	4.369	4.042	3.812	3.640	3.506	3.398	3.310	3.173	2.801	2.360
22	7.945	5.719	4.817	4.313	3.988	3.758	3.587	3.453	3.346	3.258	3.121	2.749	2.306
23	7.881	5.664	4.765	4.264	3.939	3.710	3.539	3.406	3.299	3.211	3.074	2.702	2.256
24	7.823	5.614	4.718	4.218	3.895	3.667	3.496	3.363	3.256	3.168	3.032	2.659	2.211
25	7.770	5.568	4.675	4.177	3.855	3.627	3.457	3.324	3.217	3.129	2.993	2.620	2.170
26	7.721	5.526	4.637	4.140	3.818	3.591	3.421	3.288	3.182	3.094	2.958	2.585	2.132
27	7.677	5.488	4.601	4.106	3.785	3.558	3.388	3.256	3.149	3.062	2.926	2.552	2.097
28	7.636	5.453	4.568	4.074	3.754	3.528	3.358	3.226	3.120	3.032	2.896	2.522	2.064
29	7.598	5.420	4.538	4.045	3.725	3.499	3.330	3.198	3.092	3.005	2.868	2.495	2.034

(*continued*)

Appendix 4c (*continued*)

	1	2	3	4	5	6	7	8	9	10	12	24	∞
30	7.562	5.390	4.510	4.018	3.699	3.473	3.305	3.173	3.067	2.979	2.843	2.469	2.006
35	7.419	5.268	4.396	3.908	3.592	3.368	3.200	3.069	2.963	2.876	2.740	2.364	1.891
40	7.314	5.178	4.313	3.828	3.514	3.291	3.124	2.993	2.888	2.801	2.665	2.288	1.805
50	7.171	5.057	4.199	3.720	3.408	3.186	3.020	2.890	2.785	2.698	2.563	2.183	1.683
60	7.077	4.977	4.126	3.649	3.339	3.119	2.953	2.823	2.718	2.632	2.496	2.115	1.601
80	6.963	4.881	4.036	3.563	3.255	3.036	2.871	2.742	2.637	2.551	2.415	2.032	1.494
100	6.895	4.824	3.984	3.513	3.206	2.988	2.823	2.694	2.590	2.503	2.368	1.983	1.427
120	6.851	4.787	3.949	3.480	3.174	2.956	2.792	2.663	2.559	2.472	2.336	1.950	1.381
∞	6.635	4.605	3.782	3.319	3.017	2.802	2.639	2.511	2.407	2.321	2.185	1.791	1.000

Example: $F_{(5,9,1\%)} = 6.057$ means that the probability of an *F*-value greater than 6.057 is 1% for (5, 9) df.

Appendix 4d

0.1 per cent Points of the F-distribution

Column represents degrees of freedom (ν_1) for numerator of F-test
Row represents degrees of freedom (ν_2) for denominator of F-test

	1	2	3	4	5	6	7	8	9	10	12	24	∞
1	405312	499725	540257	562668	576496	586033	593185	597954	602245	605583	610352	623703	636578
2	998.4	998.8	999.3	999.3	999.3	999.3	999.3	999.3	999.3	999.3	999.3	999.3	999.3
3	167.1	148.5	141.1	137.1	134.6	132.8	131.6	130.6	129.9	129.2	128.3	125.9	123.5
4	74.13	61.25	56.17	53.43	51.72	50.52	49.65	49.00	48.47	48.05	47.41	45.77	44.05
5	47.18	37.12	33.20	31.08	29.75	28.83	28.17	27.65	27.24	26.91	26.42	25.13	23.79
6	35.51	27.00	23.71	21.92	20.80	20.03	19.46	19.03	18.69	18.41	17.99	16.90	15.75
7	29.25	21.69	18.77	17.20	16.21	15.52	15.02	14.63	14.33	14.08	13.71	12.73	11.70
8	25.41	18.49	15.83	14.39	13.48	12.86	12.40	12.05	11.77	11.54	11.19	10.30	9.333
9	22.86	16.39	13.90	12.56	11.71	11.13	10.70	10.37	10.11	9.894	9.570	8.724	7.813
10	21.04	14.90	12.55	11.28	10.48	9.926	9.517	9.204	8.956	8.754	8.446	7.638	6.763
11	19.69	13.81	11.56	10.35	9.579	9.047	8.655	8.355	8.116	7.923	7.625	6.848	5.999
12	18.64	12.97	10.80	9.633	8.892	8.378	8.001	7.711	7.480	7.292	7.005	6.249	5.420
13	17.82	12.31	10.21	9.073	8.355	7.856	7.489	7.206	6.982	6.799	6.519	5.782	4.967
14	17.14	11.78	9.730	8.622	7.922	7.436	7.078	6.802	6.583	6.404	6.130	5.407	4.604
15	16.59	11.34	9.335	8.253	7.567	7.091	6.741	6.471	6.256	6.081	5.812	5.101	4.307
16	16.12	10.97	9.006	7.944	7.272	6.805	6.460	6.195	5.984	5.812	5.547	4.846	4.060
17	15.72	10.66	8.727	7.683	7.022	6.562	6.224	5.962	5.754	5.584	5.324	4.631	3.850
18	15.38	10.39	8.487	7.460	6.808	6.355	6.021	5.763	5.557	5.390	5.132	4.447	3.670
19	15.08	10.16	8.280	7.265	6.622	6.175	5.845	5.591	5.387	5.222	4.967	4.288	3.514
20	14.82	9.953	8.098	7.096	6.461	6.019	5.692	5.440	5.239	5.075	4.823	4.149	3.378
21	14.59	9.773	7.938	6.947	6.318	5.881	5.557	5.308	5.109	4.946	4.696	4.027	3.258
22	14.38	9.612	7.796	6.814	6.191	5.758	5.437	5.190	4.993	4.832	4.583	3.919	3.151
23	14.20	9.469	7.669	6.696	6.078	5.649	5.331	5.085	4.889	4.730	4.483	3.822	3.055
24	14.03	9.340	7.554	6.589	5.977	5.551	5.235	4.991	4.797	4.638	4.393	3.735	2.969
25	13.88	9.222	7.451	6.493	5.885	5.462	5.148	4.906	4.713	4.555	4.311	3.657	2.891
26	13.74	9.117	7.357	6.406	5.802	5.381	5.070	4.829	4.637	4.480	4.238	3.586	2.819
27	13.61	9.019	7.271	6.326	5.726	5.308	4.998	4.759	4.568	4.412	4.170	3.521	2.755
28	13.50	8.930	7.193	6.253	5.657	5.241	4.933	4.695	4.505	4.349	4.109	3.462	2.695
29	13.39	8.848	7.121	6.186	5.592	5.179	4.873	4.636	4.447	4.292	4.053	3.407	2.640

(*continued*)

0.1 PER CENT POINTS OF THE F-DISTRIBUTION

Appendix 4d (*continued*)

	1	2	3	4	5	6	7	8	9	10	12	24	∞
30	13.29	8.773	7.054	6.125	5.534	5.122	4.817	4.582	4.393	4.239	4.001	3.357	2.589
35	12.90	8.470	6.787	5.876	5.298	4.894	4.595	4.363	4.178	4.027	3.792	3.156	2.383
40	12.61	8.251	6.595	5.698	5.128	4.731	4.436	4.207	4.024	3.874	3.643	3.011	2.233
50	12.22	7.956	6.336	5.459	4.901	4.512	4.222	3.998	3.819	3.671	3.443	2.817	2.027
60	11.97	7.768	6.171	5.307	4.757	4.372	4.086	3.865	3.687	3.542	3.315	2.694	1.891
80	11.67	7.540	5.972	5.123	4.582	4.204	3.923	3.705	3.530	3.386	3.162	2.545	1.720
100	11.50	7.408	5.857	5.017	4.482	4.107	3.829	3.612	3.439	3.296	3.074	2.458	1.615
120	11.38	7.321	5.781	4.947	4.416	4.044	3.767	3.552	3.379	3.237	3.016	2.402	1.544
∞	10.83	6.908	5.422	4.617	4.103	3.743	3.474	3.266	3.098	2.959	2.742	2.132	1.015

Example: $F_{(5,9,0.1\%)} = 11.71$ means that the probability of an F-value greater than 11.71 is 0.1% for (5, 9) df.

Appendix 5

Percentage Points of the Sample Correlation Coefficient (r) when the Population Correlation Coefficient is 0 and n is the Number of X, Y Pairs

n	Top tail percentage				n	Top tail percentage			
	5	2.5	1	0.5		5	2.5	1	0.5
3	0.988	0.997	1.000	1.000	31	0.301	0.355	0.416	0.456
4	0.900	0.950	0.980	0.990	32	0.296	0.349	0.409	0.449
5	0.805	0.878	0.934	0.959	33	0.291	0.344	0.403	0.442
6	0.729	0.811	0.882	0.917	34	0.287	0.339	0.397	0.436
7	0.669	0.754	0.833	0.875	35	0.283	0.334	0.392	0.430
8	0.621	0.707	0.789	0.834	36	0.279	0.329	0.386	0.424
9	0.582	0.666	0.750	0.798	37	0.275	0.325	0.381	0.418
10	0.549	0.632	0.715	0.765	38	0.271	0.320	0.376	0.413
11	0.521	0.602	0.685	0.735	39	0.267	0.316	0.371	0.408
12	0.497	0.576	0.658	0.708	40	0.264	0.312	0.367	0.403
13	0.476	0.553	0.634	0.684	42	0.257	0.304	0.358	0.393
14	0.458	0.532	0.612	0.661	44	0.251	0.297	0.350	0.384
15	0.441	0.514	0.592	0.641	46	0.246	0.291	0.342	0.376
16	0.426	0.497	0.574	0.623	48	0.240	0.285	0.335	0.368
17	0.412	0.482	0.558	0.606	50	0.235	0.279	0.328	0.361
18	0.400	0.468	0.543	0.590	55	0.224	0.266	0.313	0.345
19	0.389	0.456	0.529	0.575	60	0.214	0.254	0.300	0.330
20	0.378	0.444	0.516	0.561	61	0.213	0.252	0.297	0.327
21	0.369	0.433	0.503	0.549	62	0.211	0.250	0.295	0.325
22	0.360	0.423	0.492	0.537	65	0.206	0.244	0.288	0.317
23	0.352	0.413	0.482	0.526	70	0.198	0.235	0.278	0.306
24	0.344	0.404	0.472	0.515	75	0.191	0.227	0.268	0.296
25	0.337	0.396	0.462	0.505	80	0.185	0.220	0.260	0.286
26	0.330	0.388	0.453	0.496	85	0.180	0.213	0.252	0.278
27	0.323	0.381	0.445	0.487	90	0.174	0.207	0.245	0.270
28	0.317	0.374	0.437	0.479	95	0.170	0.202	0.238	0.263
29	0.311	0.367	0.430	0.471	100	0.165	0.197	0.232	0.256
30	0.306	0.361	0.423	0.463					

Example: To be significant at the 5% level on a one-tailed test, r must be at least 0.378 when based on 20 X, Y pairs ($n = 20$, top tail percentage = 5).

Example: To be significant at the 5% level on a two-tailed test, r must be at least 0.444 in magnitude when based on 20 X, Y pairs ($n = 20$, top tail percentage = 2.5).

Appendix 6

5 per cent Points of the Studentized Range, for Use in Tukey and SNK Tests

Degrees of freedom	Number of means being compared								
	2	3	4	5	6	7	8	9	10
1	17.97	26.98	32.82	37.08	40.41	43.12	45.40	47.36	49.07
2	6.085	8.331	9.798	10.88	11.74	12.44	13.03	13.54	13.99
3	4.501	5.910	6.825	7.502	8.037	8.478	8.853	9.177	9.462
4	3.927	5.040	5.757	6.287	6.707	7.053	7.347	7.602	7.826
5	3.635	4.602	5.218	5.673	6.033	6.330	6.582	6.802	6.995
6	3.461	4.339	4.896	5.305	5.628	5.895	6.122	6.319	6.493
7	3.344	4.165	4.681	5.060	5.359	5.606	5.815	5.998	6.158
8	3.261	4.041	4.529	4.886	5.167	5.399	5.597	5.767	5.918
9	3.199	3.949	4.415	4.756	5.024	5.244	5.432	5.595	5.739
10	3.151	3.877	4.327	4.654	4.912	5.124	5.305	5.461	5.599
11	3.113	3.820	4.256	4.574	4.823	5.028	5.202	5.353	5.487
12	3.082	3.773	4.199	4.508	4.751	4.950	5.119	5.265	5.395
13	3.055	3.735	4.151	4.453	4.690	4.885	5.049	5.192	5.318
14	3.033	3.702	4.111	4.407	4.639	4.829	4.990	5.131	5.254
15	3.014	3.674	4.076	4.367	4.595	4.782	4.940	5.077	5.198
16	2.998	3.649	4.046	4.333	4.557	4.741	4.897	5.031	5.150
17	2.984	3.628	4.020	4.303	4.524	4.705	4.858	4.991	5.108
18	2.971	3.609	3.997	4.277	4.495	4.673	4.824	4.956	5.071
19	2.960	3.593	3.977	4.253	4.469	4.645	4.794	4.924	5.038
20	2.950	3.578	3.958	4.232	4.445	4.620	4.768	4.896	5.008
24	2.919	3.532	3.901	4.166	4.373	4.541	4.684	4.807	4.915
30	2.888	3.486	3.845	4.102	4.302	4.464	4.602	4.720	4.824
40	2.858	3.442	3.791	4.039	4.232	4.389	4.521	4.635	4.735
60	2.829	3.399	3.737	3.977	4.163	4.314	4.441	4.550	4.646
120	2.800	3.356	3.685	3.917	4.096	4.241	4.363	4.468	4.560
∞	2.772	3.314	3.633	3.858	4.030	4.170	4.286	4.387	4.474

(*continued*)

Appendix 6 (*continued*)

Degrees of freedom	\multicolumn{10}{c}{Number of means being compared}
	11	12	13	14	15	16	17	18	19	20
1	50.59	51.96	53.20	54.33	55.36	56.32	57.22	58.04	58.83	59.56
2	14.39	14.75	15.08	15.38	15.65	15.91	16.14	16.37	16.57	16.77
3	9.717	9.946	10.15	10.35	10.53	10.69	10.84	10.98	11.11	11.24
4	8.027	8.208	8.373	8.525	8.664	8.794	8.914	9.028	9.134	9.233
5	7.168	7.324	7.466	7.596	7.717	7.828	7.932	8.030	8.122	8.208
6	6.649	6.789	6.917	7.034	7.143	7.244	7.338	7.426	7.508	7.587
7	6.302	6.431	6.550	6.658	6.759	6.852	6.939	7.020	7.097	7.170
8	6.054	6.175	6.287	6.389	6.483	6.571	6.653	6.729	6.802	6.870
9	5.867	5.983	6.089	6.186	6.276	6.359	6.437	6.510	6.579	6.644
10	5.722	5.833	5.935	6.028	6.114	6.194	6.269	6.339	6.405	6.467
11	5.605	5.713	5.811	5.901	5.984	6.062	6.134	6.202	6.265	6.326
12	5.511	5.615	5.710	5.798	5.878	5.953	6.023	6.089	6.151	6.209
13	5.431	5.533	5.625	5.711	5.789	5.862	5.931	5.995	6.055	6.112
14	5.364	6.463	5.554	5.637	5.714	5.786	5.852	5.915	5.974	6.029
15	5.306	5.404	5.493	5.574	5.649	5.720	5.785	5.864	5.904	5.958
16	5.256	5.352	5.439	5.520	5.593	5.662	5.727	5.786	5.843	5.897
17	5.212	5.307	5.392	5.471	5.544	5.612	5.675	5.734	5.790	5.842
18	5.174	5.267	5.352	5.429	5.501	5.568	5.630	5.688	5.743	5.794
19	5.140	5.231	5.315	5.391	5.462	5.528	5.589	5.647	5.701	5.752
20	5.108	5.199	5.282	5.357	5.427	5.493	5.553	5.610	5.663	5.714
24	5.012	5.099	5.179	5.251	5.319	5.381	5.439	5.494	5.545	5.594
30	4.917	5.001	5.077	5.147	5.211	5.271	5.327	5.379	5.429	5.475
40	4.824	4.904	4.977	5.044	5.106	5.163	5.216	5.266	5.313	5.358
60	4.732	4.808	4.878	4.942	5.001	5.056	5.107	5.154	5.199	5.241
120	4.641	4.714	4.781	4.842	4.898	4.950	4.998	5.044	5.086	5.126
∞	4.552	4.622	4.685	4.743	4.796	4.845	4.891	4.934	4.974	5.012

Appendix 7

Percentage Points of the Chi-square Distribution

df	\multicolumn{7}{c}{Percentage in top tail}						
	10	5	2.5	1	0.5	0.1	0.05
1	2.706	3.841	5.024	6.635	7.879	10.83	12.12
2	4.605	5.991	7.378	9.210	10.60	13.82	15.20
3	6.251	7.815	9.348	11.34	12.84	16.27	17.73
4	7.779	9.488	11.14	13.28	14.86	18.47	20.00
5	9.236	11.07	12.83	15.09	16.75	20.51	22.11
6	10.64	12.59	14.45	16.81	18.55	22.46	24.10
7	12.02	14.07	16.01	18.48	20.28	24.32	26.02
8	13.36	15.51	17.53	20.09	21.95	26.12	27.87
9	14.68	16.92	19.02	21.67	23.59	27.88	29.67
10	15.99	18.31	20.48	23.21	25.19	29.59	31.42
11	17.28	19.68	21.92	24.73	26.76	31.26	33.14
12	18.55	21.03	23.34	26.22	28.30	32.91	34.82
13	19.81	22.36	24.74	27.69	29.82	34.53	36.48
14	21.06	23.68	26.12	29.14	31.32	36.12	38.11
15	22.31	25.00	27.49	30.58	32.80	37.70	39.72
16	23.54	26.30	28.85	32.00	34.27	39.25	41.31
17	24.77	27.59	30.19	33.41	35.72	40.79	42.88
18	25.99	28.87	31.53	34.81	37.16	42.31	44.43
19	27.20	30.14	32.85	36.19	38.58	43.82	45.97
20	28.41	31.41	34.17	37.57	40.00	45.31	47.50
21	29.62	32.67	35.48	38.93	41.40	46.80	49.01
22	30.81	33.92	36.78	40.29	42.80	48.27	50.51
23	32.01	35.17	38.08	41.64	44.18	49.73	52.00
24	33.20	36.42	39.36	42.98	45.56	51.18	53.48
25	34.38	37.65	40.65	44.31	46.93	52.62	54.95
26	35.56	38.89	41.92	45.64	48.29	54.05	56.41
27	36.74	40.11	43.19	46.96	49.65	55.48	57.86
28	37.92	41.34	44.46	48.28	50.99	56.89	59.30
29	39.09	42.56	45.72	49.59	52.34	58.30	60.73
30	40.26	43.77	46.98	50.89	53.67	59.70	62.16
35	46.06	49.80	53.20	57.34	60.27	66.62	69.20
40	51.81	55.76	59.34	63.69	66.77	73.40	76.10
50	63.17	67.50	71.42	76.15	79.49	86.66	89.56
60	74.40	79.08	83.30	88.38	91.95	99.61	102.7
70	85.53	90.53	95.02	100.4	104.2	112.3	115.6
80	96.58	101.9	106.6	112.3	116.3	124.8	128.3
90	107.6	113.1	118.1	124.1	128.3	137.2	140.8
100	118.5	124.3	129.6	135.8	140.2	149.4	153.2

Example: $\chi^2_{(6,5\%)} = 12.59$ means that the probability of a χ^2 value greater than 12.59 is 5% for 6 df.

Appendix 8

Probabilities of S or fewer Successes in the Binomial Distribution with n 'Trials' and p = 0.5

	S = 0	1	2	3	4	5	6	7	8	9	10
n = 4	0.0625	0.3125	0.6875	0.9375	1.0						
5	0.0313	0.1875	0.5000	0.8125	0.9688	1.0					
6	0.0156	0.1094	0.3438	0.6562	0.8906	0.9844	1.0				
7	0.0078	0.0625	0.2266	0.5000	0.7734	0.9375	0.9922	1.0			
8	0.0039	0.0352	0.1445	0.3633	0.6367	0.8555	0.9648	0.9961	1.0		
9	0.0020	0.0195	0.0898	0.2539	0.5000	0.7461	0.9102	0.9805	0.9980	1.0	
10	0.0010	0.0107	0.0547	0.1719	0.3770	0.6230	0.8281	0.9453	0.9893	0.9990	1.0
11	0.0005	0.0059	0.0327	0.1133	0.2744	0.5000	0.7256	0.8867	0.9673	0.9941	0.9995
12	0.0002	0.0032	0.0193	0.0730	0.1938	0.3872	0.6128	0.8062	0.9270	0.9807	0.9968
13	0.0001	0.0017	0.0112	0.0461	0.1334	0.2905	0.5000	0.7095	0.8666	0.9539	0.9888
14	0.0001	0.0009	0.0065	0.0287	0.0898	0.2120	0.3953	0.6047	0.7880	0.9102	0.9713
15	*	0.0005	0.0037	0.0176	0.0592	0.1509	0.3036	0.5000	0.6964	0.8491	0.9408
16	*	0.0003	0.0021	0.0106	0.0384	0.1051	0.2272	0.4018	0.5982	0.7728	0.8949
17	*	0.0001	0.0012	0.0064	0.0245	0.0717	0.1662	0.3145	0.5000	0.6855	0.8338
18	*	0.0001	0.0007	0.0038	0.0154	0.0481	0.1189	0.2403	0.4073	0.5927	0.7597
19	*	*	0.0004	0.0022	0.0096	0.0318	0.0835	0.1796	0.3238	0.5000	0.6762
20	*	*	0.0002	0.0013	0.0059	0.0207	0.0577	0.1316	0.2517	0.4119	0.5881

	S = 11	12	13	14	15	16	17	18	19	20
n = 11	1									
12	0.9998	1								
13	0.9983	0.9999	1							
14	0.9935	0.9991	0.9999	1						
15	0.9824	0.9963	0.9995	#	1					
16	0.9616	0.9894	0.9979	0.9997	#	1				
17	0.9283	0.9755	0.9936	0.9988	0.9999	#	1			
18	0.8811	0.9519	0.9846	0.9962	0.9993	0.9999	#	1		
19	0.8204	0.9165	0.9682	0.9904	0.9978	0.9996	#	#	1	
20	0.7483	0.8684	0.9423	0.9793	0.9941	0.9987	0.9998	#	#	1

* zero to four decimal places
\# 1 to four decimal places

Example: The probability of four or fewer 'successes' in 12 'trials' when $p = 0.5$ is 0.1938.

For n greater than 20 use the normal approximation. Calculate

$$z = \frac{(S + 0.5) - 0.5n}{\sqrt{0.25n}}$$

and refer to Appendix 1.

Appendix 9

Critical Values of T in the Wilcoxon Signed Rank or Matched Pairs Test

One-tailed Two-tailed	5% 10%	2.5% 5%	1% 2%	0.5% 1%	0.1% 0.2%	0.05% 0.1%
$n = 5$	0					
6	2	0				
7	3	2	0			
8	5	3	1	0		
9	8	5	3	1		
10	10	8	5	3	0	
11	13	10	7	5	1	0
12	17	13	9	7	2	1
13	21	17	12	9	4	2
14	25	21	15	12	6	4
15	30	25	19	15	8	6
16	35	29	23	19	11	9
17	41	34	27	23	14	11
18	47	40	32	27	18	14
19	53	46	37	32	21	18
20	60	52	43	37	26	21
21	67	58	49	42	30	26
22	75	65	55	48	35	30
23	83	73	62	54	40	35
24	91	81	69	61	45	40
25	100	89	76	68	51	45
26	110	98	84	75	58	51
27	119	107	92	83	64	57
28	130	116	101	91	71	64
29	140	126	110	100	79	71
30	151	137	120	109	86	78
31	163	147	130	118	94	86
32	175	159	140	128	103	94
33	187	170	151	138	112	102
34	200	182	162	148	121	111
35	213	195	173	159	131	120
36	227	208	185	171	141	130
37	241	221	198	182	151	140
38	256	235	211	194	162	150
39	271	249	224	207	173	161
40	286	264	238	220	185	172

Example: If $n = 16$ and $T = 21$ reject H_0 at the 2% level but not at the 1% level on a two-tailed test because T is less than or equal to 23 but not less than or equal to 19.

Appendix 10

Critical Values of U in the Mann–Whitney Test

To carry out a test at the 5% level, use the 5% column for a one-tailed test and the 2.5% column for a two-tailed test.
To carry out a test at the 1% level, use the 1% column for a one-tailed test and the 0.5% column for a two-tailed test.

n_1	n_2	5%	2.5%	1%	0.5%	n_1	n_2	5%	2.5%	1%	0.5%
2	4	-	-	-	-	4	16	14	11	7	5
2	5	0	-	-	-	4	17	15	11	8	6
2	6	0	-	-	-	4	18	16	12	9	6
2	7	0	-	-	-	4	19	17	13	9	7
2	8	1	0	-	-	4	20	18	14	10	8
2	9	1	0	-	-	5	5	4	2	1	0
2	10	1	0	-	-	5	6	5	3	2	1
2	11	1	0	-	-	5	7	6	5	3	1
2	12	2	1	-	-	5	8	8	6	4	2
2	13	2	1	0	-	5	9	9	7	5	3
2	14	3	1	0	-	5	10	11	8	6	4
2	15	3	1	0	-	5	11	12	9	7	5
2	16	3	1	0	-	5	12	13	11	8	6
2	17	3	2	0	-	5	13	15	12	9	7
2	18	4	2	0	-	5	14	16	13	10	7
2	19	4	2	1	0	5	15	18	14	11	8
2	20	4	2	1	0	5	16	19	15	12	9
3	3	0	-	-	-	5	17	20	17	13	10
3	4	0	-	-	-	5	18	22	18	14	11
3	5	1	0	-	-	5	19	23	19	15	12
3	6	2	1	-	-	5	20	25	20	16	13
3	7	2	1	0	-	6	6	7	5	3	2
3	8	3	2	0	-	6	7	8	6	4	3
3	9	4	2	1	0	6	8	10	8	6	4
3	10	4	3	1	0	6	9	12	10	7	5
3	11	5	3	1	0	6	10	14	11	8	6
3	12	5	4	2	1	6	11	16	13	9	7
3	13	6	4	2	1	6	12	17	14	11	9
3	14	7	5	2	1	6	13	19	16	12	10
3	15	7	5	3	2	6	14	21	17	13	11
3	16	8	6	3	2	6	15	23	19	15	12
3	17	9	6	4	2	6	16	25	21	16	13
3	18	9	7	4	2	6	17	26	22	18	15
3	19	10	7	4	3	6	18	28	24	19	16
3	20	11	8	5	3	6	19	30	25	20	17

(*continued*)

CRITICAL VALUES OF U IN THE MANN–WHITNEY TEST

Appendix 10 (*continued*)

n_1	n_2	5%	2.5%	1%	0.5%	n_1	n_2	5%	2.5%	1%	0.5%
4	4	1	0	-	-	6	20	32	27	22	18
4	5	2	1	0	-	7	7	11	8	6	4
4	6	3	2	1	0	7	8	13	10	7	6
4	7	4	3	1	0	7	9	15	12	9	7
4	8	5	4	2	1	7	10	17	14	11	9
4	9	6	4	3	1	7	11	19	16	12	10
4	10	7	5	3	2	7	12	21	18	14	12
4	11	8	6	4	2	7	13	24	20	16	13
4	12	9	7	5	3	7	14	26	22	17	15
4	13	10	8	5	3	7	15	28	24	19	16
4	14	11	9	6	4	7	16	30	26	21	18
4	15	12	10	7	5	7	17	33	28	23	19

Example: If $n_1 = 7, n_2 = 10$ and $U = 12$ reject H_0 at the 5% level but not at the 1% level on a two-tailed test because U is less than or equal 14 but not less than or equal to 9.

n_1	n_2	5%	2.5%	1%	0.5%	n_1	n_2	5%	2.5%	1%	0.5%
7	18	35	30	24	21	11	19	65	58	50	45
7	19	37	32	26	22	11	20	69	62	53	48
7	20	39	34	28	24	12	12	42	37	31	27
8	8	15	13	9	7	12	13	47	41	35	31
8	9	18	15	11	9	12	14	51	45	38	34
8	10	20	17	13	11	12	15	55	49	42	37
8	11	23	19	15	13	12	16	60	53	46	41
8	12	26	22	17	15	12	17	64	57	49	44
8	13	28	24	20	17	12	18	68	61	53	47
8	14	31	26	22	18	12	19	72	65	56	51
8	15	33	29	24	20	12	20	77	69	60	54
8	16	36	31	26	22	13	13	51	45	39	34
8	17	39	34	28	24	13	14	56	50	43	38
8	18	41	36	30	26	13	15	61	54	47	42
8	19	44	38	32	28	13	16	65	59	51	45
8	20	47	41	34	30	13	17	70	63	55	49
9	9	21	17	14	11	13	18	75	67	59	53
9	10	24	20	16	13	13	19	80	72	63	57
9	11	27	23	18	16	13	20	84	76	67	60
9	12	30	26	21	18	14	14	61	55	47	42
9	13	33	28	23	20	14	15	66	59	51	46
9	14	36	31	26	22	14	16	71	64	56	50
9	15	39	34	28	24	14	17	77	69	60	54
9	16	42	37	31	27	14	18	82	74	65	58
9	17	45	39	33	29	14	19	87	78	69	63

(*continued*)

Appendix 10 (*continued*)

n_1	n_2	5%	2.5%	1%	0.5%	n_1	n_2	5%	2.5%	1%	0.5%
9	18	48	42	36	31	14	20	92	83	73	67
9	19	51	45	38	33	15	15	72	64	56	51
9	20	54	48	40	36	15	16	77	70	61	55
10	10	27	23	19	16	15	17	83	75	66	60
10	11	31	26	22	18	15	18	88	80	70	64
10	12	34	29	24	21	15	19	94	85	75	69
10	13	37	33	27	24	15	20	100	90	80	73
10	14	41	36	30	26	16	16	83	75	66	60
10	15	44	39	33	29	16	17	89	81	71	65
10	16	48	42	36	31	16	18	95	86	76	70
10	17	51	45	38	34	16	19	101	92	82	74
10	18	55	48	41	37	16	20	107	98	87	79
10	19	58	52	44	39	17	17	96	87	77	70
10	20	62	55	47	42	17	18	102	93	82	75
11	11	34	30	25	21	17	19	109	99	88	81
11	12	38	33	28	24	17	20	115	105	93	86
11	13	42	37	31	27	18	18	109	99	88	81
11	14	46	40	34	30	18	19	116	106	94	87
11	15	50	44	37	33	18	20	123	112	100	92
11	16	54	47	41	36	19	19	123	113	101	93
11	17	57	51	44	39	19	20	130	119	107	99
11	18	61	55	47	42	20	20	138	127	114	105

References

Baker, R. J. (1980) Multiple comparison tests. *Canadian Journal of Plant Sciences* **60**, 325–327.
Carmer, S. G. and Walker, W. M. (1982) Baby Bear's dilemma: a statistical tale. *Agronomy Journal* **74**, 122–124.
Cochran, W. G. and Cox, G. M. (1957) *Experimental Designs* (2nd edn). John Wiley, New York.
Dobson, A. J. (1990) *An Introduction to Generalized Linear Models*. Chapman and Hall, London.
Draper, N. R. and Smith, H. (1998) *Applied Regression Analysis* (3rd edn). John Wiley, New York.
Dyke, G. V. (1988) *Comparative Experiments with Field Crops* (2nd edn). Griffin, London.
Fernandez, C. J. (1992) Residual analysis and data transformation: Important tools in statistical analysis. *HortScience* **27**, 297–300
Hunt, R. (1982) *Plant Growth Curves: The Functional Approach to Plant Growth Analysis*. Edward Arnold, London.
Lindley, D. V. and Scott, W. F. (1995) *New Cambridge Statistical Tables* (2nd edn). Cambridge University Press, Cambridge.
Little, T. M. (1981) Interpretation and presentation of results. *HortScience* **16**, 637–640.
McConway, K. J., Jones, M. C. and Taylor, P. C. (1999) *Statistical Modelling using Genstat*. Edward Arnold, London.
Mead, R. (1988) *The Design of Experiments*. Cambridge University Press, Cambridge.
Mead, R., Curnow, R. N. and Hasted, A. M. (1993) *Statistical Methods in Agriculture and Experimental Biology* (2nd edn). Chapman and Hall, London.
Morse, P. M. and Thompson, B. K. (1981) Presentation of experimental results. *Canadian Journal of Plant Science* **61**, 799–802.
Petersen, R. G. (1977) Use and misuse of multiple comparison procedures. *Agronomy Journal* **69**, 205–208.
Siegel, S. and Castellan, N. J. (1988) *Nonparametric Statistics for the Behavioral Sciences* (2nd edn). McGraw–Hill, New York.
Snedecor, G. W. and Cochran, W. G. (1989) *Statistical Methods* (8th edn). Iowa State University Press.
Steel, R. D. G., Torrie, J. H. and Dickey, D. A. (1996) *Principles and Procedures of Statistics A biometrical approach* (3rd edn). McGraw–Hill, New York.
Zar, J. H. (1998) *Biostatistical Analysis* (4th edn). Prentice-Hall, Upper Saddle River, NJ.

Further Reading

Gates, C. E. (1991) A user's guide to misanalysing planned experiments. *HortScience* **26**, 1262–1265

This paper deals with the abuses and misuses of statistical methods that result from an incomplete understanding. It is now very easy to do the wrong analysis due to the availability of very powerful software packages so it is important to analyse the experiment as designed. Gates gives nine examples of common abuses and misuses of planned experiments and provides a summary of nine principles of experimental design.

Nelson, L. A. (1989) A statistical editor's viewpoint of statistical usage in horticultural publications. *HortScience* **24**, 53–57

This paper is concerned with the reporting of results. Nelson gives recommendations about how to present results for publication in a journal article. He emphasises that 'before the experiment is run, one should know precisely which comparisons are going to be made when the data are available'.

Pearce, S. C., Clarke, G. M., Dyke, G. V. and Kempson, R. E. (1988) *Manual of Crop Experimentation*. Griffin, London

This book deals with the theoretical and practical aspects of experimentation. It covers contrasts, transformations, split plots, confounding, analysis of covariance, experiments over several sites, experiments over several seasons, systematic designs, nearest-neighbour methods, and intercropping.

Preece, D. A. (1982) The design and analysis of experiments: What has gone wrong? *Utilitas Mathematica* **21A**, 201–244

This is a very readable paper with no mathematics. Preece is concerned that despite the fact that the basic principles of the design and analysis of comparative experiments have been widely taught for many years, badly designed, unsatisfactorily run and unsoundly analysed agricultural experiments are still common. He discusses what has gone wrong by dividing the subject into five components as follows: (i) planning, design and layout; (ii) data recording; (iii) scrutiny and editing of the data; (iv) computational analysis; (v) interpreting and reporting the results.

Index

2^n experiments, 179
Additive model, 57, 217
Adjusted R-square, 74, 91, 96, 101
Adjusted sum of squares, 231, 264
Adjusted treatment means, 261–264, 269
Alternative hypothesis, 40, 57
Analysis of counts, 272
Analysis of covariance, 260, 271
Analysis of variance, 77, 105, 110
Analysis of variance models, 108, 138, 155, 172
Angular transformation, 223
Arcsin transformation, 223
Association, 284, 289
Assumptions, 75, 83, 140, 213–225
Asymptotic growth curve, 97

Background variation, 5, 126, 132, 145
Balanced incomplete blocks, 235
Bar chart, 21, 177
Binomial distribution, 272–274, 278, 282, 294, 295
Boxplot, 21

Calculator, 15, 104, 117, 141, 156, 174, 246
Categorical data, 10
Central limit theorem, 30, 32
Chi square
  contingency tables, 284, 286, 287
  distribution, 280, 302, 304
  goodness of fit, 280
  heterogeneity test, 290
Coefficient of variation, 14, 61, 115, 146, 155, 171, 249
Comparing means after ANOVA, 118–128, 142, 182–208
Comparing proportions, 279, 286, 287
Comparing regression lines, 256, 265
Completely randomised design, 50, 102–130, 226, 250, 260
Confidence bands, 82
Confidence intervals
  difference in means, 55, 121, 126
  difference in proportions, 279
  for a proportion, 275, 276
  mean, 10, 31–35, 121, 143
  one-sided, 191
  slope of regression line, 80
Confounding, 179
Contingency tables, 284, 286, 287
Continuity correction, 274, 281, 295, 300
Continuous data, 6, 10, 18
Contrasts, 202, 204, 208
Corrected sum of products, 68
Corrected sum of squares, 13, 68, 113
Correction factor, 13, 116
Correlation, 67, 74, 83, 253
Covariance efficiency factor, 264
Covariance; analysis of, 260, 271
Covariates, 260
Critical region, 42
Curve fitting, 87–101

Data, types of, 6, 9
Degrees of freedom, 12, 34, 54, 89, 113, 136, 139, 150, 168, 240
Dependent variable, 63
Design construction, 103, 133, 151, 161, 235, 239
Descriptive statistics, 22, 23
Discrete data, 6, 10, 16
Distribution
  Binomial, 272–274
  chi-square, 280, 302, 304
  F, 58, 60, 112, 114, 115
  Normal, 24–31, 38
  t, 34, 38–43
Dotplot, 22
Duncan's Multiple Range test, 188

Efficiency of covariate, 264
Error mean square, 111, 139
Error rate, 183
Error sum of squares, 109, 141, 231
Error variance, 110, 133
Estimation, 31, 36, 46, 277
Experimental error, 102, 110, 133
Experimental unit, 9, 46, 57, 102, 129, 253
Exponential curves, 93

Factorial experiments, 159–180, 193
Family error rate, 128

INDEX

Fisher's exact test, 288
Fisher's pairwise comparisons, 127, 128
Fisher's protected LSD, 185
Fitted values, 68–72, 92, 108, 138–140, 156, 173, 219
Fixed effects, 154
Fractional replication, 180
Frequency distribution, 16, 18
Friedman's test, 304
F-distribution, 58, 60, 112, 114, 115
F-test, 58, 114, 217

General linear model, 230, 237
Generalised linear model, 213, 225
Goodness of fit, 72, 280
Growth cabinet problem, 250

Harmonic mean, 15, 230
Heterogeneity chi square, 290
Hidden replication, 162
Histogram, 19, 83, 140, 220–222
Homogeneity of variance, 214–217
Hypothesis tests, 38
   after ANOVA, 118, 142, 182–208
   blocked data, 132–143, 304–306
   chi square heterogeneity, 290
   chi-square goodness of fit, 280
   correlation coefficient, 83
   differences in means, 47, 51–55
   equality of intercepts, 258, 269
   equality of slopes, 257, 267, 268
   equality of variances, 58, 114, 217
   fit to binomial distribution, 282
   independent samples, 51, 102, 299, 302
   intercept of regression line, 81
   LSD in ANOVA, 120–128, 142, 184, 185
   paired samples, 46, 297
   parallel lines, 257, 267
   proportions, 277, 279, 286, 287
   ratio of two variances, 58, 114, 217
   single sample, 39, 294, 296
   slope of regression line, 77–80
   trend in means, 195, 197

Incomplete block designs, 234–237
Independence of errors, 213
Independent samples t-test, 51–58
Independent variable, 63
Interaction, 150, 161, 160–165
Intercept of regression line, 66, 68
Interquartile range, 21

Kruskal–Wallis test, 302

Latin square design, 149–157

Least significant difference, 15, 120–128, 142, 145, 153, 171, 184
Least significant increase, 191
Least squares means, 230, 231, 233, 237
Least squares regression line, 67–70
Level of significance, 40, 44
Leverage, 82
Line diagram, 17
Linear trend in means, 195
Log transformation, 218–222
Logarithmic curve, 96
Logistic growth curve, 97

Main effects, 163
Main plot analysis, 246
Main plot error, 242
Mann–Whitney U test, 299
Mean, 10, 11, 15, 16, 23, 273
Mean comparisons, 182–190
Median, 11, 18, 294–305
Minimum significant difference, 186, 189
Missing values, 226–234
Mode, 18
Models, 108, 138, 155, 172, 265–270
Multiple Comparison tests, 182–190
Multiple regression, 100
Multiplicative model, 218
Multivariate analysis, 253

Non-linear curves, 93–100
Non-parametric tests, 213, 293–306
Normal approximation, 274, 295, 296, 299–301
Normal distribution, 24–31, 38
Normal probability plot, 30, 31, 84, 140, 220–222
Normality assumption, 29, 47, 51, 83, 75, 110, 214
Null hypothesis, 40, 44, 57, 110

One-sided confidence intervals, 62, 191
One-tailed test, 40, 45, 62, 120, 280
One-way analysis of variance, 102–118, 124–130, 302
Orthogonal contrasts, 204
Outlier, 20, 22, 82, 83, 219

Paired samples t-test, 46
Pairwise comparisons, 127, 182–190
Parallel lines model, 266
Parametric tests, 293
Percentage points, 29, 34
Percentage variance accounted for, 74
Percentages, 223, 224
Plot sampling, 6, 146, 214

INDEX 331

Polynomial fitting, 87, 200
Pooled variance estimate, 54, 57, 111, 119
Pooling not valid, 215
Population mean, 10
Population variance, 11
Power, 44
Power curve, 94
Precision, 36, 60, 103, 81, 122
Prediction in regression, 70, 80, 81
Predictor variable, 63
Presenting results, see Reporting results
P-value, 42, 52, 78, 82, 106, 114, 137, 170

Quadratic regression, 89–93
Quadratic trend in means, 197
Quartiles, 21, 23

Random effects, 153
Random numbers, 104, 133, 151
Randomisation, 102–105, 133, 151, 236
Randomised blocks, 47, 132–146, 160, 229, 238, 304
Range, 19, 21
Rank correlation, 76
Ranks, 104, 133, 151, 293
Regression, 63–101, 195–202
Repeated measures, 252
Replication, 60, 102, 128, 144, 162, 277
Reporting results, 123, 144, 172, 176, 193, 201, 245
Residual degrees of freedom, 111, 112, 139
Residual diagnostics, 83, 84, 140, 219–222
Residual mean square, 77, 111, 119, 125, 139
Residual plots, 84, 140, 220–222
Residual standard deviation, 127
Residual sum of squares, 69, 71, 109, 111, 139, 141
Residuals, 68, 71, 108, 109, 139
Response variable, 63
R-square, 66, 73, 91, 95, 101, 196, 197, 200

Sample mean, 11
Sample size determination, 36, 60, 128, 144, 277
Sample variance, 12
Sampling, 3, 6, 146, 214
Sampling distribution of mean, 32, 38
Satellite mapping, 5
Sequential sum of squares, 90, 231, 267, 268
Serial measurements, 252
Sign test, 294
Significance level, 41, 44
Significance testing, See hypothesis testing
Single sample t-test, 39
Site, choice of, 3

Skeleton ANOVA tables, 150, 160, 162, 178, 240–242
Skewness, 11, 20, 22
Slope of regression line, 66, 68
Soil testing, 4
Split plot experiments, 238–253
Split split plot experiments, 252
Split unit designs, See split plot designs
Square root transformation, 220, 221
Standard deviation, 14, 17
Standard error
  difference in adjusted means, 270
  difference in intercepts, 259
  difference in means, 55, 57, 119, 125, 142, 170, 247–249
  difference in slopes, 257, 268
  intercept of regression line, 81
  mean, 23, 32, 40, 41, 121, 143
  adjusted treatment mean, 269, 270
  contrast, 205, 208
  predicted Y mean, 80
  predicted Y-value, 81
  slope of regression line, 79
Standard normal distribution, 25, 38
Standardised residuals, 71
Stem and leaf plot, 21, 22
Strata, 153, 245
Student–Newman–Keuls test, 187
Sub plot analysis, 247
Sub plot error, 242
Sum of squares, 13
Sxx, 13, 68, 107, 111, 117
Sxy, 68
Systems of experimentation, 1, 2
Syy, 72, 78

t-distribution, 34, 38–43, 57, 59
Total sum of squares, 78, 108, 109, 141
Total yield estimation, 36
Transformation, 213, 219–225
Treatment mean square, 111
Treatment sum of squares, 109, 111, 141
Trimmed mean, 23
t-test
  correlation coefficient, 83
  in one-way ANOVA, 119
  independent samples, 50, 51
  intercept of regression line, 81
  paired samples, 46
  single sample, 39
  slope of regression line, 79
Tukey simultaneous tests, 233
Tukey's test, 128, 185–187
Two-tailed test, 40
Type I error, 43, 44, 129, 183

Type I sums of squares, 231
Type II error, 44, 183
Type III sums of squares, 231
Types of variable, 9

Underscore method, 184
Unequal replication, 124
Unequal sample variances, 59, 215

Variables, 6, 10, 63
Variance, 11, 12, 17, 273

Variance ratio, 58, 77, 90, 112, 170, 142
Variation, background, 5, 126, 132, 145

Waller–Duncan's Bayes test, 189
Weighted mean, 15
Wilcoxon matched pairs test, 297
Wilcoxon single sample test, 296

Yates' correction, 289